Variational
Methods
In Optimization

Variational Methods In Optimization

DONALD R. SMITH

Department of Mathematics
University of California, San Diego

Prentice-Hall, Inc., Englewood Cliffs, N. J.

Library of Congress Cataloging in Publication Data

SMITH, DONALD R.
Variational methods in optimization.

Includes bibliographical references.
1. Mathematical optimization. 2. Calculus of variations. I. Title.
QA402.5.S55 515'.64 72-9884
ISBN 0-13-940627-1

10 9 8 7 6 5 4 3 2 1

Printed in the United States of America.

PRENTICE-HALL INTERNATIONAL, INC., *London*
PRENTICE-HALL OF AUSTRALIA, PTY. LTD., *Sydney*
PRENTICE-HALL OF CANADA, LTD., *Toronto*
PRENTICE-HALL OF INDIA PRIVATE LIMITED, *New Delhi*
PRENTICE-HALL OF JAPAN, INC., *Tokyo*

To Andrew J. Galambos

in appreciation for the theory of primary property

Preface

Almost every text on calculus shows how to use differentiation to find greatest or least values of a function, and most calculus texts show how to use this method to solve various maximum and minimum problems that arise in certain practical situations of interest. The main purpose of this book is to give an elementary exposition of an extension of this method which leads to an increase in both the number and type of such problems that can be solved. It is in fact possible to solve a remarkably wide range of optimization problems arising in such fields as engineering, astronautics, mathematics, physics, economics, and operations research using only certain abstractions of what are today considered to be very simple ideas from elementary calculus.

The book considers many of the classical problems of the calculus of variations, including problems with fixed endpoints, variable endpoints, isoperimetric constraints, and certain types of global inequality constraints, along with certain other optimization problems customarily handled by the methods of optimal control theory. A unified theory is presented that requires only the fundamentals of elementary calculus as a prerequisite and permits the solution of a strikingly wide variety of optimization problems. The essential tool used time and time again is an abstraction of the Euler–Lagrange multiplier theorem of differential calculus. (Indeed, this book might have been subtitled "Variations on a Theme of Euler!") Endpoint conditions and boundary conditions are handled explicitly as constraints on an equal basis with isoperimetric constraints; this leads to a considerable simplification in the handling of problems with *variable* boundary conditions. The book is built around various specific applications arising in economics, business management, engineering, and the physical sciences, which serve to motivate the discussion throughout. The mathematical ideas are always developed in connection with these concrete examples.

This book originated as lecture notes for a one-quarter undergraduate course in variational calculus taught by the author in 1967 at the University of California at San Diego; the students in this class were mostly third- and fourth-year undergraduate students of engineering, physical science, and applied mathematics. A later version of the manuscript was used in 1971 in an evening course taught by the author through the University of California Extension (San Diego) to a class made up of engineers and applied scientists from local industry. In 1972 I used the final version of the manuscript in a one-quarter "supplementary enrichment course" for a selected group of highly motivated first-year undergraduate students who were at that time just completing their first year course in elementary calculus.

Readers who have completed or are completing a first course in calculus may find this book useful for several reasons. The inherent interest associated with the various optimization problems considered makes the text quite useful as a vehicle on which the reader may obtain a broader appreciation of such calculus topics as the chain rule of differentiation, the fundamental theorem of calculus, integration by parts, parametric curves, line integrals, double integrals, and elementary differential equations. The material of the book exhibits the great power of successful abstraction in solving practical problems; moreover, the book can provide to those readers not already familiar with these topics a first glimpse at some of the interesting and nontrivial practical uses of the inverse function theorem, linear algebra, vector spaces, linear and nonlinear functionals, and certain aspects of applied analysis. At the same time it furnishes a brief and elementary first introduction to the theory and applications of optimal control theory and the calculus of variations. For these reasons the book should be particularly useful to students of engineering, applied mathematics, and the physical sciences.

The book is designed so that it can be used either by itself or as a supplementary text in a variety of courses. The essential material which must be covered in any course which uses this text is contained within the first three chapters, and consists of Sections 1.1, 1.2, 1.4, 2.1–2.4, 3.1, 3.3, 3.4, and 3.6. The later chapters are independent of each other, so that any combination of any of the later chapters may be covered in any order once this essential material has been covered. For example, a course emphasizing applications in economics and business management would probably also include Sections 2.5, 3.7, 3.8, 3.10, 4.1, 4.4, and parts of Chapters 5 and 6. On the other hand, a traditional course in the calculus of variations for engineering and science students would include much of Chapter 4 as a minimum, along with selected material from the remaining chapters, depending on the needs and length of the particular course. Any course touching on optimal control theory would include Chapter 6. The author covered the essential material along with Sections 4.1, 4.4, 4.6, and Chapter 6 in a "supplementary enrichment course" for terminal calculus students. The proof of the Euler–Lagrange multiplier theorem found in Section 3.5 may safely be skipped in any course if desired, although one would want to cover the geometrical idea behind the proof as indicated in Section 3.4. Enough examples and applications have been included in the text so that those readers who are primarily interested in engineering or scientific applications may skip the examples dealing with economics or business management if desired, and vice-versa. The book may provide a useful supplementary text for an applied course in differential equations or a course in applied advanced calculus.

An elementary text on variational methods in optimization intended to be readable and understandable to students in a short course cannot include all the important topics in the subject. I have stressed the fundamental necessary condition of the theory, which involves only the first variation of a functional. At the same time I emphasize the importance of checking whether or not the necessary condition actually yields the desired optimum solution sought in any given problem, and I show how to do this directly in most of the examples considered. I give only a brief discussion in Chapter 8 of the use of the second variation in optimization theory. I omit any detailed consideration of the maximum principle of optimal control theory, although I do consider many problems which have customarily been solved by the maximum principle. I also omit any detailed consideration of numerical aspects of optimization, although I do discuss briefly certain direct methods of optimization in Chapter 7. I should also mention that it has been my experience that a first course in calculus is usually adequate to enable the reader to grasp the essential content and plausibility of the inverse function theorem in its simplest form, and I have felt free to use this theorem, without proof, in the proof of the Euler–Lagrange multiplier theorem.

DONALD R. SMITH

Acknowledgements

This book is my presentation of various works of many different authors. It is a pleasure for me to acknowledge their work here. Leonhard Euler produced the first general theory of variational calculus, based on certain earlier work of James Bernoulli and John Bernoulli. Joseph Lagrange was instrumental in the early development of the theory as we know it today. Both Euler and Lagrange were instrumental in providing us with the "multiplier theorem" which is so very basic to this book. Throughout the text I have acknowledged these and other authors who are responsible for various aspects of the work discussed here. My general dependence on other authors, even to the limited extent to which it is known, is too wide for complete citation here.

I learned the importance of physics and history from Andrew J. Galambos. His influence is largely responsible for the historical approach I have taken in this book.

A public lecture given by Clifford Truesdell led me to a fuller appreciation of the work of Leonhard Euler. Christina Zemansky provided me with English translations from the Latin of certain pages of Euler's work relating to the multiplier theorem.

Jerry L. Kazdan read the original and final versions of the manuscript and gave me good advice that led to many substantial improvements in the book, including neater proofs for several results, a simpler presentation due to a greater emphasis on the Gâteaux variation as compared with the Fréchet differential, the inclusion of several particularly helpful examples, and the inclusion of the material on the variational description of Sturm–Liouville eigenvalues. Several other individuals, including Karl Keating and William Avrin, read parts of the manuscript and gave me helpful advice. Richard Schoonover and Carolyn G. Smith each helped with several of the illustrations.

Linda Van Note typed about two-thirds of the manuscript, while Lillian C. Johnson, Lola Mitchell, and Dora A. Sanders typed the remaining third. They all did excellent work. The Prentice-Hall staff did an outstanding job in seeing the book through press; I wish especially to thank Arthur H. Wester and Judy Ann Burke.

Finally, I wish to thank my wife Carolyn for her understanding and encouragement during the months while I was preoccupied with the writing of this book.

D.R.S.

Contents

Variational Methods In Optimization

1. Functionals

In this chapter we shall give examples of maximum and minimum problems which lead naturally to the consideration of functionals defined on subsets of vector spaces. The properties of continuity and linearity will be introduced for functionals. Many of the functionals arising in applications are continuous, but many are not linear.

1.1. Introduction; Examples of Optimization Problems

Many of the problems which have concerned scientists since the seventeenth century have dealt with finding the largest or smallest values of varying quantities. One such problem studied by Isaac Newton (1642–1729), Christian Huygens (1629–1695), and Leonhard Euler (1707–1789) was to find the **greatest projectile range** that can be achieved by varying the initial elevation angle of motion of a projectile which is hurled up from the surface of the earth while taking into account the effect of the resistance of the air on the projectile. Another problem considered by Newton in 1687 was to find the **least water resistance** that can be achieved by varying the shape of an object propelled through the water. A problem discussed originally by Galileo Galilei (1564–1642) in his *Dialogues* (1632, 1638) and later solved by various intuitive and geometric methods by several eminent mathematicians [including John

Bernoulli (1667–1748), his older brother James Bernoulli (1654–1705), the Marquis de L'Hôpital (1661–1704), Gottfried von Leibniz (1646–1716) and Newton] was to find the **shortest time of descent** that can be achieved by varying the shape of a wire down which a small bead slides under gravity from one point to a nearby lower point, as shown in Figure 1. One might think that

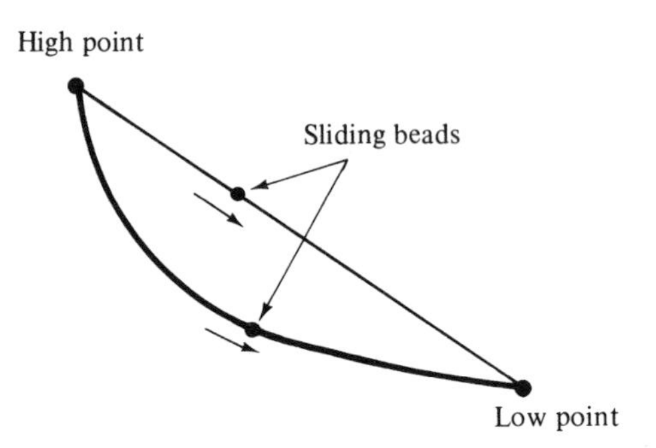

Figure 1

the quickest motion of the bead will be along the straight line joining the two points, but already Galileo had noticed that some other curves give shorter times. John Bernoulli was responsible for stimulating general interest in the calculus of variations when in 1696 he challenged the scholars of his day to find the solution of this problem. He called the curve giving the quickest descent the **brachistochrone** ($\beta\rho\alpha\chi\acute{\iota}\sigma\tau o\varsigma$ = shortest, $\chi\rho\acute{o}\nu o\varsigma$ = time); it is a *cycloid* in this case (see Figure 6 in Section 4.2).

A related brachistochrone problem which seems to have occurred to Newton† is to find the **shortest transit time** that can be achieved by varying the path of a tunnel through the interior of the earth through which a bead slides under gravity from one fixed point on the earth's surface to another, as shown in Figure 2. Again one might think that the quickest time will be achieved through a straight tunnel connecting the two points. In fact, however, the shortest transit time is obtained by having the tunnel follow the path of a *hypocycloid* (see Figure 9 in Section 4.2). The resulting tunnel is longer than a

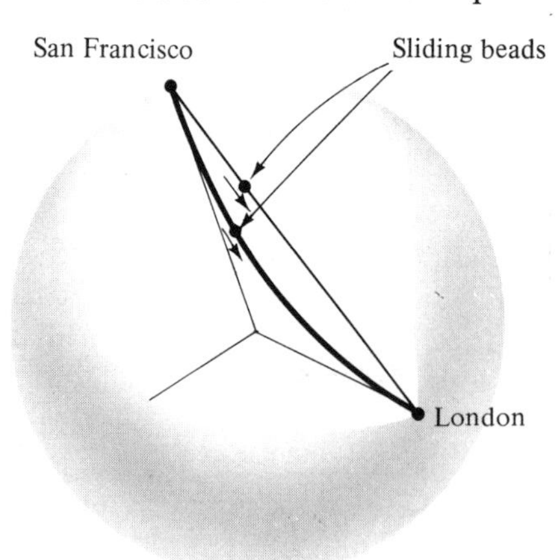

Figure 2

† See Morris Kline, *Calculus: An Intuitive and Physical Approach* (New York: John Wiley & Sons, Inc., 1967), p. 263.

straight tunnel between the points, but it starts out directed toward the center of the earth and this results in an initial gain in acceleration that more than offsets the increased length. For example, a straight tunnel through the earth connecting San Diego and San Francisco would be about 500 miles long with a maximum depth below the earth's surface of 8 miles at its midpoint, while the brachistochrone hypocycloid tunnel would in this case be about 685 miles long and reach a maximum depth of about 160 miles beneath the earth's surface. The bead could slide through the longer hypocycloid tunnel in only 12 minutes, while it would take about 42 minutes to slide through the straight tunnel. Of course this altogether neglects the effects of friction.

Another problem involving the shape of a curve is to find the **greatest area** that can be enclosed between a given straight line and an arbitrary curve of *fixed length* joining two points on the line when it is permitted to vary the shape of the curve but not its length, as indicated in Figure 3. Legend has it that this problem was solved intuitively by Queen Dido of Carthage in about 850 B.C. after she persuaded a North African chieftan to give her as much land as she could enclose within the hide of a bull (Virgil's *Aeneid*, Book I, line 367). She is reported to have cut the bull's hide into many thin strips, which she laid out along a *semicircle*, using the Mediterranean coast as a supplementary boundary. In this way she got the largest possible area in which to establish the state of Carthage.

A modern version of Queen Dido's problem was posed and solved by the Russian mathematician S. A. Chaplygin,† who considered the problem of finding the **greatest area** that can be encircled in a *given time* by varying the closed path of an airplane which flies at constant natural speed relative to the surrounding air while a constant wind blows. Chaplygin showed that the airplane should navigate against the wind so as to fly along an *ellipse* with its major axis perpendicular to the wind direction and its eccentricity given by the ratio of the wind speed to the aircraft speed, as indicated in Figure 4. The resulting elliptical path reduces, as expected, to a circular path if there is no wind blowing.

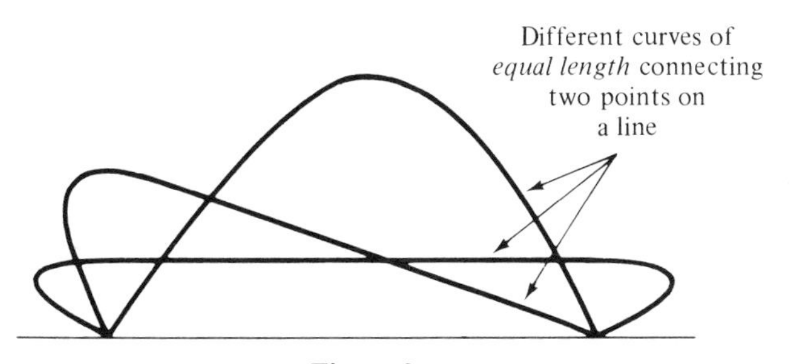

Figure 3

†See Naum I. Akhiezer, *The Calculus of Variations*, trans. Aline H. Frink (Waltham, Mass.: Ginn/Blaisdell, 1962), p. 206.

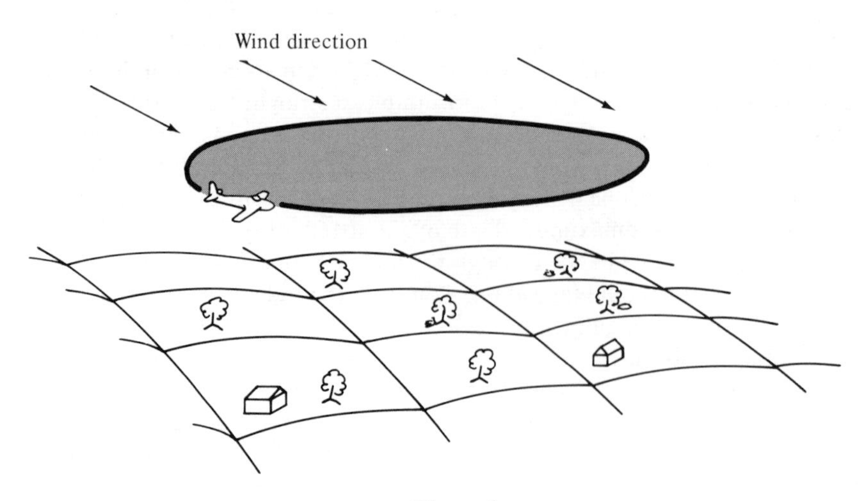

Figure 4

Another navigational problem is to find the **minimum transit time** that can be achieved by varying the path of a boat crossing an expanse of water from a given starting point to a given goal. It is assumed that the velocity of the water current is everywhere known, and the boat is assumed to have a constant source of motive power which allows it to move with a constant natural speed relative to the surrounding water. The German mathematician Ernst Zermelo (1871–1953) solved this problem in 1931.

Optimization problems play a prominent role in the fields of astronautics and rocket control. For example, if a rocket is launched from the surface of the earth, a rocket engineer may wish to find the **minimum time** required for the rocket to attain an altitude of 1 mile above the earth's surface if the rocket has a known fixed quantity of fuel and if the rocket experiences some resistance from the atmosphere of the earth. The problem is to find the best thrust rate at each instant of time so that the rocket will consume the fuel at the proper rate in order to minimize the total flight time. Or in another situation the rocket engineer may wish to find the **minimum amount of fuel** required for the rocket to achieve an altitude of 1 mile, where in this case the time of flight is unimportant.

Similar optimization problems also arise today in the fields of economics and business management. For example, a manufacturing company may wish to find the **smallest cost** that can be achieved by varying its production rate if the cost is known to depend variously on the expenses of maintaining a large inventory and on the expenses of making rapid changes in the production rate which might be required if the inventory level is low.† Or an individual may

†The author learned of this problem from C. L. Hwang, L. T. Fan, and L. E. Erickson, "Optimum Production Planning by the Maximum Principle," *Management Science*, **13** (May 1967), 751–755.

wish to find the **largest average consumption** that he can achieve over the next 10 years from his salary and his savings (which earn a given interest return) by varying his weekly consumption rate if he requires that a specified level of savings remain at the end of 10 years and if he assumes that inflation will occur at a constant rate so that the true value of paper exchange will decrease with each passing year. The individual may wish to consume little and save much during the early years so as to increase his savings and get a net increase in interest return over the 10-year period. In this way he may think that he can achieve a higher average consumption by consuming at a greater rate during the later years. On the other hand, if the inflation rate is high, the individual may think that he should consume his resources as quickly as possible.

Most of the problems discussed above cannot be solved using the tools and techniques of elementary calculus alone. In fact only the first-mentioned problem concerning the greatest range of a projectile can be solved by elementary calculus. In that case the range of the projectile can be represented as a real-valued function of the initial elevation angle, and every text on calculus shows how to use differentiation to find the greatest or smallest values of a function, based on the fact that the derivative of a smooth function gives the slope of the tangent line to the graph of the function, as illustrated in Figure 5. Since the graph must have a horizontal tangent line at any interior maximum or minimum point, it follows that the derivative of the function must vanish there. Of course the derivative also vanishes at horizontal inflection points, which may or may not be of interest. In practice one locates *all* points at which the derivative vanishes and then tests these points further along with any boundary points to find the desired extreme points. (See

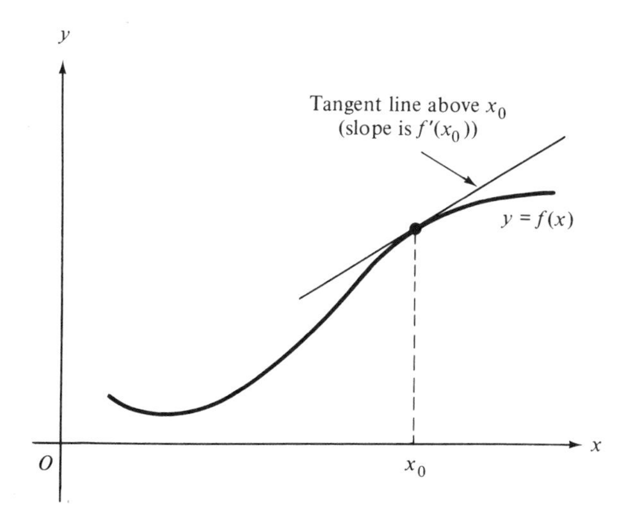

Figure 5

almost any text on calculus.) *We shall see in this book how to extend this familiar method so as to handle all the other problems mentioned above along with many other such optimization problems.*

1.2. Vector Spaces

A typical real-valued function f of the type usually encountered in elementary calculus is defined on some subset D of the set of all real numbers $\mathcal{R}$. In this case D is called the **domain** of f, and it is sometimes written as $D = D(f)$ to indicate its relation to f. If x is any number in D, then the expression $f(x)$ is used to denote the numerical value of the function f at x. For example, if f is the trigonometric *sine function*, then its value at x may be written as $f(x) = \sin x$, and in this case the domain D consists of the entire set of numbers $\mathcal{R}$. The value of the *natural logarithm function* at x may be written as $f(x) = \log x$, and its domain D consists of all *positive* numbers.

Although such functions f whose domains $D(f)$ are contained in the set of real numbers are adequate in dealing with the problems considered in elementary calculus, this is no longer true with most of the problems that we shall want to consider in this book. We shall need to consider real-valued functions defined on sets of objects *other* than numbers. For example, the *time of descent* of the bead in John Bernoulli's brachistochrone problem mentioned in Section 1.1 is presumably a function of the shape of the entire path followed by the bead, and this shape cannot be described by any single number.

In general all our functions will be defined on subsets of various **vector spaces** (or **linear spaces**, as they are sometimes called). The most common vector space is simply the set of all real numbers $\mathcal{R}$, so that the situation obtaining in elementary calculus will be included as a special case within our more general framework.

By a **vector space** (over the set of real numbers $\mathcal{R}$) we mean a set $\mathfrak{X}$ of elements $x, y, z, \ldots$ *of any kind*, referred to as **vectors**, for which the operations of addition of vectors and multiplication of vectors by real numbers $a, b, c, \ldots$ are defined and obey the following rules:

1. The sum $x + y$ is a vector in $\mathfrak{X}$ for every pair of vectors x and y in $\mathfrak{X}$.
2. The product ax is a vector in $\mathfrak{X}$ for every vector x in $\mathfrak{X}$ and for every number a in $\mathcal{R}$.
3. $x + y = y + x$ for any two vectors x and y in $\mathfrak{X}$.
4. $(x + y) + z = x + (y + z)$ for any vectors x, y, and z in $\mathfrak{X}$.
5. $\mathfrak{X}$ contains an element 0, called the *zero vector*, such that $x + 0 = x$ for every vector x in $\mathfrak{X}$.
6. $\mathfrak{X}$ contains, for every vector x, a vector $-x$ such that $x + (-x) = 0$.
7. $a(bx) = (ab)x$ for any a, b in $\mathcal{R}$ and any x in $\mathfrak{X}$.

8. $a(x + y) = ax + ay$ for any a in $\mathcal{R}$ and any x, y in $\mathcal{X}$.
9. $(a + b)x = ax + bx$ for any a, b in $\mathcal{R}$ and any x in $\mathcal{X}$.
10. $1x = x$ for all x in $\mathcal{X}$.

It is clear that $\mathcal{R}$ is a vector space with the usual definitions of addition and multiplication. Similarly, *n-dimensional Euclidean space* $\mathcal{R}_n$, consisting of all ordered n-tuples of real numbers $x = (x_1, x_2, \ldots, x_n)$, is a vector space with addition defined as

$$(x_1, x_2, \ldots, x_n) + (y_1, y_2, \ldots, y_n) = (x_1 + y_1, x_2 + y_2, \ldots, x_n + y_n)$$

for every $x = (x_1, x_2, \ldots, x_n)$ and $y = (y_1, y_2, \ldots, y_n)$ in $\mathcal{R}_n$, and with multiplication by numbers defined as

$$a(x_1, x_2, \ldots, x_n) = (ax_1, ax_2, \ldots, ax_n)$$

for any vector $x = (x_1, x_2, \ldots, x_n)$ and any number a.

Another example of a vector space is the set $\mathcal{X}$ of all real-valued functions defined on some fixed interval I of numbers. For any two such functions ϕ and ψ we define their sum $\phi + \psi$ by the formula

$$(\phi + \psi)(x) = \phi(x) + \psi(x) \tag{1.2.1}$$

for any x in I, while for any number a we define the product $a\phi$ by the formula

$$(a\phi)(x) = a\phi(x) \tag{1.2.2}$$

for any x in I. The addition $\phi(x) + \psi(x)$ and multiplication $a\phi(x)$ appearing on the right-hand sides of formulas (1. 2. 1) and (1. 2. 2) are the usual addition and multiplication of real numbers. Clearly the sum $\phi + \psi$ and the product $a\phi$ so defined are real-valued functions on I (and hence vectors in $\mathcal{X}$) for any vectors ϕ and ψ in $\mathcal{X}$ and for any number a. Moreover, one checks easily that all the rules listed above for vector spaces are satisfied in this case. The zero vector is the *zero function*, which is the function which has the value 0 everywhere on I.

If $\mathcal{X}$ is any fixed vector space and if $\mathcal{Y}$ is a subset of $\mathcal{X}$ such that $x + y$ and ax are in $\mathcal{Y}$ for every x and y in $\mathcal{Y}$ and for every number a, then it is clear that $\mathcal{Y}$ is itself a vector space with the same operations of addition and multiplication by numbers as inherited from $\mathcal{X}$. In this case $\mathcal{Y}$ is said to be a **subspace** of $\mathcal{X}$.

For example, the set of all n-tuples of numbers $x = (x_1, x_2, \ldots, x_n)$ with $x_1 = 0$ is a subspace of n-dimensional Euclidean space $\mathcal{R}_n$. An important subspace of the vector space of all real-valued functions on some fixed interval I [see formulas (1.2.1) and (1.2.2) for the definitions of addition and multiplication by numbers] is given by the set of all such functions which have continuous derivatives of all orders up to and including kth order. Here k may be any fixed nonnegative integer, and the resulting subspace is denoted as $\mathcal{C}^k(I)$ or $\mathcal{C}^k[a, b]$ if the underlying interval is $I = [a, b]$. We shall often refer to vectors in $\mathcal{C}^k(I)$ as **functions of class $\mathcal{C}^k$ on I.**

Note that any vector space $\mathfrak{X}$ must have a relation of **equality** (=), which we have taken for granted above. For example, any useful construction of the real numbers $\mathcal{R}$ (from the positive integers, say) must include adequate instructions which tell how to show whether two numbers are equal.† (For example, the numbers $\frac{3}{2}$ and $\frac{6}{4}$ are equal, as we all know.) In $\mathcal{R}_n$ two vectors $x = (x_1, \ldots, x_n)$ and $y = (y_1, \ldots, y_n)$ are equal whenever their *coordinates* x_i and y_i are equal for each i. Whenever we say that two vectors ϕ and ψ in $\mathcal{C}^k(I)$ are equal, we shall mean that the values $\phi(x)$ and $\psi(x)$ of the two functions are equal for each x in I.

Exercises

1. Verify that the set of all real-valued functions on the interval $a \leq x \leq b$ is a vector space with the operations defined by formulas (1.2.1) and (1.2.2).

2. Let $\mathcal{Y}$ be a subset of a given vector space $\mathfrak{X}$ with the property that $x + y$ and ax are in $\mathcal{Y}$ for every pair of vectors x and y in $\mathcal{Y}$ and for every number a. Prove that $\mathcal{Y}$ is itself a vector space with the same operations of addition of vectors and multiplication of vectors by numbers as inherited from $\mathfrak{X}$.

3. Let $x_1, x_2, \ldots, x_k$ be any given k vectors in a vector space $\mathfrak{X}$, and let $a_1, a_2, \ldots, a_k$ be any k real numbers. Does the expression $a_1x_1 + a_2x_2 + \cdots + a_kx_k$ necessarily give a vector in $\mathfrak{X}$?

4. Let $x_1, x_2, \ldots, x_k$ be any given k vectors in a vector space $\mathfrak{X}$, and let $\mathcal{Y}$ be the set of *all* vectors x of the form $x = a_1x_1 + a_2x_2 + \cdots + a_kx_k$, where $a_1, a_2, \ldots, a_k$ may be any arbitrary real numbers. Show that $\mathcal{Y}$ is a subspace of $\mathfrak{X}$.

5. Which of the following subsets of $\mathcal{R}_3$ are subspaces?
(a) The set of all vectors $x = (x_1, x_2, x_3)$ with $x_2 + x_3 = 0$.
(b) The set of all vectors $x = (x_1, x_2, x_3)$ with $x_1 = 0$.
(c) The set of all vectors with $x_2 + x_3 = 0$ *and* $x_1 = 0$.
(d) The vectors with $x_1 = 1$.
(e) The vectors with $x_3 = 2x_1 - x_2$.
(f) The vectors with $x_1 = x_2x_3$.

6. Let $\mathcal{C}^0[a, b]$ be the vector space of all continuous real-valued functions defined on the interval $a \leq x \leq b$. Which of the following subsets of $\mathcal{C}^0[a, b]$ are subspaces?
(a) The set of all functions ϕ such that $\phi(a) = 0$ and $\phi(b) = 0$.
(b) The set of all functions ϕ such that $\phi(a) = 0$ and $\phi(b) = 1$.

† The author has learned much about the nature of equality from Errett Bishop, *Foundations of Constructive Analysis* (New York: McGraw-Hill Book Company, 1967). Bishop gives a beautiful development of the real numbers in Chapter 2 of this reference.

(c) The functions with $\phi(a) + 2\phi(b) = 0$.
(d) The functions with $\phi(a) - \phi(b) = 1$.

1.3. Functionals

Throughout the rest of this book we shall be considering various real-valued functions J whose domains $D(J)$ will be certain specified subsets of certain given vector spaces. Such functions are called **functionals**.

For example, let D be the set of all *positive-valued* continuous functions $\phi = \phi(x)$ on the interval $0 \leq x \leq \pi/2$, and define the functional J by

$$J(\phi) = \int_0^{\pi/2} \sqrt{\phi(x) \sin x}\, dx$$

for any ϕ in D. Here D is a subset of the vector space $\mathcal{C}^0[0, \pi/2]$ of *all* continuous functions ϕ on $[0, \pi/2]$.

A Brachistochrone Functional. As another example, consider again the time of descent of a bead along a wire joining two nearby fixed points, as mentioned briefly in Section 1.1 in connection with John Bernoulli's brachistochrone problem. In this case we can represent the wire as a smooth curve γ in the (x, y)-plane joining two fixed points $P_0 = (x_0, y_0)$ and $P_1 = (x_1, y_1)$, as illustrated in Figure 6. Then the *time* T required for a bead to move from P_0 down to P_1 along γ is given by the line integral

$$T = \int_0^T dt = \int_\gamma \frac{ds}{v}, \tag{1.3.1}$$

where s measures the *arc length* along γ, ds/dt is the rate of change of arc length with respect to time t, and the *speed of motion* is

$$v = \frac{ds}{dt}.$$

For definiteness we assume that the earth's gravitational force acts *down* along the negative y-direction, as indicated in Figure 6, with the x-component zero and the y-component given by $-g$, where g is the acceleration due to gravity, assumed to be constant. Then a bead located at the position (x, y) and sliding along γ under the force of gravity (assume that $y_0 > y_1$) will have kinetic energy of motion given as

$$\text{kinetic energy} = \tfrac{1}{2}mv^2$$

and potential energy given as

$$\text{potential energy} = mgy,$$

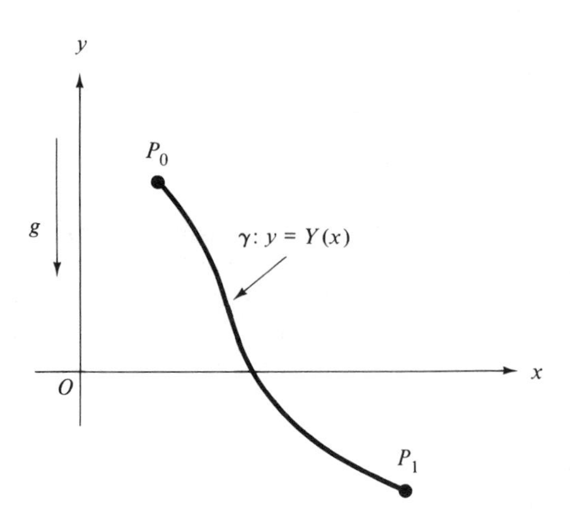

Figure 6

where m is the mass of the bead. *Conservation of energy*† requires that the sum of the kinetic and potential energies remains constant during the motion (we neglect friction), so that if the bead starts from *rest* at P_0 with zero initial kinetic energy and initial potential energy equal to mgy_0, it follows that during the motion the relation

$$\tfrac{1}{2}mv^2 + mgy = mgy_0 \tag{1.3.2}$$

must hold. If the curve γ is represented parametrically as

$$\gamma: \quad y = Y(x), \qquad x_0 \leq x \leq x_1,$$

for some suitable function $Y(x)$ relating x and y along γ, then equation (1.3.2) can be solved for the speed along γ as

$$v = \sqrt{2g[y_0 - Y(x)]},$$

while the differential element of arc length along γ is then given as‡

$$ds = \sqrt{1 + Y'(x)^2}\, dx,$$

where $Y'(x)$ denotes the derivative of the function $Y(x)$. Hence the integral appearing in equation (1.3.1) which gives the time of motion for the bead to

†The conservation of energy for such motions in the presence of conservative force fields follows easily from Newton's law of motion and the definitions of kinetic and potential energy. See Section 4.8.

‡Sherman K. Stein, *Calculus in the First Three Dimensions* (New York: McGraw-Hill Book Company, 1967), p. 208.

slide from P_0 down to P_1 can be written as

$$\int_\gamma \frac{ds}{v} = \int_{x_0}^{x_1} \sqrt{\frac{1 + Y'(x)^2}{2g[y_0 - Y(x)]}}\, dx,$$

which may be considered to be a functional with domain D given by the set of all continuously differentiable functions $Y = Y(x)$ on the interval $[x_0, x_1]$ satisfying the constraints $Y(x_0) = y_0$ and $Y(x_1) = y_1$. The value $T = T(Y)$ of this brachistochrone functional is then

$$T(Y) = \int_{x_0}^{x_1} \sqrt{\frac{1 + Y'(x)^2}{2g[y_0 - Y(x)]}}\, dx \tag{1.3.3}$$

for any Y in $D = D(T)$. Note that the domain $D(T)$ is a subset of the vector space $\mathcal{C}^1[x_0, x_1]$ of *all* continuously differentiable functions on $[x_0, x_1]$.

An Area Functional. As another example, consider Chaplygin's problem on the greatest area that can be encircled in a *given* time T by varying the closed path γ flown by an airplane at *constant natural speed* v_0 while a constant wind blows. The airplane is assumed to be flying in a fixed horizontal plane surface above the surface of the earth which we take to be the (x, y)-coordinate plane. We represent the closed path of the airplane in the fixed coordinate plane parametrically as

$$\gamma: \begin{cases} x = X(t) \\ y = Y(t) \qquad \text{for } 0 \leq t \leq T \end{cases}$$

for suitable functions $X(t)$ and $Y(t)$, where the parameter t is *time*, and where in order to close the path in time T we require that (see Figure 7)

$$X(0) = X(T), \qquad Y(0) = Y(T). \tag{1.3.4}$$

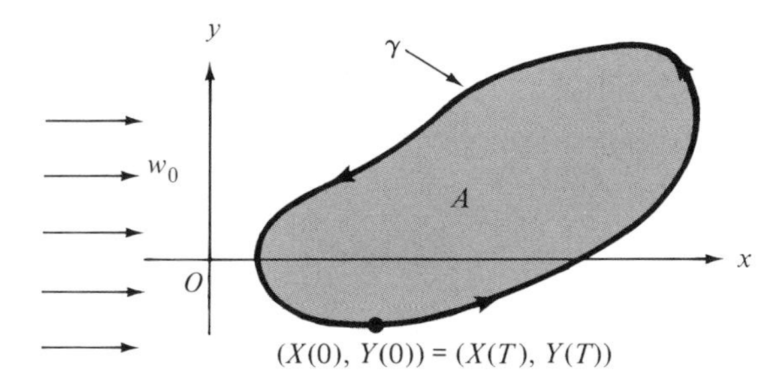

Figure 7

The *area A* enclosed by the path of the airplane is given by†

$$A = \frac{1}{2}\int_0^T [X(t)Y'(t) - Y(t)X'(t)]\,dt. \tag{1.3.5}$$

The origin of coordinates O may be taken to be some fixed point, and for simplicity we suppose that the coordinate plane has been rotated about this point once and for all so that the x-direction coincides with the fixed wind direction. Finally we let w_0 be the *constant wind speed*, with $0 \leq w_0 \leq v_0$, where the airplane flies with constant natural speed v_0 relative to the surrounding air. We let $\alpha = \alpha(t)$ be the *steering angle* between the positive x-direction and the direction of the axis of the airplane, as indicated in Figure 8. The velocity of the airplane *relative to the surrounding air* will be given by the vector whose x-component equals $v_0 \cos\alpha$ and y-component equals $v_0 \sin\alpha$. The *absolute velocity* of the airplane relative to the ground will be obtained by adding the wind velocity to the airplane's velocity relative to the air. In this way we find that

$$\begin{aligned} X'(t) &= v_0 \cos\alpha(t) + w_0 \\ Y'(t) &= v_0 \sin\alpha(t), \end{aligned} \tag{1.3.6}$$

where the velocity vector $(X'(t), Y'(t))$ is obtained by differentiating the position vector $(x, y) = (X(t), Y(t))$ along γ.

We may integrate both sides of the equations of (1.3.6) with respect to time to find that

$$\begin{aligned} X(t) &= x_0 + v_0 \int_0^t \cos\alpha(\tau)\,d\tau + w_0 t \\ Y(t) &= y_0 + v_0 \int_0^t \sin\alpha(\tau)\,d\tau, \end{aligned} \tag{1.3.7}$$

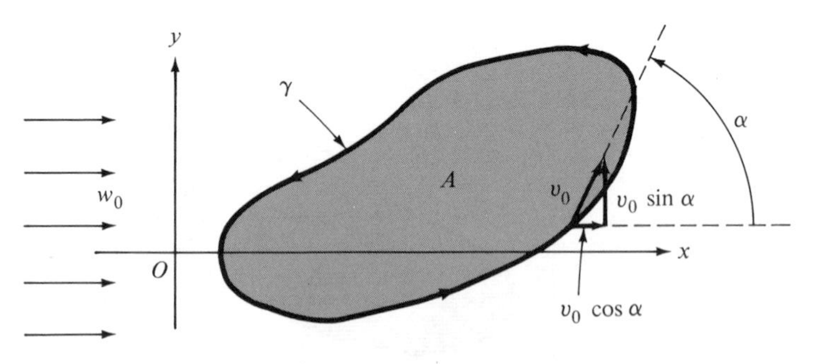

Figure 8

† *Ibid.*, p. 433.

where we have set

$$X(0) = x_0, \qquad Y(0) = y_0. \tag{1.3.8}$$

Then the requirement (1.3.4) with (1.3.7) and (1.3.8) implies the following conditions (explain):

$$\begin{aligned} \int_0^T \cos \alpha(t)\, dt &= -\frac{w_0}{v_0} T \\ \int_0^T \sin \alpha(t)\, dt &= 0. \end{aligned} \tag{1.3.9}$$

These conditions (1.3.9) are *constraints* which must be satisfied by all *admissible* steering control angles $\alpha = \alpha(t)$. We might also impose the additional constraint that the *initial* steering angle $\alpha(0)$ be specified as

$$\alpha(0) = \alpha_0 \tag{1.3.10}$$

for some given initial angle α_0.

If we now substitute equations (1.3.6) and (1.3.7) into (1.3.5), for the area we find that

$$\begin{aligned} A(\alpha) = \frac{1}{2} \int_0^T \Big\{ & v_0 \sin \alpha(t) \Big[x_0 + w_0 t + v_0 \int_0^t \cos \alpha(\tau)\, d\tau \Big] \\ & - [v_0 \cos \alpha(t) + w_0] \Big[y_0 + v_0 \int_0^t \sin \alpha(\tau)\, d\tau \Big] \Big\}\, dt, \end{aligned} \tag{1.3.11}$$

where we have written $A(\alpha)$ to indicate that the area is a functional of the steering control angle α. The domain $D = D(A)$ of the functional defined by (1.3.11) might be taken to be the set of all continuous functions $\alpha = \alpha(t)$ on the interval $0 \leq t \leq T$ satisfying the constraints of (1.3.9) and (1.3.10). Hence $D(A)$ is a subset of the vector space $\mathcal{C}^0[0, T]$.

A Transit Time Functional. Another example of a functional is given by the transit time of a boat crossing a river from a fixed initial point on one bank to a specified terminal point on the other bank.† For simplicity we assume that the river has parallel banks, as shown in Figure 9, where we let the y-axis coincide with the left bank. The river is l units wide so that the right bank coincides with the line $x = l$. We consider a river without cross currents so that the current velocity is everywhere directed downstream along the y-direction. Moreover, we assume that this *downstream current speed* w depends only on x as

$$w = w(x),$$

†The author first learned of this problem from Hans Sagan, *Introduction to the Calculus of Variations* (New York: McGraw-Hill Book Company, 1969), p. 6.

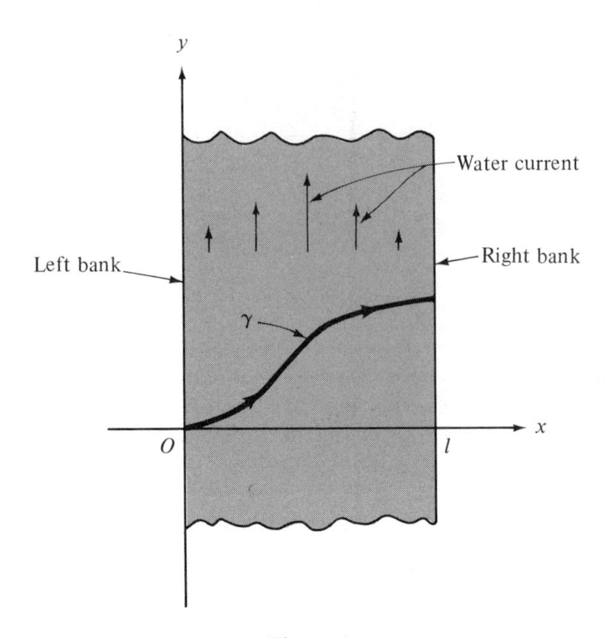

Figure 9

where $w(x)$ is a known function defined for $0 \leq x \leq l$. The boat travels at a *constant natural speed* v_0 relative to the surrounding water. If the path of the boat is represented by a curve γ given parametrically as

$$\gamma: \begin{cases} x = \xi(t) \\ y = \eta(t) \qquad \text{for } 0 \leq t \leq T \end{cases}$$

for suitable functions $\xi(t)$ and $\eta(t)$, it then follows that the components of the absolute velocity of the boat will be given along γ as [compare with (1.3.6)]

$$\begin{aligned} \frac{dx}{dt} &= \xi'(t) = v_0 \cos \alpha(t) \\ \frac{dy}{dt} &= \eta'(t) = v_0 \sin \alpha(t) + w(\xi(t)), \end{aligned} \tag{1.3.12}$$

where $\alpha = \alpha(t)$ is the *steering angle* of the boat measured between the positive x-direction and the direction of the axis of the boat, as indicated in Figure 10.

The *time of transit* T of the boat may be given by the integral

$$T = \int_0^T dt = \int_0^l \frac{dt}{dx}\, dx,$$

where $dx/dt = \xi'(t)$ is given by the first equation of (1.3.12). In this way for

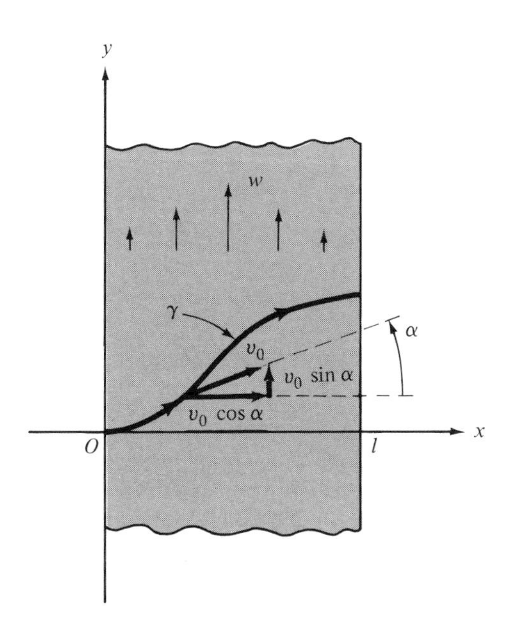

Figure 10

the time T we find that

$$T = \int_0^l \frac{dx}{v_0 \cos \alpha}, \tag{1.3.13}$$

where now we shall use x as the parameter along γ as

$$\gamma: \quad y = Y(x) \qquad \text{for } 0 \leq x \leq l, \tag{1.3.14}$$

where the function $Y(x)$ is obtained from the previous functions $x = \xi(t)$ and $y = \eta(t)$ as $Y(x) = \eta(\xi^{-1}(x))$, where ξ^{-1} is the inverse function of ξ. Along γ the chain rule of differentiation† gives

$$Y'(x) = \frac{\eta'(t)}{\xi'(t)} \qquad \left(\text{i.e., } \frac{dy}{dx} = \frac{dy/dt}{dx/dt}\right),$$

so that the equations of (1.3.12) imply that

$$Y'(x) = \frac{\sin \alpha + e(x)}{\cos \alpha},$$

where we have set

$$e(x) = \frac{w(x)}{v_0}. \tag{1.3.15}$$

†Stein, *op. cit.*, p. 103.

Hence along γ we have

$$\begin{aligned} Y' \cos \alpha &= e + \sin \alpha \\ &= e \pm \sqrt{1 - \cos^2 \alpha}, \end{aligned}$$

which can be solved for $\cos \alpha$ to give

$$\cos \alpha = \frac{1 - e(x)^2}{\sqrt{1 - e(x)^2 + Y'(x)^2} - e(x)Y'(x)}. \tag{1.3.16}$$

Finally, then, for the transit time, equations (1.3.13) and (1.3.16) give

$$T(Y) = \frac{1}{v_0} \int_0^l \frac{\sqrt{1 - e(x)^2 + Y'(x)^2} - e(x)Y'(x)}{1 - e(x)^2} \, dx, \tag{1.3.17}$$

where we have written $T = T(Y)$ to indicate that the transit time is a functional depending on the particular curve γ given in terms of the function $Y = Y(x)$ by (1.3.14). The domain of the functional defined by (1.3.17) may by taken to be the vector space $\mathcal{C}^1[0, l]$ consisting of all continuously differentiable functions $Y(x)$ on the fixed interval $0 \leq x \leq l$.

A Cost Functional. As a final example, we consider a problem in production planning that has application to certain manufacturing companies. We consider a company that manufactures and sells some particular product. We assume that the company has on hand sufficient long-term orders for its product so that it can predict its future *sales rate* $\mathcal{S}$ with certainty (at least over some fixed time period). In this case the sales rate $\mathcal{S}$ is a given function of time t, say

$$\mathcal{S} = \mathcal{S}(t), \tag{1.3.18}$$

for some known function $\mathcal{S}(t)$.

If the product is completely durable, it is natural to assume that the *production rate* $P(t)$ and the finished product *inventory level* $I(t)$ are related by the differential equation

$$\dot{I} = P - \mathcal{S},$$

which simply states that the time *rate of change* of the inventory level ($\dot{I} = dI/dt$) is equal to the difference of the production and sales rates. We allow for some spoilage of the inventory by considering instead the equation

$$\dot{I} = P - (\mathcal{S} + \alpha I), \tag{1.3.19}$$

which states that the production is depleted not only by sales but also by certain spoilage which is proportional to the inventory level on hand. For simplicity we assume that the *spoilage proportionality factor* α is a given *constant*.

Finally, we assume that on the basis of the known sales forecast $\mathcal{S}$ of formula (1.3.18) the company has decided on a *desired inventory level* $\mathcal{I}(t)$, resulting in a corresponding *desired production rate* $\mathcal{P}(t)$, obtained from equation (1.3.19) as

$$\mathcal{P} = \dot{\mathcal{I}} + \mathcal{S} + \alpha\mathcal{I}. \tag{1.3.20}$$

For example, the company might choose $\mathcal{I} = 0$ with $\mathcal{P} = \mathcal{S}$; i.e., the company might wish to have production equal sales.

We now suppose that the inventory level (but not the sales rate $\mathcal{S}$) has been *disturbed* away from its desired level $\mathcal{I}$ (perhaps by a malfunction at the manufacturing plant), and we seek a *new production rate* $P = P(t)$ that will guide the inventory level back toward the desired level in a specified time period T so as to minimize a given *cost functional* C, which might depend, for example, on the *deviations* of the inventory level I and the production rate P away from their known desired levels $\mathcal{I}$ and $\mathcal{P}$. In fact, for definiteness and simplicity we shall take the cost functional C to be defined by

$$C = \int_0^T \{\beta^2[I(t) - \mathcal{I}(t)]^2 + [P(t) - \mathcal{P}(t)]^2\}\, dt, \tag{1.3.21}$$

where β is a fixed constant which the company might want to specify so as to give different relative weights to the *unwanted deviations* of the inventory level and production rate away from their known desired levels.

Note that the absolute minimum cost obtainable with formula (1.3.21) is $C = 0$, obtained by taking $I = \mathcal{I}$ and $P = \mathcal{P}$. The company's difficulty arises from the fact that the actual inventory level has already been disturbed initially away from $\mathcal{I}$, so that at time $t = 0$

$$I(0) = I_0 \tag{1.3.22}$$

holds for some given nonnegative constant I_0 with

$$I_0 \neq \mathcal{I}_0 = \mathcal{I}(0).$$

The difference $I_0 - \mathcal{I}_0$ furnishes a measure of the initial disturbance away from the desired state. In this situation every admissible new production rate $P = P(t)$ leads with equations (1.3.19) and (1.3.22) to a *positive* cost C in formula (1.3.21) which the company hopes to minimize. In fact, equation (1.3.19) can be integrated† with (1.3.22) to give

$$I(t) = e^{-\alpha t}\left(I_0 + \int_0^t e^{\alpha\tau}[P(\tau) - \mathcal{S}(\tau)]\, d\tau\right),$$

† See, for example, Chapter 20 of George B. Thomas, Jr., *Calculus and Analytic Geometry* (Reading, Mass.: Addison-Wesley Publishing Company, Inc., 1968), or any elementary text on differential equations.

so that the inventory level $I = I(t)$ is already determined once the future production rate $P = P(t)$ is specified. In this case it is convenient to write $I = I_P$ to indicate this dependency of I on P, where now

$$I_P(t) = e^{-\alpha t}\left(I_0 + \int_0^t e^{\alpha\tau}[P(\tau) - \mathcal{S}(\tau)]\, d\tau\right). \tag{1.3.23}$$

The cost functional C given by formula (1.3.21) can now be thought of as a functional of the production rate P, denoted as $C = C(P)$, where

$$C(P) = \int_0^T \{\beta^2[I_P(t) - \mathcal{I}(t)]^2 + [P(t) - \mathcal{P}(t)]^2\}\, dt, \tag{1.3.24}$$

and where I_P is given by formula (1.3.23). The domain of the functional $C = C(P)$ can be taken to be the vector space $\mathcal{C}^0[0, T]$ consisting of all continuous functions $P = P(t)$ on the interval $0 \leq t \leq T$, or some specified subset of $\mathcal{C}^0[0, T]$ if we wish to exclude certain production rates. The problem of the manufacturing company would be to choose the particular production rate P that will minimize the cost functional $C = C(P)$.

1.4. Normed Vector Spaces

We shall repeatedly be interested in the problem of maximizing or minimizing the value of some given functional defined on some specified subset of a suitable vector space. The special case in which the vector space is the set of real numbers is studied in elementary differential calculus, where an important role is played in such extremum problems by the notion of **distance** between numbers. In that case, distance is defined in terms of the **absolute value function** $|\cdot|$, which is defined for any number x by

$$|x| = \begin{cases} x & \text{if } x \geq 0 \\ -x & \text{if } x < 0. \end{cases} \tag{1.4.1}$$

The *distance between any two numbers x and y* is then given by the *absolute value of their difference*:

$$\text{distance between } x \text{ and } y = |x - y|. \tag{1.4.2}$$

We shall need some such similar notion of **distance between vectors** in a more general vector space in order to solve the optimization problems considered in this book. Such a distance concept can be defined in terms of a **norm function** defined on the vector space, which takes the place of the absolute value function of formula (1.4.1). We shall use the notation $\|\cdot\|$ to denote such a norm function, much as we use the notation $|\cdot|$ to denote the absolute value function on $\mathcal{R}$.

A vector space $\mathfrak{X}$ is said to be a **normed vector space** whenever there is a real-valued norm function $||\cdot||$ defined on $\mathfrak{X}$ which assigns the number $||x||$ (called the **norm of** x, or the **length of** x) to the vector x in $\mathfrak{X}$ such that

1. $||x|| \geq 0$ for all vectors x in $\mathfrak{X}$, and $||x|| = 0$ if and only if x is the zero vector in $\mathfrak{X}$.
2. $||ax|| = |a|\,||x||$ for every x in $\mathfrak{X}$ and every number a in $\mathcal{R}$.
3. $||x + y|| \leq ||x|| + ||y||$ for every pair of vectors x and y in $\mathfrak{X}$.

The first condition here simply states that the length or norm of every vector in $\mathfrak{X}$ is positive except for the zero vector in $\mathfrak{X}$ which has zero length. The second condition ensures, among other things, that the length of the vector $-x$ is the same as the length of x. The last condition asserts that the length of the sum $x + y$ can never exceed the sum of the lengths of x and y and is called the triangle inequality (*the length of one side of a triangle is always less than or equal to the sum of the lengths of the other two sides*; see Figure 11).

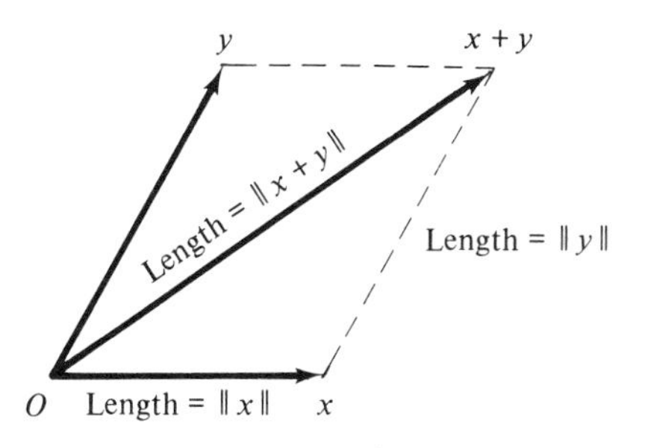

Figure 11

We now define the **distance between any two vectors** x **and** y of a normed vector space $\mathfrak{X}$ to be the *length of their difference* [compare with formula (1.4.2)]:

$$\text{distance between } x \text{ and } y = ||x - y||. \tag{1.4.3}$$

It is easy to check that the set of all numbers $\mathcal{R}$ is a normed vector space with norm given by the absolute value function of formula (1.4.1); i.e., $||x|| = |x|$ for any number x. Similarly, n-dimensional Euclidean space $\mathcal{R}_n$ is a normed vector space with norm function defined by

$$||x|| = \sqrt{x_1^2 + x_2^2 + \cdots + x_n^2} \tag{1.4.4}$$

for any vector $x = (x_1, x_2, \ldots, x_n)$ in $\mathcal{R}_n$. The triangle inequality in this case can be seen to follow from **Cauchy's inequality**,

$$\left(\sum_{i=1}^{n} x_i y_i\right)^2 \leq \left(\sum_{i=1}^{n} x_i^2\right)\left(\sum_{j=1}^{n} y_j^2\right), \tag{1.4.5}$$

which holds for all n-tuples $x = (x_1, x_2, \ldots, x_n)$ and $y = (y_1, y_2, \ldots, y_n)$ in $\mathcal{R}_n$. A proof of Cauchy's inequality (1.4.5) is given in Section A1 of the Appendix.

The vector space $\mathcal{C}^0(I)$ consisting of all continuous real-valued functions ϕ defined on some fixed interval $I = [a, b]$ with addition and multiplication by numbers defined by formulas (1.2.1) and (1.2.2) can be made into a normed vector space with norm defined by

$$\|\phi\| = \sqrt{\int_a^b |\phi(x)|^2 \, dx} \tag{1.4.6}$$

for any vector ϕ in $\mathcal{C}^0(I)$. The triangle inequality can be proved using **Schwarz's inequality**,

$$\left(\int_a^b \phi(x)\psi(x) \, dx\right)^2 \leq \int_a^b \phi(x)^2 \, dx \int_a^b \psi(x)^2 \, dx, \tag{1.4.7}$$

which is proved in Section A1 of the Appendix. The norm (1.4.6) is usually referred to as the $\boldsymbol{L_2}$ **norm** on $\mathcal{C}^0(I)$.

Another candidate for a norm function on $\mathcal{C}^0(I) = \mathcal{C}^0[a, b]$ is given by

$$\|\phi\| = \max_{a \leq x \leq b} |\phi(x)| \tag{1.4.8}$$

for any ϕ. One can check that (1.4.8) satisfies all the conditions to be a norm, so that formulas (1.4.6) and (1.4.8) give *two distinct norms* for the vector space $\mathcal{C}^0[a, b]$. It is customary to refer to the latter norm (1.4.8) as the **uniform norm** on $\mathcal{C}^0$. We give an example in Section A2 of the Appendix which shows how radically different are the two normed vector spaces induced on the vector space $\mathcal{C}^0[a, b]$ by these two norms. Hence a given vector space $\mathfrak{X}$ may lead to more than one distinct normed vector space since there may be more than one norm on $\mathfrak{X}$. It is customary to distinguish between such distinct normed vector spaces arising from a single vector space by labeling them separately with different names. For example, we might use $\mathfrak{X}_1$ to denote the normed vector space given by the vector space $\mathcal{C}^0[a, b]$ equipped with the L_2 norm (1.4.6) and then use $\mathfrak{X}_2$ to denote the normed vector space given by the same vector space $\mathcal{C}^0[a, b]$ equipped with the uniform norm (1.4.8). Whenever we speak of $\mathcal{C}^0[a, b]$ as a *normed* vector space, we shall always be careful to specify the particular norm being used, so that no confusion should result.

A useful norm function on the vector space $\mathcal{C}^k(I)$ for any fixed interval $I = [a, b]$ and any fixed nonnegative integer k (see the second to last paragraph of Section 1.2 for the definition of this vector space) is given by

$$\|\phi\| = \max_{\text{all } x \text{ in } I} |\phi(x)| + \max_{\text{all } x \text{ in } I} |\phi'(x)| + \cdots + \max_{\text{all } x \text{ in } I} |\phi^{(k)}(x)| \tag{1.4.9}$$

for any function ϕ of class $\mathcal{C}^k(I)$, where $\phi'(x) = d\phi(x)/dx$, and $\phi^{(i)}(x) = d^i\phi(x)/dx^i$ for $i = 2, 3, \ldots, k$. Note that formula (1.4.9) reduces to formula (1.4.8) in the case $k = 0$. Note also that the *distance* between two vectors ϕ and ψ in $\mathcal{C}^k(I)$, measured as $\|\phi - \psi\|$ with the norm of (1.4.9) [see (1.4.3)], is *small* whenever the values of the functions $\phi = \phi(x)$ and $\psi = \psi(x)$ are uniformly close to one another for all x in $I = [a, b]$ along with the values of their derivatives through order k.

We note in general that *any subspace* $\mathcal{Y}$ *of any given normed vector space* $\mathcal{X}$ *is itself a normed vector space with the same norm as used on* $\mathcal{X}$. For example, let $\mathcal{Y}$ be the set of all functions of class $\mathcal{C}^0$ on I which vanish at some fixed specified point c in I. Then $\mathcal{Y}$ is clearly a subspace of the vector space $\mathcal{C}^0(I)$ (why?), and we can consider $\mathcal{Y}$ to be a normed vector space with any norm function already available on $\mathcal{C}^0(I)$ (as, for example, the L_2 norm function or the uniform norm function).

Finally, it is convenient to introduce the notion of a **ball** in a normed vector space, which extends to a general normed vector space something like the notion of an **interval** from the special case in which the vector space is the set of real numbers. Since an open interval of numbers $I = (a, b)$ can be thought of as being the set of all numbers having distance less than its *radius* $(b - a)/2$ from its *center* $(a + b)/2$ (see Figure 12), it is then natural to define for any positive number ρ and any vector x in a normed vector space $\mathcal{X}$ the **ball of radius ρ centered at** x to be the set of all vectors y in $\mathcal{X}$ having distance from x less than ρ. If we denote such a ball as $B_\rho(x)$, we have then (by definition)

$$B_\rho(x) = \{\text{set of all vectors } y \text{ in } \mathcal{X} \text{ satisfying } \|y - x\| < \rho\}.$$

For example, the ball $B_\rho(0)$ of radius ρ centered at the zero vector in the normed vector space $\mathcal{C}^0[a, b]$ with the uniform norm consists of the set of all continuous functions ϕ on the interval $[a, b]$ satisfying

$$\max_{a \leq x \leq b} |\phi(x)| < \rho;$$

such a function is shown, for example, in Figure 13.

A subset D of a normed vector space $\mathcal{X}$ is said to be **open** in $\mathcal{X}$ whenever D contains along with each of its elements x some ball $B_\rho(x)$ in $\mathcal{X}$ centered at x for some positive ρ which may depend on x. Note that a subspace $\mathcal{Y}$ of a normed vector space $\mathcal{X}$ is *not* open in $\mathcal{X}$ unless $\mathcal{Y}$ is actually all of $\mathcal{X}$. For example, the subspace of $\mathcal{R}_2$ consisting of all 2-tuples $x = (x_1, x_2)$ with $x_2 = 0$

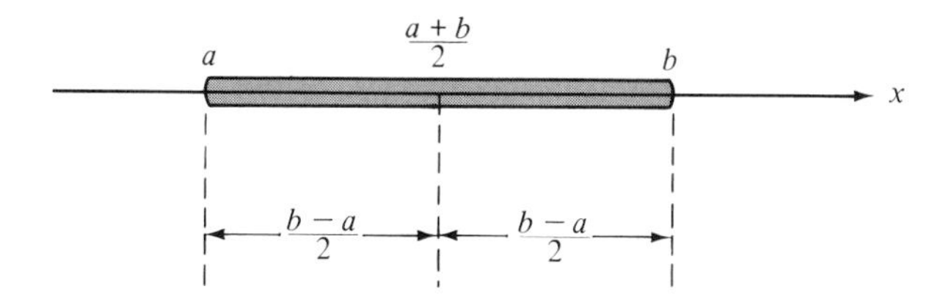

Figure 12

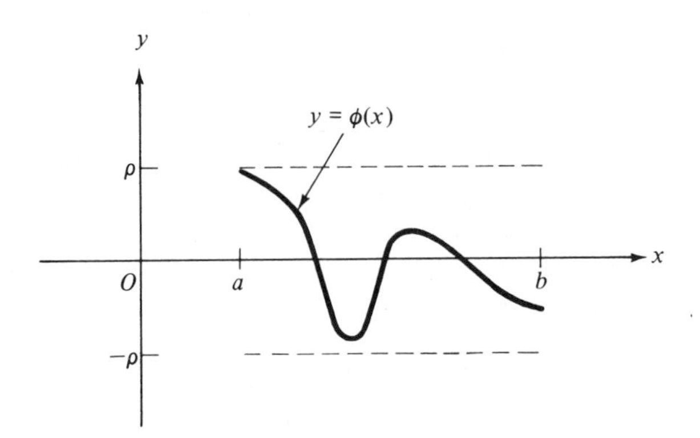

Figure 13

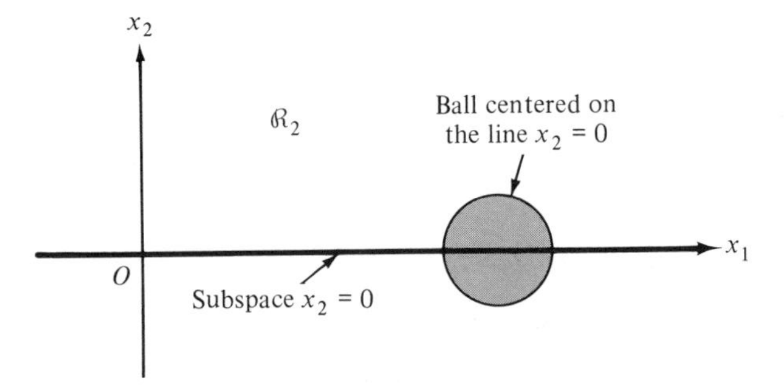

Figure 14

is clearly not open in $\mathcal{R}_2$ since no ball in $\mathcal{R}_2$ centered on the line $x_2 = 0$ will be contained in that line (see Figure 14). However, any such subspace $\mathcal{Y}$ may always be considered to be a normed vector space itself with the same norm as used in $\mathcal{X}$, and then $\mathcal{Y}$ *is an open subset of itself* (as is every normed vector space).

Exercises

1. Let the function $\|\cdot\|$ be defined on the set of all real numbers $\mathcal{R}$ by the formula $\|x\| = x^2$, any x in $\mathcal{R}$. Determine whether or not this function is a norm on $\mathcal{R}$.
2. Verify that the function $\|\cdot\|$ defined by formula (1.4.4) satisfies all the properties required of a norm function on $\mathcal{R}_n$.
3. Verify that each of the functions $\|\cdot\|$ defined by formulas (1.4.6) and (1.4.8) satisfies all the properties of a norm function on $\mathcal{C}^0[a, b]$.
4. Prove the inequality $|\|x\| - \|y\|| \leq \|x - y\|$ for any two vectors x and y in a normed vector space.

for any continuous function f on $[0, T]$ [see Section A3 of the Appendix for a proof of (1.5.9)], we may replace the estimate (1.5.8) with the larger estimate (explain)

$$|C(P) - C(Q)| \leq \beta^2 \int_0^T |I_P - I_Q||I_P + I_Q - 2\mathcal{I}|\,dt + \int_0^T |P - Q||P + Q - 2\mathcal{P}|\,dt. \tag{1.5.10}$$

The triangle inequality for numbers can be used again to give

$$|I_P + I_Q - 2\mathcal{I}| = |2(I_P - \mathcal{I}) + (I_Q - I_P)| \leq 2|I_P - \mathcal{I}| + |I_Q - I_P|$$

and, similarly,

$$|P + Q - 2\mathcal{P}| \leq 2|P - \mathcal{P}| + |P - Q|,$$

so that the inequality (1.5.10) implies in turn that

$$|C(P) - C(Q)| \leq \beta^2 \int_0^T |I_P(t) - I_Q(t)|(2|I_P(t) - \mathcal{I}(t)| + |I_P(t) - I_Q(t)|)\,dt + \int_0^T |P(t) - Q(t)|(2|P(t) - \mathcal{P}(t)| + |P(t) - Q(t)|)\,dt. \tag{1.5.11}$$

Now since formula (1.5.5) implies that

$$|P(t) - Q(t)| \leq ||P - Q|| \tag{1.5.12}$$

for every t in the interval $0 \leq t \leq T$, and, similarly, that

$$|P(t) - \mathcal{P}(t)| \leq ||P - \mathcal{P}||,$$

it follows that the last integral on the right-hand side of inequality (1.5.11) can be estimated above as

$$\begin{aligned}
&\int_0^T |P(t) - Q(t)|(2|P(t) - \mathcal{P}(t)| + |P(t) - Q(t)|)\,dt \\
&\quad \leq \int_0^T ||P - Q||(2||P - \mathcal{P}|| + ||P - Q||)\,dt \\
&\quad = ||P - Q||(2||P - \mathcal{P}|| + ||P - Q||)T.
\end{aligned} \tag{1.5.13}$$

We can get a similar estimate for the first integral on the right-hand side of inequality (1.5.11) by using formulas (1.3.23) and (1.5.7) to write

$$I_P(t) - I_Q(t) = e^{-\alpha t} \int_0^t e^{\alpha\tau}[P(\tau) - Q(\tau)]\,d\tau,$$

from which we find as before [using the inequality (1.5.9) and the fact that the exponential function is positive]

$$|I_P(t) - I_Q(t)| \leq e^{-\alpha t} \int_0^t e^{\alpha\tau} |P(\tau) - Q(\tau)| \, d\tau. \tag{1.5.14}$$

The inequality (1.5.12) along with the result

$$e^{\alpha\tau} \leq e^{\alpha t},$$

which is valid for $\tau \leq t$, now gives with (1.5.14) the estimate

$$|I_P(t) - I_Q(t)| \leq ||P - Q|| T$$

for all t in the interval $[0, T]$. Hence the first integral on the right-hand side of the inequality (1.5.11) can be estimated above as

$$\begin{aligned} &\beta^2 \int_0^T |I_P(t) - I_Q(t)|(2|I_P(t) - \mathcal{I}(t)| + |I_P(t) - I_Q(t)|) \, dt \\ &\quad \leq \beta^2 ||P - Q|| T(2||I_P - \mathcal{I}|| + ||P - Q|| T)T, \end{aligned} \tag{1.5.15}$$

where we also used the inequality

$$|I_P(t) - \mathcal{I}(t)| \leq ||I_P - \mathcal{I}||,$$

which follows directly from the definition of the norm (1.4.8).

We may now use the inequalities (1.5.13) and (1.5.15) to estimate the right-hand side of inequality (1.5.11) and finally obtain

$$\begin{aligned} |C(P) - C(Q)| \leq ||P - Q|| T\{&\beta^2 T(2||I_P - \mathcal{I}|| \\ &+ ||P - Q|| T) + 2||P - \mathcal{P}|| + ||P - Q||\} \end{aligned} \tag{1.5.16}$$

for any vector Q in $\mathcal{C}^0[0, T]$. This result shows that the given cost functional C is indeed continuous at P since the estimate (1.5.16) guarantees that $C(Q)$ will be close to $C(P)$ for all vectors Q which are close to P in $\mathcal{C}^0[0, T]$. In fact, for any given positive number ϵ it follows from (1.5.16) that *the desired result* (1.5.3) *will hold if Q satisfies*

$$||P - Q|| T\{\beta^2 T(2||I_P - \mathcal{I}|| + ||P - Q|| T) + 2||P - \mathcal{P}|| + ||P - Q||\} < \epsilon \tag{1.5.17}$$

since (1.5.17) will guarantee that the right-hand side of (1.5.16) (hence also the left-hand side) is less than ϵ. We can simplify the condition (1.5.17) somewhat if we arbitrarily restrict consideration to those vectors Q contained in the ball $B_1(P)$ of radius 1 centered at P. (The reader should convince himself that this is permissible.) Then (1.5.17) will certainly hold for all such vectors Q in $B_1(P)$

which also satisfy (explain)

$$\|P - Q\| T\{\beta^2 T(2\|I_P - \mathscr{I}\| + T) + 2\|P - \mathscr{P}\| + 1\} < \epsilon,$$

and this will hold if

$$\|P - Q\| < \epsilon T^{-1}\{\beta^2 T(2\|I_P - \mathscr{I}\| + T) + 2\|P - \mathscr{P}\| + 1\}^{-1}.$$

Hence, finally, the desired inequality (1.5.3) will certainly hold for all Q satisfying condition (1.5.4) *if we define* ρ *to be the smaller of the two possible numbers*, 1 and

$$\epsilon T^{-1}\{\beta^2 T(2\|I_P - \mathscr{I}\| + T) + 2\|P - \mathscr{P}\| + 1\}^{-1}.$$

Note that this choice of ρ depends only on ϵ and P and on the fixed expressions $\mathscr{I}$, $\mathscr{P}$, $\mathscr{S}$, I_0, α, β, and T, which are involved in the definition of the cost functional C, *but not on* Q. We have therefore shown that the cost functional C is continuous at P in the normed vector space $\mathscr{C}^0[0, T]$ equipped with the uniform norm. Symbolically we have shown the result [see (1.5.2)]

$$\underset{Q \to P \text{ in } \mathscr{C}^0[0,T]}{\text{limit}} C(Q) = C(P).$$

Moreover, since this is true for *any* P, we have actually shown that this functional C is continuous on $\mathscr{C}^0[0, T]$ (relative to the uniform norm).

Exercises

1. Show that any norm function defined on a vector space $\mathscr{X}$ is a continuous functional on $\mathscr{X}$ relative to the norm itself. *Hint:* See Exercise 4 of Section 1.4.

2. Let a functional J be defined on the vector space $\mathscr{C}^0[0, 1]$ by the formula $J(\phi) = \int_0^1 (\sin x)\,[\phi(x)]^3\,dx$ for any function ϕ of class $\mathscr{C}^0$ on $0 \leq x \leq 1$. Show that J is continuous on the normed vector space $\mathscr{C}^0[0, 1]$ equipped with the uniform norm, with $\|\phi\| = \max_{0 \leq x \leq 1} |\phi(x)|$.

3. Let a functional J be defined on an open set D in a normed vector space $\mathscr{X}$ as the product $J(x) = K(x)L(x)$ for any x in D, where K and L are continuous functionals on D. Show that J is continuous on D.

4. Let $\mathscr{X}$ be the normed vector space consisting of all continuous functions $\phi(t)$ on the interval $0 \leq t \leq 1$, equipped with the L_2 norm given as $\|\phi\| = (\int_0^1 [\phi(t)]^2\,dt)^{1/2}$ for any vector ϕ in $\mathscr{X}$, and let the functional J be defined by $J(\phi) = \phi(0)$ for any ϕ in $\mathscr{X}$. Show that J fails to be continuous at ϕ for each vector ϕ in $\mathscr{X}$. *Hint:* Construct vectors ψ in $\mathscr{X}$ which are of small distance from ϕ in $\mathscr{X}$ but for which the difference $J(\psi) - J(\phi) = \psi(0) - \phi(0)$ fails to be small. (Try $\psi = \phi + \chi$ for suitable vectors χ of small norm.)

5. Let $\mathscr{Y}$ be the normed vector space consisting of all continuous functions $\phi(t)$

on the interval $0 \leq t \leq 1$ equipped with the uniform norm given as $\|\phi\| = \max_{0 \leq t \leq 1} |\phi(t)|$ for any vector ϕ in $\mathcal{Y}$, and let J be as in Exercise 4, with $J(\phi) = \phi(0)$ for any ϕ in $\mathcal{Y}$. Show that J is continuous at each vector ϕ in $\mathcal{Y}$.

Remark. The results of Exercises 4 and 5 show that the continuity or lack of continuity of a given functional defined on a vector space may depend on the particular norm used. This can happen, however, only in vector spaces such as $\mathcal{C}^0[a, b]$, which are not *finite dimensional* since all norms are *equivalent* on any finite dimensional space such as $\mathcal{R}_n$.†

1.6. Linear Functionals

There is an important special class of functionals for which continuity is often easy to prove, namely the class of *linear* functionals. A functional J is said to be **linear** if the domain of J consists of some entire vector space $\mathcal{X}$ and if J satisfies the *linearity relation*

$$J(ax + by) = aJ(x) + bJ(y) \tag{1.6.1}$$

for all numbers a and b in $\mathcal{R}$ and for all vectors x and y in $\mathcal{X}$. For example, the functional $K = K(f)$ defined on the vector space $\mathcal{C}^0[0, 1]$ by

$$K(f) = \int_0^1 f(t)\,dt \qquad \text{for any continuous function } f = f(t)$$

clearly satisfies the condition $K(af + bg) = aK(f) + bK(g)$ for all numbers a and b and for all continuous functions f and g on the interval $[0, 1]$. Hence this functional K is linear.

On the other hand, the cost functional C considered in Section 1.5 is *not* linear since the required linearity relation $C(aP + bQ) = aC(P) + bC(Q)$ does *not* hold in general for all numbers a and b and for all vectors P and Q in the vector space $\mathcal{C}^0[0, T]$. This assertion can be easily checked using formulas (1.3.23) and (1.3.24). Such functionals which are not linear are called **nonlinear.**

It is easy to check from equation (1.6.1) that *every linear functional J vanishes at the zero vector* in its domain $\mathcal{X}$, with $J(0) = 0$.

We now verify the useful result that **a linear functional is continuous on its domain $\mathcal{X}$ if and only if it is continuous at the zero vector in $\mathcal{X}$.** Indeed, if a linear functional J is continuous on $\mathcal{X}$, then by definition J is continuous at each vector in $\mathcal{X}$ and hence at the zero vector. Conversely, suppose that J is continuous at the zero vector in $\mathcal{X}$, so that equation (1.5.2) implies that

† Richard E. Williamson, Richard H. Crowell, and Hale F. Trotter, *Calculus of Vector Functions* (Englewood Cliffs, N.J.: Prentice-Hall, Inc., 1968), pp. 407–410.

[recall that $J(0) = 0$]

$$\operatorname*{limit}_{y\to 0 \text{ in } \mathfrak{X}} J(y) = 0.$$

Hence given any $\epsilon > 0$ there is a number $\rho > 0$ such that

$$|J(y)| < \epsilon \tag{1.6.2}$$

for all vectors y satisfying $\|y\| < \rho$. If x is *any* fixed vector in $\mathfrak{X}$, then by the linearity relation (1.6.1)

$$J(z) = J(z - x + x) = J(z - x) + J(x)$$

holds for all vectors z in $\mathfrak{X}$, and so (explain)

$$|J(z) - J(x)| = |J(z - x)|. \tag{1.6.3}$$

If we now take $y = z - x$ in the inequality (1.6.2), we then find with equation (1.6.3) that

$$|J(z) - J(x)| < \epsilon$$

for all vectors z satisfying $\|z - x\| < \rho$, which shows that J is continuous at x. Since x was arbitrary, this shows that J is continuous everywhere on $\mathfrak{X}$ and completes the proof of the asserted result.

We remark finally that **a linear functional J is continuous at the zero vector in its domain $\mathfrak{X}$ if an estimate of the type**

$$|J(x)| \leq \text{constant } \|x\| \tag{1.6.4}$$

holds for all vectors x in $\mathfrak{X}$ and for some *fixed constant* depending only on the functional J but not depending on x. (The reader should be able to supply a short proof of this assertion based directly on the definition of continuity for functionals given in the second paragraph of Section 1.5.) Hence it follows from these results that *we need only prove an estimate of the form* (1.6.4) *for a linear functional J on a normed vector space $\mathfrak{X}$ in order to conclude that J is continuous everywhere on $\mathfrak{X}$.*

Exercises

1. Verify that any linear functional J on a normed vector space $\mathfrak{X}$ is continuous on $\mathfrak{X}$ if an estimate of the type given by inequality (1.6.4) holds.
2. Determine which of the following formulas define linear functionals on the vector space $\mathcal{C}^0[0, 1]$.
   (a) $J(\phi) = \phi(0)$.

(b) $J(\phi) = \phi(1) - \phi(0)$.
(c) $J(\phi) = [\phi(1) - \phi(0)]^2$.
(d) $J(\phi) = \int_0^1 x^2\phi(x)\,dx$.
(e) $J(\phi) = \int_0^1 x\phi(x)^2\,dx$.
(f) $J(\phi) = \int_{1/2}^1 (\sin x)\phi(x)\,dx$.

3. Let $e_1, e_2, \ldots, e_n$ be the *unit* vectors in $\mathcal{R}_n$ given as $e_1 = (1, 0, \ldots, 0)$, $e_2 = (0, 1, \ldots, 0), \ldots, e_n = (0, 0, \ldots, 1)$, where e_i has its ith component equal to 1 and all other components equal to 0. Show that any vector $x = (x_1, x_2, \ldots, x_n)$ in $\mathcal{R}_n$ can be written as a *linear combination* of these unit vectors as $x = \sum_{i=1}^n x_ie_i$ and derive the result $J(x) = \sum_{i=1}^n x_iJ(e_i)$ for any linear functional J on $\mathcal{R}_n$. Use the latter result to prove that *every linear functional on $\mathcal{R}_n$ is continuous.*† *Hint:* Use Cauchy's inequality (1.4.5) along with the norm given by formula (1.4.4).

4. Determine whether or not the functional defined by $J(\phi) = \phi'(0) = (d\phi/dx)|_{x=0}$ is a continuous linear functional on the normed vector space $\mathcal{C}^1[-1, 1]$ equipped with the uniform norm defined as $\|\phi\| = \max_{-1\le x\le 1}|\phi(x)|$ for any vector ϕ in $\mathcal{C}^1[-1, 1]$. *Hint:* Consider the vectors $\phi_1, \phi_2, \phi_3, \ldots, \phi_k, \ldots$ in $\mathcal{C}^1[-1, 1]$ defined by the formulas $\phi_1(x) = \sin x$, $\phi_2(x) = (\sin 2x)/\sqrt{2}$, $\phi_3(x) = (\sin 3x)/\sqrt{3}, \ldots, \phi_k(x) = (\sin kx)/\sqrt{k}, \ldots$.

5. Let a functional J be defined on some ball $B_\rho(0)$ centered at the zero vector in a normed vector space $\mathfrak{X}$ and assume that J satisfies the linearity relation $J(ax + by) = aJ(x) + bJ(y)$ for all vectors x and y in $B_\rho(0)$ and for all (small) numbers a and b such that $ax + by$ remains in $B_\rho(0)$. Show how to extend the domain of J to be all of $\mathfrak{X}$ in such a way as to obtain a linear functional on $\mathfrak{X}$. (This explains why it is "natural" to take the domain of a *linear* functional to be an entire vector space.)

† This result is true not only for $\mathcal{R}_n$ but also for *any* finite dimensional vector space $\mathfrak{X}$. Exercise 4 shows, however, that it need not be true for *every* normed vector space. Hence the notion of continuity is of interest for linear functionals mainly in vector spaces which are *not* finite dimensional.

2. A Fundamental Necessary Condition for an Extremum

In this chapter we shall introduce the Gâteaux variation of a functional. We shall prove that the variation must vanish at a local maximum or minimum vector, and we shall show how to use this result to solve certain extremum problems. We shall also discuss briefly the Fréchet differential of a functional.

2.1. Introduction

Let f be any typical real-valued function as is commonly studied in elementary calculus, with domain D consisting of all numbers x in some given open interval $D = (a, b)$. We assume that f is differentiable on D so that the graph of f has a well-defined tangent line at each point, as indicated in Figure 1. The slope of the tangent line at any point x is understood to be given by the limit

$$\operatorname*{limit}_{\epsilon \to 0} \frac{f(x + \epsilon) - f(x)}{\epsilon} = f'(x) \tag{2.1.1}$$

which is also the value of the derivative of f at x.

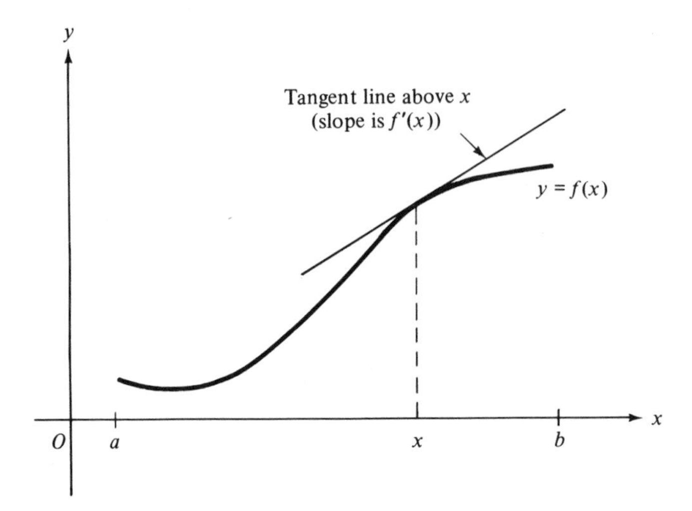

Figure 1

If f has a local minimum or maximum value at a point x^* in D, then the tangent line to the graph of f will be *horizontal* at x^* with zero slope, as shown in Figure 2, and the condition

$$f'(x^*) = 0 \tag{2.1.2}$$

will hold at any such interior local maximum or minimum point x^*. Almost every calculus text shows how to use the condition (2.1.2) to solve certain minimum and maximum problems arising in various situations. (See the

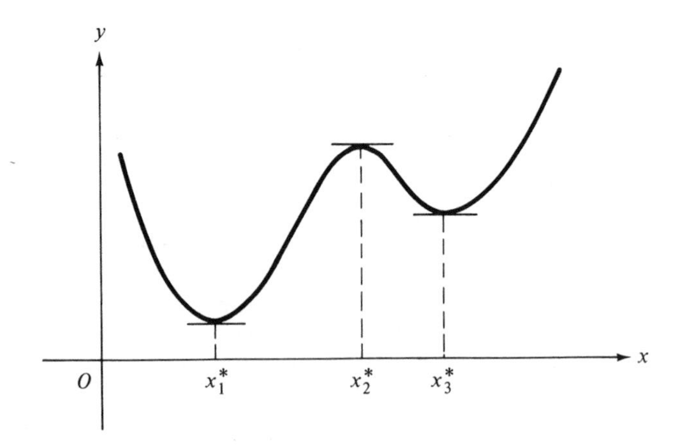

Figure 2

Exercises at the end of this section.) We wish to generalize this familiar approach so as to obtain a similar method which can be used to solve the optimization and extremum problems considered in this book.

Exercises

1. Find the two positive numbers whose sum is 50 and whose product is a maximum.
2. A soup can is to be made in the form of a right circular cylinder to contain one pint. Find the dimensions of the can that will require the least amount of metal.
3. The experimental values $a_1, a_2, a_3, \ldots, a_n$ are obtained in an experiment repeated n times, and the experimenter wishes to find a number x that will minimize the sum $(x - a_1)^2 + (x - a_2)^2 + \cdots + (x - a_n)^2$. (That is, he wishes to minimize the sum of the squares of the *deviations*.) Find x.
4. An apple orchard has 30 trees per acre at present, and the average yield is 400 apples per tree. If new trees are planted, it is estimated that the average yield per tree will drop 7 apples for each new tree added per acre. What is the largest possible apple crop per acre? Compare the largest possible crop with the present crop.†
5. An apartment complex has 55 units. It has been observed that every unit will remain occupied when the rent of each unit is $150 per month, while one additional unit will be vacated and remain empty for every $5 increase in the rent above $150. The owner must pay $6000 per month in fixed operating expenses (including insurance payments and a monthly payment on the original indebtedness incurred during construction), and there is an additional monthly expense of $15 per *occupied* unit for services and repairs. What is the maximum possible monthly profit, and what rent should be charged in order to attain it? Compare the maximum profit with the profit obtained when all units are occupied at $150 per unit per month.

2.2. *A Fundamental Necessary Condition for an Extremum*

Let D be a fixed nonempty subset of a normed vector space $\mathfrak{X}$, and let J be a functional defined on D. A vector x^* in D is said to be a **maximum vector**

†Exercises 4 and 5 involve discrete, not continuous, variables. Indeed, the number of apple trees in Exercise 4 must be a nonnegative *integer*, and the same is true of the number of occupied apartments in Exercise 5. Nevertheless, the reader should be able to convince himself that these problems can be solved using the calculus by first allowing the discrete variables to take on any numerical value and then later restricting consideration to integer values.

in D for J if $J(x) \leq J(x^*)$ for all vectors x in D. The vector x^* in D is a *local* maximum vector in D for J if there is some ball $B_\rho(x^*)$ in $\mathfrak{X}$ centered at x^* such that $J(x) \leq J(x^*)$ for all vectors x that are simultaneously in D and in $B_\rho(x^*)$. [If D is an *open* subset of $\mathfrak{X}$, we require the ball $B_\rho(x^*)$ to be contained in D.] A local *minimum* vector in D for J is defined similarly, using $J(x) \geq J(x^*)$. For brevity we shall say that x^* is a local extremum vector in D for J if x^* is either a local maximum vector or a local minimum vector, and in this case we say that J has a local *extremum* at x^*. The functional value $J(x^*)$ is said to be a local extreme value for J in D.

We now consider the case of a functional J defined on an *open* subset D of the normed vector space $\mathfrak{X}$. If x^* is a local minimum vector in D for J and if h is any fixed vector in $\mathfrak{X}$, then the inequality

$$J(x^* + \epsilon h) - J(x^*) \geq 0$$

will hold for all sufficiently small numbers ϵ. (Note that the vector $x^* + \epsilon h$ will be in the domain of J for all small numbers ϵ since x^* is a vector in the open set D.) Hence

$$\frac{J(x^* + \epsilon h) - J(x^*)}{\epsilon} \geq 0 \tag{2.2.1}$$

for all small *positive* numbers ϵ, while

$$\frac{J(x^* + \epsilon h) - J(x^*)}{\epsilon} \leq 0 \tag{2.2.2}$$

for all small *negative* numbers $\epsilon < 0$. If we let ϵ tend toward zero in (2.2.1), we find that

$$\operatorname*{limit}_{\substack{\epsilon \to 0 \\ \epsilon > 0}} \frac{J(x^* + \epsilon h) - J(x^*)}{\epsilon} \geq 0,$$

and similarly from (2.2.2) we find that

$$\operatorname*{limit}_{\substack{\epsilon \to 0 \\ \epsilon < 0}} \frac{J(x^* + \epsilon h) - J(x^*)}{\epsilon} \leq 0,$$

provided, of course, that these limits exist. (These limits are to be understood in the usual sense of elementary calculus.) We can now conclude from these last results that the condition

$$\operatorname*{limit}_{\epsilon \to 0} \frac{J(x^* + \epsilon h) - J(x^*)}{\epsilon} = 0 \tag{2.2.3}$$

must hold at any such local minimum vector x^* in D for J, *provided that this* *limit exists.* It is clear that this same condition must hold similarly at any local *maximum* vector. We shall see that this condition (2.2.3) is the desired generalization of the familiar condition (2.1.2) from elementary calculus.

A functional J defined on an open subset D of a normed vector space $\mathfrak{X}$ is said to have a **Gâteaux variation** at a vector x in D whenever there is a functional $\delta J(x)$ with values $\delta J(x; h)$ defined for all vectors h in $\mathfrak{X}$ and such that

$$\operatorname*{limit}_{\epsilon \to 0} \frac{J(x + \epsilon h) - J(x)}{\epsilon} = \delta J(x; h)$$

holds [compare with (2.1.1)] for every vector h in $\mathfrak{X}$. The functional $\delta J(x)$ is called the *Gâteaux variation* of J at x, or simply the *variation* of J at x.†

The result of the calculation leading up to equation (2.2.3) can now be summarized in the following basic theorem.

Theorem. *If a functional J defined on an open set D contained in a normed vector space $\mathfrak{X}$ has a local extremum at a vector x^* in D, and if J has a variation at x^*, then the variation of J at x^* must vanish; that is,*

$$\delta J(x^*; h) = 0 \qquad \textit{for all vectors } h \textit{ in } \mathfrak{X} \tag{2.2.4}$$

must hold.

Hence the vanishing of the variation is a necessary condition which must hold at any local extremum vector x^*. We shall see that this necessary condition can be used to solve a wide variety of extremum problems. Indeed, in practice it is often possible to eliminate the arbitrary vector h from equation (2.2.4) so as to obtain a simpler equation which involves only the extremum vector x^* and which can be solved to give the desired extremum vector. The elimination of the vector h will always hinge on the fact that equation (2.2.4) is required to hold for *every* vector h in $\mathfrak{X}$ and will always involve making some suitable special choice (or choices) of h, as is done, for example, in Exercise 3 at the end of this section to obtain the condition (2.2.6) which involves only x^*.

We should mention that even if we find a vector x^* in D which satisfies the necessary condition (2.2.4), we must still check whether or not $J(x^*)$ is actually a local extreme value for J in D. Indeed, equation (2.2.4) may hold also at certain *non*extremum vectors x^* such as *saddle points* or certain

†The concept of the variation for certain special functionals has its roots in the work of Joseph Louis Lagrange (1736–1813). The general theory of functionals came later through the work of Vito Volterra (1860–1940), Jacques Hadamard (1865–1963), and others, while the variation as we know it today was introduced and studied in the early twentieth century by R. Gâteaux, who was killed in September 1914 early in World War I. Gâteaux's work on the variation was published by the French Mathematical Society after his death.

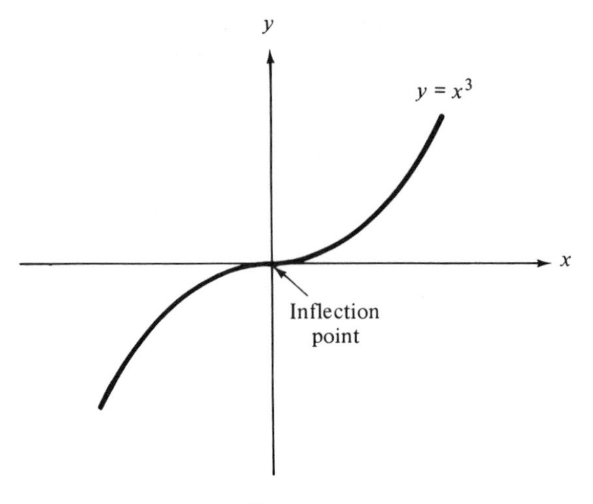

Figure 3

inflection points of J in D. For example, the function J defined on $\mathcal{R} = \mathcal{R}_1$ by

$$J(x) = x^3 \qquad \text{for any number } x \text{ in } \mathcal{R}$$

has a vanishing derivative at $x = 0$ (see Figure 3), and so the variation of J vanishes also at the point $x^* = 0$ (see Exercise 1). However, the point $x^* = 0$ is not a local extremum vector in $\mathcal{R}$ for J but is rather a horizontal inflection point. (See Exercise 6 of Section 2.5 for a related example.) Another example is given by the function K defined on $\mathcal{R}_2$ by

$$K(x) = x_2^2 - x_1^2$$

for any point $x = (x_1, x_2)$ in $\mathcal{R}_2$. This function has a variation given as [see

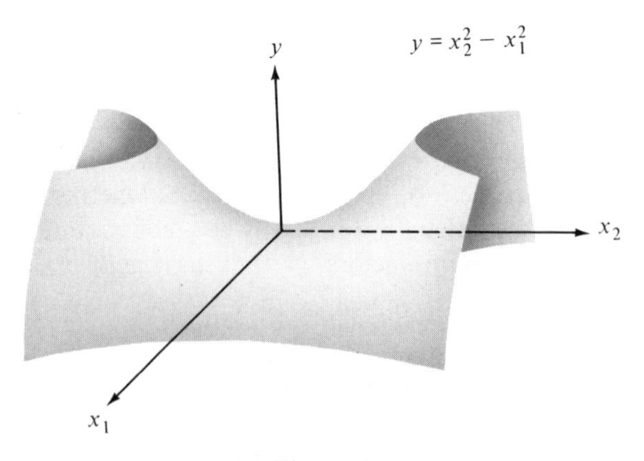

Figure 4

equation (2.2.5) in Exercise 3]

$$\begin{aligned}\delta K(x; h) &= \frac{\partial K(x)}{\partial x_1} h_1 + \frac{\partial K(x)}{\partial x_2} h_2 \\ &= -2x_1 h_1 + 2x_2 h_2\end{aligned}$$

for any vector $h = (h_1, h_2)$ in $\mathcal{R}_2$, and so the variation clearly vanishes at $x^* = (0, 0)$. However, the point $x^* = (0, 0)$ is not a local extremum point for K in $\mathcal{R}_2$ but is rather a saddle point (see Figure 4).

Exercises

1. Let J be a typical real-valued function as is commonly studied in elementary calculus, with domain D consisting of all numbers x in some given open interval $D = (a, b)$, and assume that J is differentiable at x. In this case show that J has a variation at x given as $\delta J(x; h) = J'(x)h$ for any number h in $\mathcal{R}$, where $J'(x)$ denotes the derivative of J at x.

2. Let J be as in Exercise 1, and assume that J has a local extremum at a point x^* in D. Show that the necessary condition (2.2.4) is equivalent to the following condition, which is commonly used in elementary calculus: $J'(x^*) = 0$. *Hints:* The last condition clearly implies (2.2.4). The converse is also true since (2.2.4) is required to hold for *every* number h.

3. Let $J = J(x)$ be a real-valued function defined for all n-tuples $x = (x_1, x_2, \ldots, x_n)$ in some given open region D in $\mathcal{R}_n$, and assume that J has continuous first-order partial derivatives at x denoted as $J_{x_i}(x) = \partial J(x)/\partial x_i$ for $i = 1, 2, \ldots, n$. Here the partial derivative J_{x_i} is just the derivative of J considered as a function of the ith variable x_i alone, with all other variables $x_1, \ldots, x_{i-1}, x_{i+1}, \ldots, x_n$ held fixed. Hence, for example, we have the definition $J_{x_1}(x) = \text{limit}_{\epsilon_1 \to 0} \{[J(x_1 + \epsilon_1, x_2, \ldots, x_n) - J(x_1, x_2, \ldots, x_n)]/\epsilon_1\}$ and similar definitions for $J_{x_2}(x), \ldots, J_{x_n}(x)$. It can be shown in this case that J has a variation at x given as

$$\delta J(x; h) = \sum_{i=1}^{n} J_{x_i}(x) h_i \tag{2.2.5}$$

for any vector $h = (h_1, h_2, \ldots, h_n)$ in $\mathcal{R}_n$. [See Exercise 4 of Section 2.3 for an indication of the derivation of (2.2.5).] If J has a local extremum at a vector $x^* = (x_1^*, x_2^*, \ldots, x_n^*)$ in D, show that the necessary condition (2.2.4) is equivalent to the following condition, which is commonly used in elementary calculus:

$$\frac{\partial J(x^*)}{\partial x_i} = 0 \qquad \text{for } i = 1, 2, \ldots, n. \tag{2.2.6}$$

Hints: This last condition on the vanishing of all the first-order partial derivatives along with (2.2.5) clearly implies (2.2.4) in this case. The converse can be shown to be true by taking a suitable special choice for the arbitrary vector $h = (h_1, h_2, \ldots, h_n)$ in (2.2.4). Indeed, if we assume that (2.2.4) holds and if we then take in particular $h_i = J_{x_i}(x^*)$ for $i = 1, 2, \ldots, n$, we find with (2.2.5) that $\sum_{i=1}^{n} [J_{x_i}(x^*)]^2 = 0$, from which (2.2.6) follows directly.

4. Let the functional $J = J(\phi)$ be defined on the vector space $\mathcal{C}^0[0, 1]$ by $J(\phi) = \int_0^1 [\phi(x)^2 + \phi(x)^3]\, dx$ for any function ϕ of class $\mathcal{C}^0$ on $0 \leq x \leq 1$. Show that the variation of J at an arbitrary fixed vector $\phi = \phi(x)$ in $\mathcal{C}^0[0, 1]$ is given by $\delta J(\phi; \psi) = \int_0^1 [2\phi(x) + 3\phi(x)^2]\psi(x)\, dx$ for any vector ψ in $\mathcal{C}^0[0, 1]$.

5. Find all possible local extremum vectors ϕ^* in $\mathcal{C}^0[0, 1]$ for the functional J of Exercise 4. *Hint:* Use the following condition [see (2.2.4)], $\delta J(\phi^*; \psi) = 0$ for all vectors ψ in $\mathcal{C}^0[0, 1]$, which must necessarily hold if ϕ^* is a local extremum vector. If you make the special choice $\psi(x) = 2\phi^*(x) + 3\phi^*(x)^2$ in the last equation, you should be able to deduce the result $\phi^*(x)[2 + 3\phi^*(x)] = 0$ for $0 \leq x \leq 1$, from which you should be able to find all possible local extremum vectors.

6. Let the functional $K = K(\phi)$ be defined on the vector space $\mathcal{C}^0[0, 2]$ by $K(\phi) = \int_0^2 [3x^5 - 12x^2\phi(x) + 3x\phi(x)^2 + 2\phi(x)^3]\, dx$ for any function ϕ of class $\mathcal{C}^0$ on $0 \leq x \leq 2$. Calculate the variation of K at an arbitrary vector $\phi = \phi(x)$ in $\mathcal{C}^0[0, 2]$ and find all possible candidates for local extremum vectors ϕ^* in $\mathcal{C}^0[0, 2]$ for K.

7. Let the functional $L = L(\phi)$ be defined on the vector space $\mathcal{C}^0[0, \pi/2]$ by $L(\phi) = \int_0^{\pi/2} [2\phi(x)^3 + 9 (\sin x)\phi(x)^2 + 12 (\sin^2 x)\phi(x) - \cos x]\, dx$ for any function ϕ of class $\mathcal{C}^0$ on $0 \leq x \leq \pi/2$. Calculate the variation of L at an arbitrary vector $\phi = \phi(x)$ in $\mathcal{C}^0[0, \pi/2]$ and find all possible candidates for local extremum vectors ϕ^* in $\mathcal{C}^0[0, \pi/2]$ for L.

2.3. Some Remarks on the Gâteaux Variation

We have noted in the preceding section that a functional J defined on an open subset D of a normed vector space $\mathfrak{X}$ is said to have a variation at x in D whenever the following limit exists for every vector h in $\mathfrak{X}$:

$$\operatorname*{limit}_{\epsilon \to 0} \frac{J(x + \epsilon h) - J(x)}{\epsilon} = \delta J(x; h). \tag{2.3.1}$$

Here ϵ is a *number*, and the limit is understood to be taken in the usual sense of elementary calculus. Since the limit of a function is unique if it exists, it follows from (2.3.1) that *a functional can have at most one variation at x*.

The value of the variation $\delta J(x; h)$ is a *directional derivative of J at x in the direction of the vector h* and is a generalization of the directional derivative of multivariable calculus.† On the other hand, if we compare equation (2.3.1) with equation (2.1.1), we find that *the value of the variation is just the ordinary derivative of the function $J(x + \epsilon h)$ considered as a function of the real num-*

†See John G. Hocking, *Calculus with an Introduction to Linear Algebra* (New York: Holt, Rinehart and Winston, Inc., 1970), p. 645.

ber ϵ and evaluated at $\epsilon = 0$; i.e.,

$$\delta J(x; h) = \frac{d}{d\epsilon} J(x + \epsilon h)\bigg|_{\epsilon=0}. \tag{2.3.2}$$

Hence the result of the preceding section on the vanishing of the variation of J at a local extremum vector x^* is actually a direct corollary of the corresponding result from elementary calculus according to which the derivative of the ordinary function $f(\epsilon) = J(x^* + \epsilon h)$ must vanish at $\epsilon = 0$ if the number 0 is a local minimum or a local maximum point for $f(\epsilon)$.

It follows that the vanishing of the variation at x^* is actually independent of the particular norm that is used on the vector space $\mathfrak{X}$. Strictly speaking, we require only that all vectors of the form $x^* + \epsilon h$ be in the domain D for every fixed vector h in $\mathfrak{X}$ and for all sufficiently small numbers ϵ and that the number $\epsilon = 0$ be a local extremum point for the function $J(x^* + \epsilon h)$ considered as a function of ϵ. Of course, this will automatically be the case if D is an open set in $\mathfrak{X}$ relative to some norm and if x^* is a local extremum vector in D for J relative to this same norm. Hence we have found it convenient (although not essential) to use the notions of *open set* and *local extremum vector*. (These notions *do* depend on the particular norm being used; see the example in Section A2 of the Appendix.)

If J has a variation at x, then $\delta J(x; 0) = 0$ necessarily holds, as the reader should verify. Moreover, the variation must satisfy the *homogeneity* relation

$$\delta J(x; ah) = a\, \delta J(x; h) \tag{2.3.3}$$

for any number a, since

$$\delta J(x; ah) = \frac{d}{d\epsilon} J(x + \epsilon ah)\bigg|_{\epsilon=0} = a\frac{d}{d\sigma} J(x + \sigma h)\bigg|_{\sigma=0} = a\delta J(x; h),$$

where we have put $\sigma = \epsilon a$ and made use of the chain rule of differentiation to find $dJ/d\epsilon = a\, dJ/d\sigma$.

Finally, we mention that we shall often use the symbol Δx (read "delta x") rather than h to denote the second argument in the variation $\delta J(x; \Delta x)$, in which case

$$\delta J(x; \Delta x) = \operatorname*{limit}_{\epsilon \to 0} \frac{J(x + \epsilon \Delta x) - J(x)}{\epsilon} = \frac{d}{d\epsilon} J(x + \epsilon \Delta x)\bigg|_{\epsilon=0} \tag{2.3.4}$$

will hold [see (2.3.1) and (2.3.2)] for any vector x in the domain of J and for any vector Δx in $\mathfrak{X}$. The use here of the symbol Δx to represent an arbitrary vector in $\mathfrak{X}$ is only a matter of notation and has no fundamental significance, although it will later perform a useful bookkeeping service.

Exercises

1. Let the functional $J = J(\phi)$ be defined on the vector space $\mathcal{C}^0[0, 1]$ by $J(\phi) = \int_0^1 [\phi(x)^2 + 2(x - 1)\phi(x) - 2e^x\phi(x)]\, dx$ for any function ϕ of class $\mathcal{C}^0$ on $0 \leq x \leq 1$. Show that the variation of J at an arbitrary fixed vector $\phi = \phi(x)$ in $\mathcal{C}^0[0, 1]$ is given by $\delta J(\phi; \Delta\phi) = 2 \int_0^1 [\phi(x) + x - 1 - e^x]\, \Delta\phi(x)\, dx$ for any vector $\Delta\phi$ in $\mathcal{C}^0[0, 1]$. *Hint:* Calculate $J(\phi + \epsilon\, \Delta\phi)$ and use (2.3.4).

2. Use the theorem of Section 2.2 to find a minimum vector ϕ^* in $\mathcal{C}^0[0, 1]$ for the functional J of the preceding exercise. *Hint:* Make the special choice $\Delta\phi(x) = \phi^*(x) + x - 1 - e^x$ in $\delta J(\phi^*; \Delta\phi) = 0$.

3. Let the functional $K = K(\phi)$ be defined on the vector space $\mathcal{C}^0[0, 1]$ by $K(\phi) = \int_0^1 [2 \tan x(1 + \cos x)\phi(x) - \phi(x)^2]\, dx$ for any function ϕ of class $\mathcal{C}^0$ on $0 \leq x \leq 1$. Calculate the variation of K at an arbitrary vector $\phi = \phi(x)$ in $\mathcal{C}^0[0, 1]$ and find a maximum vector ϕ^* in $\mathcal{C}^0[0, 1]$ for K.

4. Let $F = F(x)$ be a real-valued function defined for all n-tuples $x = (x_1, x_2, \ldots, x_n)$ in some given open set D in $\mathcal{R}_n$, and assume that F has continuous first-order partial derivatives at x denoted as $F_{x_i}(x) = \partial F(x)/\partial x_i$ for $i = 1, 2, \ldots, n$. In this case show that F has a variation at x given as

$$\delta F(x; h) = \sum_{i=1}^{n} F_{x_i}(x)h_i \tag{2.3.5}$$

for any vector $h = (h_1, h_2, \ldots, h_n)$ in $\mathcal{R}_n$. *Hints:* Use (2.3.2) to find $\delta F(x; h) = (d/d\epsilon)F(x + \epsilon h)|_{\epsilon=0}$, and then use the chain rule of differential calculus† to calculate the derivative $dF(x + \epsilon h)/d\epsilon$. For later reference we note here that (2.3.5) implies that

$$\lim_{\epsilon \to 0} \frac{F(x + \epsilon h) - F(x)}{\epsilon} = \sum_{i=1}^{n} F_{x_i}(x)h_i \tag{2.3.6}$$

for any n-tuple $h = (h_1, \ldots, h_n)$, where F may be any smooth function defined on an open subset of $\mathcal{R}_n$.

5. Let J be the functional defined in Exercise 4 of Section 1.5. Show that J has a Gâteaux variation at each vector ϕ in $\mathfrak{X}$, even though J is not continuous at ϕ.

6. Let J be defined for any vector $x = (x_1, x_2)$ in two-dimensional Euclidean space $\mathcal{R}_2$ by

$$J(x) = \begin{cases} \dfrac{x_1 x_2^2}{x_1^2 + x_2^4} & \text{if } x_1 \neq 0 \\ 0 & \text{if } x_1 = 0. \end{cases}$$

Show that J has a Gâteaux variation at the origin $x = (0, 0)$ with (finite) value

†See Richard E. Williamson, Richard H. Crowell, and Hale F. Trotter, *Calculus of Vector Functions* (Englewood Cliffs, N.J.: Prentice-Hall, Inc., 1968), pp. 160–161.

at $h = (h_1, h_2)$ given as

$$\delta J(0; h) = \begin{cases} \dfrac{h_2^2}{h_1} & \text{if } h_1 \neq 0 \\ 0 & \text{if } h_1 = 0. \end{cases}$$

Show that J is *not* continuous at $x = (0, 0)$ by considering the values of J along the parabola $x_1 = x_2^2$.

7. Show that a functional J which has a variation at x is *continuous along each fixed direction at* x; i.e., show that $\text{limit}_{\epsilon \to 0} J(x + \epsilon h) = J(x)$ for each fixed vector h in $\mathfrak{X}$.

8. Verify directly that the functional J of Exercise 6 is continuous along each fixed direction at the origin $x = (0, 0)$.

9. Verify directly that the functional J of Exercise 5 (which is the same as the functional of Exercise 4 of Section 1.5) is continuous along each fixed direction at each vector ϕ in $\mathfrak{X}$.

10. Let the functional J be defined on an open set D in a normed vector space $\mathfrak{X}$ by $J(x) = \sin K(x)$ for any vector x in D, where K is a given functional on D which is known to have a variation at a vector x_0 in D. Prove that J has a variation at x_0 given by $\delta J(x_0; h) = \cos K(x_0)\, \delta K(x_0; h)$ for any vector h in $\mathfrak{X}$.

11. Let the functional J be defined on an open set D in a normed vector space $\mathfrak{X}$ by the product formula $J(x) = K(x)L(x)$ for any vector x in D, where K and L are given functionals on D which are known to have variations at a vector x_0 in D. Show that J has a variation at x_0 given as $\delta J(x_0; h) = K(x_0)\, \delta L(x_0; h) + \delta K(x_0; h) L(x_0)$ for any vector h in $\mathfrak{X}$.

12. Let K and L be as in Exercise 11 and define J by $J(x) = K(x)/L(x)$ for any x in D for which $L(x) \neq 0$. If K and L are known to have variations at a vector x_0 in D at which $L(x_0) \neq 0$, show that J has a variation at x_0 given as $\delta J(x_0; h) = [L(x_0)\, \delta K(x_0; h) - K(x_0)\, \delta L(x_0; h)]/L(x_0)^2$ for any vector h in $\mathfrak{X}$.

2.4. Examples on the Calculation of Gâteaux Variations

In this section we shall calculate the variations of the functionals considered in Section 1.3. Specifically, we shall consider the brachistochrone functional T defined by formula (1.3.3), which gives the time of descent of a bead sliding down a wire connecting two points; the area functional A defined by formula (1.3.11), which gives the area enclosed by the path of an airplane; the time functional T defined by formula (1.3.17), which gives the transit time of a boat crossing a river; and the cost functional C defined by formulas (1.3.23) and (1.3.24).

Cost Functional. We first consider the cost functional C defined by (1.3.23) and (1.3.24) on the vector space $\mathcal{C}^0[0, T]$ consisting of all functions (production rates) P of class $\mathcal{C}^0$ on the fixed interval $[0, T]$. From (1.3.24) we calculate

$$C(P + \epsilon\,\Delta P) = \int_0^T \{\beta^2[I_{P+\epsilon\Delta P}(t) - \mathcal{I}(t)]^2 + [P(t) + \epsilon\,\Delta P(t) - \mathcal{P}(t)]^2\}\,dt \tag{2.4.1}$$

for any number ϵ and any vectors P and ΔP in the vector space $\mathcal{C}^0[0, T]$, where $I_{P+\epsilon\Delta P}$ is found from (1.3.23) to be given as (explain)

$$\begin{aligned} I_{P+\epsilon\Delta P}(t) &= e^{-\alpha t}\Big(I_0 + \int_0^t e^{\alpha\tau}[P(\tau) + \epsilon\,\Delta P(\tau) - \mathcal{S}(\tau)]\,d\tau\Big) \\ &= I_P(t) + \epsilon e^{-\alpha t}\int_0^t e^{\alpha\tau}\Delta P(\tau)\,d\tau. \end{aligned}$$

If we insert this last result back into equation (2.4.1), after minor simplification there follows

$$\begin{aligned} C(P + \epsilon\,\Delta P) = C(P) &+ 2\epsilon\int_0^T \Big\{\beta^2[I_P(t) - \mathcal{I}(t)]e^{-\alpha t}\int_0^t e^{\alpha\tau}\,\Delta P(\tau)\,d\tau \\ &+ [P(t) - \mathcal{P}(t)]\,\Delta P(t)\Big\}\,dt \\ &+ \epsilon^2\int_0^T \Big\{\beta^2\Big[e^{-\alpha t}\int_0^t e^{\alpha\tau}\,\Delta P(\tau)\,d\tau\Big]^2 + [\Delta P(t)]^2\Big\}\,dt, \end{aligned}$$

where we again used formula (1.3.24). If we now differentiate the last equation with respect to the numerical variable ϵ, we easily find that

$$\begin{aligned} \frac{d}{d\epsilon}C(P + \epsilon\,\Delta P) = 2&\int_0^T \Big\{\beta^2[I_P(t) - \mathcal{I}(t)]e^{-\alpha t}\int_0^t e^{\alpha\tau}\,\Delta P(\tau)\,d\tau \\ &+ [P(t) - \mathcal{P}(t)]\,\Delta P(t)\Big\}\,dt \\ &+ 2\epsilon\int_0^T \Big\{\beta^2\Big[e^{-\alpha t}\int_0^t e^{\alpha\tau}\,\Delta P(\tau)\,d\tau\Big]^2 + [\Delta P(t)]^2\Big\}\,dt, \end{aligned}$$

and if we now evaluate this equation at $\epsilon = 0$, we find with (2.3.4) that

$$\begin{aligned} \delta C(P; \Delta P) = 2&\int_0^T \Big\{\beta^2[I_P(t) - \mathcal{I}(t)]e^{-\alpha t}\int_0^t e^{\alpha\tau}\,\Delta P(\tau)\,d\tau \\ &+ [P(t) - \mathcal{P}(t)]\,\Delta P(t)\Big\}\,dt \end{aligned} \tag{2.4.2}$$

for any vector ΔP in the vector space $\mathcal{C}^0[0, T]$. Hence the cost functional C has a variation at each vector P in $\mathcal{C}^0[0, T]$, and this variation may be given by formula (2.4.2).

Area Functional. We now turn to a consideration of the area functional A defined by formula (1.3.11) on the vector space $\mathcal{C}^0[0, T]$ consisting of all functions (steering controls) $\alpha = \alpha(t)$ of class $\mathcal{C}^0$ on the fixed interval $[0, T]$. We neglect for the moment any consideration of the constraints given by equations (1.3.9) and (1.3.10) which appear in Chaplygin's problem for the functional A. We shall see in Chapter 3 how to account for such constraints in the maximum problem through the use of Euler-Lagrange multipliers.

To derive the variation of the functional A, we first calculate from formula (1.3.11) that

$$\begin{aligned} A(\alpha + \epsilon\,\Delta\alpha) = \frac{1}{2}\int_0^T \Big\{ & v_0 \sin(\alpha(t) + \epsilon\,\Delta\alpha(t))\Big[x_0 + w_0 t \\ & + v_0 \int_0^t \cos(\alpha(\tau) + \epsilon\,\Delta\alpha(\tau))\,d\tau\Big] \\ & - [w_0 + v_0 \cos(\alpha(t) + \epsilon\,\Delta\alpha(t))] \\ & \times \Big[y_0 + v_0 \int_0^t \sin(\alpha(\tau) + \epsilon\,\Delta\alpha(\tau))\,d\tau\Big]\Big\}\,dt \end{aligned}$$

for any number ϵ and any vectors α and $\Delta\alpha$ in the vector space $\mathcal{C}^0[0, T]$. For brevity we rewrite this result as

$$A(\alpha + \epsilon\,\Delta\alpha) = \int_a^b f(t, \epsilon)\,dt \tag{2.4.3}$$

with $a = 0$ and $b = T$ and where the function $f = f(t, \epsilon)$ is defined by

$$\begin{aligned} f(t, \epsilon) = {} & \frac{1}{2} v_0 \sin[\alpha(t) + \epsilon\,\Delta\alpha(t)] \\ & \times \Big\{x_0 + w_0 t + v_0 \int_0^t \cos[\alpha(\tau) + \epsilon\,\Delta\alpha(\tau)]\,d\tau\Big\} \\ & - \frac{1}{2}\{w_0 + v_0 \cos[\alpha(t) + \epsilon\,\Delta\alpha(t)]\} \\ & \times \Big\{y_0 + v_0 \int_0^t \sin[\alpha(\tau) + \epsilon\,\Delta\alpha(\tau)]\,d\tau\Big\} \end{aligned} \tag{2.4.4}$$

for $0 \leq t \leq T$, for any number ϵ, and for any two fixed vectors α and $\Delta\alpha$ in $\mathcal{C}^0[0, T]$. In particular $f = f(t, \epsilon)$ is a continuous function of t for each fixed ϵ, and it is a continuously differentiable function of ϵ for each fixed t. Moreover, the partial derivative of f with respect to ϵ is a continuous function of t, and it is easy to check that the difference quotient

$$\frac{f(t, \epsilon + h) - f(t, \epsilon)}{h}$$

tends toward the derivative $\partial f(t, \epsilon)/\partial\epsilon$ *uniformly* for all $0 \leq t \leq T$ as $h =$

$\Delta\epsilon \longrightarrow 0$. It then follows from a well-known result in advanced calculus that†

$$\lim_{h\to 0} \int_a^b \frac{f(t, \epsilon + h) - f(t, \epsilon)}{h} dt = \int_a^b \frac{\partial f(t, \epsilon)}{\partial \epsilon} dt$$

holds, or that

$$\frac{d}{d\epsilon} \int_a^b f(t, \epsilon)\, dt = \int_a^b \frac{\partial f(t, \epsilon)}{\partial \epsilon} dt, \tag{2.4.5}$$

where $a = 0$ and $b = T$ in the present case. It follows from equations (2.4.3) and (2.4.5) that the expression $A(\alpha + \epsilon\, \Delta\alpha)$ is differentiable with respect to the parameter ϵ, and its derivative is given as

$$\frac{d}{d\epsilon} A(\alpha + \epsilon\, \Delta\alpha) = \int_0^T \frac{\partial f(t, \epsilon)}{\partial \epsilon} dt, \tag{2.4.6}$$

where the function f is defined by formula (2.4.4). We may use the techniques of elementary calculus along with equations (2.4.4) and (2.4.5) to calculate the required derivative $\partial f(t, \epsilon)/\partial \epsilon$ for each *fixed* t, and we find that

$$\begin{aligned}
\frac{\partial f(t, \epsilon)}{\partial \epsilon} = \frac{v_0\, \Delta\alpha(t)}{2} \Big\{ &\cos\,(\alpha(t) + \epsilon\, \Delta\alpha(t)) \Big[x_0 + w_0 t \\
&+ v_0 \int_0^t \cos\,(\alpha(\tau) + \epsilon\, \Delta\alpha(\tau))\, d\tau \Big] + \sin\,(\alpha(t) + \epsilon\, \Delta\alpha(t)) \Big[y_0 \\
&+ v_0 \int_0^t \sin\,(\alpha(\tau) + \epsilon\, \Delta\alpha(\tau))\, d\tau \Big]\Big\} + \frac{v_0}{2} \sin\,[\alpha(t) + \epsilon\, \Delta\alpha(t)] \\
&\times \Big\{ -v_0 \int_0^t \sin\,[\alpha(\tau) + \epsilon\, \Delta\alpha(\tau)]\, \Delta\alpha(\tau)\, d\tau \Big\} \\
&- \Big\{ \frac{w_0}{2} + \frac{v_0}{2} \cos\,[\alpha(t) + \epsilon\, \Delta\alpha(t)] \Big\} \\
&\times \Big\{ v_0 \int_0^t \cos\,[\alpha(\tau) + \epsilon\, \Delta\alpha(\tau)]\, \Delta\alpha(\tau)\, d\tau \Big\}.
\end{aligned}$$

If we now set $\epsilon = 0$ in the last equation, we find that

$$\begin{aligned}
\left.\frac{\partial f(t, \epsilon)}{\partial \epsilon}\right|_{\epsilon=0} = \frac{v_0\, \Delta\alpha(t)}{2} \Big\{ &\cos \alpha(t) \Big[x_0 + w_0 t + v_0 \int_0^t \cos \alpha(\tau)\, d\tau \Big] \\
&+ \sin \alpha(t) \Big[y_0 + v_0 \int_0^t \sin \alpha(\tau)\, d\tau \Big] \Big\} \\
&+ \frac{v_0}{2} \sin \alpha(t) \Big[-v_0 \int_0^t \Delta\alpha(\tau) \sin \alpha(\tau)\, d\tau \Big] \\
&- \Big[\frac{w_0}{2} + \frac{v_0}{2} \cos \alpha(t) \Big] \Big[v_0 \int_0^t \Delta\alpha(\tau) \cos \alpha(\tau)\, d\tau \Big],
\end{aligned}$$

†Richard Courant and Fritz John, *Introduction to Calculus and Analysis* [New York: John Wiley & Sons, Inc. (Interscience Division), 1965], p. 538.

which can be simplified using the trigonometric identity

$$\cos \alpha \cos \beta + \sin \alpha \sin \beta = \cos (\alpha - \beta)$$

to give

$$\begin{aligned}\left.\frac{\partial f(t, \epsilon)}{\partial \epsilon}\right|_{\epsilon=0} &= \frac{v_0^2}{2} \int_0^t \cos [\alpha(t) - \alpha(\tau)][\Delta\alpha(t) - \Delta\alpha(\tau)]\, d\tau \\ &\quad - \frac{v_0 w_0}{2} \int_0^t \Delta\alpha(\tau) \cos \alpha(\tau)\, d\tau \\ &\quad + \frac{v_0}{2}[(x_0 + w_0 t) \cos \alpha(t) + y_0 \sin \alpha(t)]\, \Delta\alpha(t). \qquad (2.4.7)\end{aligned}$$

It follows finally from equations (2.3.4), (2.4.6), and (2.4.7) that the functional A has a variation at each vector α in $\mathcal{C}^0[0, T]$ given by

$$\begin{aligned}\delta A(\alpha; \Delta\alpha) &= \frac{v_0^2}{2} \int_0^T \int_0^t \cos [\alpha(t) - \alpha(\tau)][\Delta\alpha(t) - \Delta\alpha(\tau)]\, d\tau\, dt \\ &\quad - \frac{v_0 w_0}{2} \int_0^T \int_0^t \cos \alpha(\tau)\, \Delta\alpha(\tau)\, d\tau\, dt \\ &\quad + \frac{v_0}{2} \int_0^T \Delta\alpha(t)[(x_0 + w_0 t) \cos \alpha(t) + y_0 \sin \alpha(t)]\, dt \qquad (2.4.8)\end{aligned}$$

for any vector $\Delta\alpha$ in the vector space $\mathcal{C}^0[0, T]$.

Some Other Functionals. We turn finally to a consideration of the time functionals $T = T(Y)$ defined, respectively, by formulas (1.3.3) and (1.3.17), which give, respectively, the time of descent of a bead sliding down a wire connecting two points and the transit time of a boat crossing a river. Since formulas (1.3.3) and (1.3.17) involve the derivative $Y'(x)$ of the function $Y = Y(x)$, it is natural to take as the domain of these functionals the vector space $\mathcal{C}^1(I)$ consisting of all functions $Y = Y(x)$ of class $\mathcal{C}^1$ on the interval I, where $I = [x_0, x_1]$ for (1.3.3) and $I = [0, l]$ for (1.3.17). We neglect to consider the constraints $Y(x_0) = y_0$ and $Y(x_1) = y_1$ which appear in John Bernoulli's brachistochrone problem for the functional (1.3.3). Rather we shall see in Chapter 3 how to account for such constraints through the use of Euler-Lagrange multipliers.

The functionals T under consideration are examples of a wider class of functionals which have the general form

$$J(Y) = \int_{x_0}^{x_1} F(x, Y(x), Y'(x))\, dx, \qquad (2.4.9)$$

where in any particular case the function $F = F(x, y, z)$ is a specified given function defined for all points (x, y, z) in some open set in three-dimensional

Euclidean space $\mathcal{R}_3$. The brachistochrone functional given by formula (1.3.3) is of the form (2.4.9), with

$$F(x, y, z) = \sqrt{\frac{1 + z^2}{2g(y_0 - y)}}, \tag{2.4.10}$$

while the functional given by (1.3.17) is of the same form, with $x_0 = 0$, $x_1 = l$, and

$$F(x, y, z) = \frac{\sqrt{1 - e(x)^2 + z^2} - e(x)z}{v_0[1 - e(x)^2]}, \tag{2.4.11}$$

where here $e(x)$ is a known function given by formula (1.3.15). It is understood in formula (2.4.9) that $F(x, y, z)$ is evaluated at $y = Y(x)$ and $z = Y'(x)$ for any x in the interval $[x_0, x_1]$, where $Y = Y(x)$ may be any suitable function of class $\mathcal{C}^1$ on $[x_0, x_1]$. We can use any suitable norm on $\mathcal{C}^1[x_0, x_1]$, as, for example, the norm given by formula (1.4.9) with $k = 1$,

$$\|Y\| = \max_{x_0 \leq x \leq x_1} |Y(x)| + \max_{x_0 \leq x \leq x_1} |Y'(x)| \tag{2.4.12}$$

for any vector Y in $\mathcal{C}^1[x_0, x_1]$. We assume that the functional J is defined by (2.4.9) for all vectors Y in some open subset D of the normed vector space $\mathcal{C}^1[x_0, x_1]$.

Hence, rather than consider separately the two functionals given by formulas (1.3.3) and (1.3.17), we shall consider the single more general functional defined by formula (2.4.9) for a given fixed function F depending on three real variables x, y, and z. We can later specialize our results to the functionals of (1.3.3) and (1.3.17) by taking F to be given, respectively, by (2.4.10) and (2.4.11).

To obtain the variation of the functional J at any fixed vector Y in its domain D, we use (2.4.9) to calculate

$$J(Y + \epsilon\,\Delta Y) = \int_{x_0}^{x_1} F(x, Y(x) + \epsilon\,\Delta Y(x), Y'(x) + \epsilon\,\Delta Y'(x))\,dx \tag{2.4.13}$$

for any vector ΔY in the vector space $\mathcal{C}^1[x_0, x_1]$ and for any small number ϵ. We shall assume that the function $F = F(x, y, z)$ is continuous with respect to all its variables and has continuous first-order partial derivatives with respect to y and z. In this case we find from (2.4.13) that [compare with (2.4.5)]

$$\begin{aligned}\frac{d}{d\epsilon}J(Y + \epsilon\,\Delta Y) &= \frac{d}{d\epsilon}\int_{x_0}^{x_1} F(x, Y(x) + \epsilon\,\Delta Y(x), Y'(x) + \epsilon\,\Delta Y'(x))\,dx \\ &= \int_{x_0}^{x_1} \frac{\partial}{\partial\epsilon} F(x, Y(x) + \epsilon\,\Delta Y(x), Y'(x) + \epsilon\,\Delta Y'(x))\,dx,\end{aligned} \tag{2.4.14}$$

where the derivative $\partial F/\partial\epsilon$ appearing in the last integral can be calculated by the chain rule of differential calculus as

$$\frac{\partial}{\partial\epsilon}F(x, Y(x) + \epsilon\,\Delta Y(x), Y'(x) + \epsilon\,\Delta Y'(x)) = F_y(x, Y(x) + \epsilon\,\Delta Y(x), Y'(x) + \epsilon\,\Delta Y'(x))\,\Delta Y(x) + F_z(x, Y(x) + \epsilon\,\Delta Y(x), Y'(x) + \epsilon\,\Delta Y'(x))\,\Delta Y'(x).$$

Here $F_y(x, y, z) = \partial F(x, y, z)/\partial y$ and $F_z(x, y, z) = \partial F(x, y, z)/\partial z$, and we have put $y = Y(x) + \epsilon\,\Delta Y(x)$ and $z = Y'(x) + \epsilon\,\Delta Y'(x)$. If we evaluate this equation at $\epsilon = 0$, we get for each fixed x

$$\frac{\partial}{\partial\epsilon}F(x, Y(x) + \epsilon\,\Delta Y(x), Y'(x) + \epsilon\,\Delta Y'(x))\Big|_{\epsilon=0} = F_Y(x, Y(x), Y'(x))\,\Delta Y(x) + F_{Y'}(x, Y(x), Y'(x))\,\Delta Y'(x), \tag{2.4.15}$$

where

$$F_Y(x, Y(x), Y'(x)) = \frac{\partial F(x, y, z)}{\partial y}\Big|_{\substack{y=Y(x)\\ z=Y'(x)}}$$
$$F_{Y'}(x, Y(x), Y'(x)) = \frac{\partial F(x, y, z)}{\partial z}\Big|_{\substack{y=Y(x)\\ z=Y'(x)}} \tag{2.4.16}$$

These last results (2.4.14) and (2.4.15) along with (2.3.4) now give for the variation of J

$$\delta J(Y; \Delta Y) = \int_{x_0}^{x_1} [F_Y(x, Y(x), Y'(x))\,\Delta Y(x) + F_{Y'}(x, Y(x), Y'(x))\,\Delta Y'(x)]\,dx \tag{2.4.17}$$

for any vector $Y = Y(x)$ in the domain D of J and for any vector $\Delta Y = \Delta Y(x)$ in the vector space $\mathcal{C}^1[x_0, x_1]$.

If we take F to be given by (2.4.10) as in John Bernoulli's brachistochrone functional, we find that

$$\frac{\partial F(x, y, z)}{\partial y} = \frac{1}{2(y_0 - y)}\sqrt{\frac{1 + z^2}{2g(y_0 - y)}}$$

and that

$$\frac{\partial F(x, y, z)}{\partial z} = \frac{z}{\sqrt{2g(y_0 - y)(1 + z^2)}},$$

so that equations (2.4.16) and (2.4.17) give in this case

$$\delta T(Y; \Delta Y) = \int_{x_0}^{x_1}\left\{\frac{1}{2[y_0 - Y(x)]}\sqrt{\frac{1 + Y'(x)^2}{2g[y_0 - Y(x)]}}\,\Delta Y(x) + \frac{Y'(x)\,\Delta Y'(x)}{\sqrt{2g[y_0 - Y(x)][1 + Y'(x)^2]}}\right\}dx$$

for the value at any ΔY of the variation of the brachistochrone functional T of (1.3.3). Similarly, if we take F to be given by (2.4.11), we find that $\partial F(x, y, z)/\partial y = 0$ and that

$$\frac{\partial F(x, y, z)}{\partial z} = \frac{z - e(x)\sqrt{1 - e(x)^2 + z^2}}{v_0[1 - e(x)^2]\sqrt{1 - e(x)^2 + z^2}},$$

so that in this case (2.4.15) and (2.4.17) give

$$\delta T(Y; \Delta Y) = \int_0^l \frac{Y'(x) - e(x)\sqrt{1 - e(x)^2 + Y'(x)^2}}{v_0[1 - e(x)^2]\sqrt{1 - e(x)^2 + Y'(x)^2}} \Delta Y'(x)\, dx$$

for the variation of the time functional T of formula (1.3.17), which gives the transit time of a boat crossing a river.

Exercises

1. Calculate the variation of the functional J defined on the normed vector space $\mathcal{C}^1[x_0, x_1]$ by $J(Y) = \int_{x_0}^{x_1} [Y(x)^2 + Y'(x)^2 - 2Y(x)\sin x]\, dx$.

2. The functional $J(Y) = \left[\int_{x_0}^{x_1} x\sqrt{1 + Y'(x)^2}\, dx\right] / \left[\int_{x_0}^{x_1} \sqrt{1 + Y'(x)^2}\, dx\right]$ gives the x-coordinate of the center of mass of a smooth homogeneous wire represented by the curve γ given parametrically as $\gamma: y = Y(x)$ for $x_0 \leq x \leq x_1$. Find the variation of J at an arbitrary fixed vector $Y = Y(x)$ in $\mathcal{C}^1[x_0, x_1]$. *Hint:* Exercise 12 of Section 2.3 may be helpful.

3. The functional $A(Y) = 2\pi \int_{x_0}^{x_1} Y(x)\sqrt{1 + Y'(x)^2}\, dx$ gives the area of the surface of revolution obtained by rotating the curve γ about the x-axis, where γ is given as $\gamma: y = Y(x)$ for $x_0 \leq x \leq x_1$. Find the variation of A at an arbitrary fixed vector $Y = Y(x)$ in $\mathcal{C}^1[x_0, x_1]$.

4. Calculate the variation of the functional J defined on the normed vector space $\mathcal{C}^1[x_0, x_1]$ by $J(Y) = \int_{x_0}^{x_1} x^2 Y'(x)^2\, dx$.

2.5. An Optimization Problem in Production Planning

We shall now solve the optimization problem described in the last few pages of Section 1.3 for the cost functional $C = C(P)$ considered there and in Section 2.4. The functional C is defined on the vector space $\mathcal{C}^0[0, T]$ consisting of all continuous functions (production rates) P on the fixed interval $[0, T]$. We refer the reader to Section 1.3 for a complete statement of the problem.

We seek an extremum vector P^* in $\mathcal{C}^0[0, T]$ that will furnish a *minimum* value to the cost functional C. It was shown in Section 2.4 that this functional has a variation at each vector P in $\mathcal{C}^0[0, T]$. Hence if P^* is the desired extremum vector, we can use the theorem of Section 2.2 to conclude that [see equation (2.2.4)]

$$\delta C(P^*; \Delta P) = 0$$

for every vector ΔP in $\mathcal{C}^0[0, T]$. We can use (2.4.2) to rewrite this equation as

$$\int_0^T \left\{\beta^2[I_{P^*}(t) - \mathcal{I}(t)]e^{-\alpha t} \int_0^t e^{\alpha\tau}\, \Delta P(\tau)\, d\tau + [P^*(t) - \mathcal{P}(t)]\, \Delta P(t)\right\} dt = 0, \tag{2.5.1}$$

which must now hold for *every* continuous function ΔP if P^* is the desired minimum vector.

We seek to eliminate the arbitrary vector ΔP from equation (2.5.1) so as to obtain a simpler equation which will involve only the extremum vector P^* and which can be solved to give the desired extremum vector. For this purpose we use the fact that the order of the repeated (iterated) integrals appearing in (2.5.1) can be interchanged as (see Exercise 1)

$$\begin{aligned}\int_0^T [I_{P^*}(t) - \mathcal{I}(t)]e^{-\alpha t} \int_0^t e^{\alpha\tau}\, \Delta P(\tau)\, d\tau\, dt \\ = \int_0^T e^{\alpha t}\, \Delta P(t) \int_t^T e^{-\alpha\tau}[I_{P^*}(\tau) - \mathcal{I}(\tau)]\, d\tau\, dt,\end{aligned} \tag{2.5.2}$$

so that (2.5.1) is equivalent to

$$\int_0^T \left\{P^*(t) - \mathcal{P}(t) + \beta^2 e^{\alpha t} \int_t^T e^{-\alpha\tau}[I_{P^*}(\tau) - \mathcal{I}(\tau)]\, d\tau\right\} \Delta P(t)\, dt = 0,$$

which must still hold for every continuous function ΔP in $\mathcal{C}^0[0, T]$. If in particular we take ΔP to be the function in braces in the last equation, defined by the formula

$$\Delta P(t) = P^*(t) - \mathcal{P}(t) + \beta^2 e^{\alpha t} \int_t^T e^{-\alpha\tau}[I_{P^*}(\tau) - \mathcal{I}(\tau)]\, d\tau \qquad \text{for } 0 \leq t \leq T,$$

then we find the condition

$$\int_0^T \left\{P^*(t) - \mathcal{P}(t) + \beta^2 e^{\alpha t} \int_t^T e^{-\alpha\tau}[I_{P^*}(\tau) - \mathcal{I}(\tau)]\, d\tau\right\}^2 dt = 0,$$

which must be satisfied by the extremum function P^*. But the only way that such an integral of a continuous *nonnegative* function can vanish is for the integrand function to vanish. Hence if P^* is the desired extremum vector, we conclude that

$$P^*(t) - \mathcal{P}(t) + \beta^2 e^{\alpha t} \int_t^T e^{-\alpha\tau}[I_{P^*}(\tau) - \mathcal{I}(\tau)]\, d\tau = 0 \qquad \text{for } 0 \leq t \leq T \tag{2.5.3}$$

must hold. Equation (2.5.3) is the desired simpler equation involving only the extremum vector P^*.

We can most readily solve equation (2.5.3) for P^* by first replacing this *integral equation* with a certain related *differential equation* which will be easy to solve.† In fact, if we differentiate both sides of equation (2.5.3) with respect to t and use the fundamental theorem of calculus in the form

$$\frac{d}{dt}\int_t^T h(\tau)\,d\tau = -h(t) \qquad \text{for any continuous function } h,$$

we find with equation (2.5.3) after some simplification that

$$\frac{d}{dt}[P^*(t) - \mathcal{P}(t)] = \alpha[P^*(t) - \mathcal{P}(t)] + \beta^2[I_{P^*}(t) - \mathcal{I}(t)], \tag{2.5.4}$$

which must hold for all t in the interval $0 < t < T$. If we now differentiate equation (2.5.4) and use it again to simplify the result, we find that

$$\frac{d^2}{dt^2}[P^*(t) - \mathcal{P}(t)] = \alpha^2[P^*(t) - \mathcal{P}(t)] + \beta^2\left\{\frac{d}{dt}[I_{P^*}(t) - \mathcal{I}(t)] + \alpha[I_{P^*}(t) - \mathcal{I}(t)]\right\} \tag{2.5.5}$$

for all t. On the other hand, (1.3.19) and (1.3.20) imply that

$$\frac{d}{dt}[I_{P^*}(t) - \mathcal{I}(t)] + \alpha[I_{P^*}(t) - \mathcal{I}(t)] = P^*(t) - \mathcal{P}(t), \tag{2.5.6}$$

so that the differential equation (2.5.5) can be written more simply as

$$\frac{d^2}{dt^2}[P^*(t) - \mathcal{P}(t)] = (\alpha^2 + \beta^2)[P^*(t) - \mathcal{P}(t)], \tag{2.5.7}$$

which must be satisfied by the extremum production rate $P^* = P^*(t)$ *for all* t *in the interval* $0 < t < T$.

The most general solution of equation (2.5.7) has the form‡

$$P^*(t) - \mathcal{P}(t) = Ae^{\gamma t} + Be^{-\gamma t}, \qquad \gamma = \sqrt{\alpha^2 + \beta^2}, \tag{2.5.8}$$

for arbitrary constants A and B, which may be specified in the present case by requiring P^* to satisfy certain additional conditions obtained from the

†The details of the following calculation need *not* be followed carefully in a first reading.

‡See, for example, Chapter 20 of George B. Thomas, Jr., *Calculus and Analytic Geometry* (Reading, Mass.: Addison-Wesley Publishing Company, Inc., 1968), or any elementary text on differential equations.

integral equation (2.5.3). For example, if we evaluate equation (2.5.3) at $t = T$, we find the necessary condition

$$P^*(t) - \mathcal{P}(t) = 0 \qquad \text{at } t = T, \tag{2.5.9}$$

while if we evaluate equation (2.5.4) at $t = 0$, we get the condition

$$\frac{d}{dt}[P^*(t) - \mathcal{P}(t)] - \alpha[P^*(t) - \mathcal{P}(t)] = \beta^2(I_0 - \mathcal{I}_0) \qquad \text{at } t = 0, \tag{2.5.10}$$

where $I_0 = I_{P^*}(0)$ and $\mathcal{I}_0 = \mathcal{I}(0)$ are known constants. If we now require that the function P^* given by equation (2.5.8) must satisfy the conditions (2.5.9) and (2.5.10) we find necessarily that the constants A and B must take on the values

$$\begin{aligned} A &= \frac{\beta^2(I_0 - \mathcal{I}_0)e^{-\gamma T}}{(\gamma + \alpha)e^{\gamma T} + (\gamma - \alpha)e^{-\gamma T}} \\ B &= \frac{-\beta^2(I_0 - \mathcal{I}_0)e^{\gamma T}}{(\gamma + \alpha)e^{\gamma T} + (\gamma - \alpha)e^{-\gamma T}}, \end{aligned} \tag{2.5.11}$$

so that the extremum function P^* is given by [see equations (2.5.8) and (2.5.11)]

$$P^*(t) = \mathcal{P}(t) + \beta^2(\mathcal{I}_0 - I_0)\frac{e^{\gamma(T-t)} - e^{-\gamma(T-t)}}{(\gamma + \alpha)e^{\gamma T} + (\gamma - \alpha)e^{-\gamma T}} \tag{2.5.12}$$

for any t in the interval $[0, T]$, where $\gamma = \sqrt{\alpha^2 + \beta^2}$. If we insert the result of equation (2.5.12) into (2.5.4), we find the corresponding optimum inventory level I_{P^*} to be given as

$$I_{P^*}(t) = \mathcal{I}(t) + (I_0 - \mathcal{I}_0)\frac{(\gamma + \alpha)e^{\gamma(T-t)} + (\gamma - \alpha)e^{-\gamma(T-t)}}{(\gamma + \alpha)e^{\gamma T} + (\gamma - \alpha)e^{-\gamma T}} \tag{2.5.13}$$

One checks easily that the production rate P^* given by equation (2.5.12), along with the corresponding inventory level I_{P^*} given by (2.5.13), does, in fact, satisfy the integral equation (2.5.3) for $0 \leq t \leq T$. Moreover, our calculation leading to (2.5.12) shows that *this function P^* is the only function for which equation* (2.5.3) *holds*. Hence (2.5.12) gives the only function P^* of class $\mathcal{C}^0$ on $[0, T]$ which satisfies the necessary condition (2.5.1) of the theorem of Section 2.2. This condition (2.5.1) is a necessary condition which must be satisfied by any possible local extremum vector for the cost functional C. As in elementary calculus, however, *we must still check whether P^* gives a maximum or a minimum value to the cost functional C, or whether perhaps P^* represents some sort of inflection point or saddle point for C* (see the examples near the end of Section 2.2 and Exercise 6 in this section). In fact, it is easy to show in the present case that P^* furnishes a *minimum* value to the cost functional C since (2.5.12), (2.5.13), and the calculations following (2.4.1) imply that

$$C(P^* + Q) - C(P^*) = \int_0^T \left\{ \beta^2 \left[e^{-\alpha t} \int_0^t e^{\alpha \tau} Q(\tau)\, d\tau \right]^2 + Q(t)^2 \right\} dt \qquad (2.5.14)$$

for any function Q of class $\mathcal{C}^0$ on $[0, T]$. Since the right-hand side of (2.5.14) is always *nonnegative* for any Q, it follows that

$$C(P^* + Q) - C(P^*) \geq 0$$

for every Q in the vector space $\mathcal{C}^0[0, T]$, which proves that *the cost functional C has a minimum in $\mathcal{C}^0[0, T]$ at the vector P^* defined by formula* (2.5.12). The resulting minimum value of C may be found by inserting the results of equations (2.5.12) and (2.5.13) back into formula (1.3.24); if this is done, we find that

$$\begin{aligned} C(P^*) &= C_{\text{minimum}} \\ &= \beta^2 (I_0 - \mathcal{I}_0)^2 \frac{e^{\gamma T} - e^{-\gamma T}}{(\gamma + \alpha) e^{\gamma T} + (\gamma - \alpha) e^{-\gamma T}}, \end{aligned} \qquad (2.5.15)$$

which value is *positive* if $\beta^2 (I_0 - \mathcal{I}_0)^2 > 0$.

Formulas (2.5.12) and (2.5.13) show that the optimum production rate P^* gives corresponding *deviations* $P^* - \mathcal{P}$ and $I_{P^*} - \mathcal{I}$ which depend linearly on the *initial disturbance* $I_0 - \mathcal{I}_0$. Moreover, these deviations will be small if the initial disturbance is small. For example, if we use the uniform norm [see (1.4.8)], we find that

$$\| P^* - \mathcal{P} \| = \beta^2 | I_0 - \mathcal{I}_0 | \frac{e^{\gamma T} - e^{-\gamma T}}{(\gamma + \alpha) e^{\gamma T} + (\gamma - \alpha) e^{-\gamma T}}$$

$$\| I_{P^*} - \mathcal{I} \| = | I_0 - \mathcal{I}_0 |.$$

If the initial disturbance is such that $I_0 - \mathcal{I}_0$ is *negative* (as might be the case following a mechanical breakdown which temporarily halts production at the manufacturing plant), then the optimum new production rate P^* will (naturally) *exceed* $\mathcal{P}$, while the corresponding inventory level I_{P^*} will be lower than $\mathcal{I}$, as indicated in Figure 5.

We note also that if the company chooses to *decrease* β^2 in formula (1.3.21) so as to place relatively *more importance* on the minimization of $P - \mathcal{P}$ as compared with $I - \mathcal{I}$, then formula (2.5.12) shows that the optimum production rate P^* will indeed be closer to $\mathcal{P}$, while formula (2.5.13) shows in this case that the corresponding inventory level I_{P^*} will be farther from $\mathcal{I}$ for $t > 0$. Similarly, one can study the effects obtained in formulas (2.5.12), (2.5.13), and (2.5.15) by varying the spoilage factor α.

Finally, we note that in every case the optimum production rate $P^* = P^*(t)$ of (2.5.12) agrees with the desired rate $\mathcal{P}$ at the terminal time $t = T$,

$$P^*(T) = \mathcal{P}(T). \qquad (2.5.16)$$

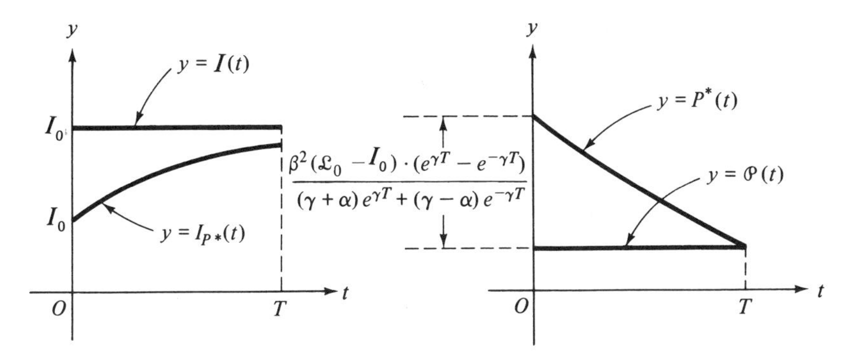

Optimum solution for minimum cost

Figure 5

In fact, this is just the condition given by equation (2.5.9) which arises upon evaluating the integral equation (2.5.3) at $t = T$. This boundary condition (2.5.16) has a somewhat different nature than the condition

$$I(0) = I_0$$

[see equation (1.3.22)] which is imposed initially at $t = 0$. Indeed, the condition (2.5.16) is *nowhere explicitly imposed* in the statement of the problem; rather it arises automatically or *naturally* in the solution of the optimization problem as a consequence of the required vanishing of the variation at any extremum vector [see equations (2.2.4) and (2.5.1)]. Such conditions as (2.5.16) are known as **natural boundary conditions** in contrast with *imposed* boundary conditions such as (1.3.22). We shall see that natural boundary conditions appear in the solutions of many (but not all) optimization or extremum problems.

Before leaving this section, we should mention that in one important respect the extremum problem just considered for the cost functional C is *not* typical of many extremum problems which arise in practice. Indeed, most extremum problems arising in practice involve various *constraints* of one kind or another. For example, we saw in Section 1.3 that Chaplygin's problem on maximizing the area encircled by the path of an airplane flying with constant natural speed while a constant wind is blowing is equivalent to the problem of maximizing the value of the area functional A defined by formula (1.3.11) subject to the constraints given by equations (1.3.9) and (1.3.10). Moreover, the minimum problem just considered for the cost functional C may be unrealistic since the manufacturing company may *not* have adequate labor and capital resources to enable it to increase production up to the level required by the optimum production rate P^* given by formula (2.5.12). In this case a realistic approach to the production planning problem must include considerations of any relevant production constraints. We shall see in

Chapter 3 how to handle such extremum problems involving constraints using the method of Euler-Lagrange multipliers.

Exercises

1. Let $g = g(t, \tau)$ be any function defined and continuous on the triangular region $0 \leq \tau \leq t \leq T$, as shown in Figure 6. Verify that the two iterated integrals $\int_0^T \{\int_0^t g(t, \tau)\, d\tau\}\, dt$ and $\int_0^T \{\int_\tau^T g(t, \tau)\, dt\}\, d\tau$ represent the same integral of $g(t, \tau)$ over the same triangular region in the (t, τ)-plane, with only the orders of integration interchanged, so that† $\int_0^T \{\int_0^t g(t, \tau)\, d\tau\}\, dt = \int_0^T \{\int_\tau^T g(t, \tau)\, dt\}\, d\tau$ holds. Use this equation to prove the related result

$$\int_0^T \left\{\int_0^t g(t, \tau)\, d\tau\right\} dt = \int_0^T \left\{\int_t^T g(\tau, t)\, d\tau\right\} dt \tag{2.5.17}$$

for any such continuous function g. Show how to use (2.5.17) to prove the result (2.5.2).

2. Show that equations (2.5.1) and (2.5.3) are equivalent in the sense that either implies the other.

3. Carry through the steps in the derivation of the differential equation (2.5.7) from the integral equation (2.5.3).

4. Show that the production rate P^* given by equation (2.5.12) satisfies equation (2.5.1) for all vectors ΔP in the vector space $\mathcal{C}^0[0, T]$.

5. Let the functional K be defined for any vector P in the vector space $\mathcal{C}^0[0, T]$ by $K(P) = \int_0^T [P(t) - \mathcal{P}(t)]^3\, dt$, where $\mathcal{P} = \mathcal{P}(t)$ is a given known function of

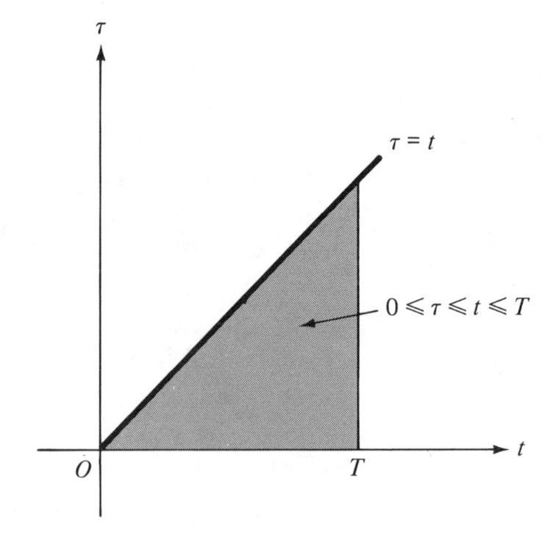

Figure 6

†See Chapter 9 of Sherman K. Stein, *Calculus in the First Three Dimensions* (New York: McGraw-Hill Book Company, 1967), or Chapter 12 of Hocking, *Op. cit.*

class $\mathcal{C}^0$ on $[0, T]$. Show that K has a variation at any P in $\mathcal{C}^0[0, T]$, with $\delta K(P; \Delta P) = 3 \int_0^T [P(t) - \mathcal{P}(t)]^2 \, \Delta P(t) \, dt$ for any vector ΔP in $\mathcal{C}^0[0, T]$.

6. Let $K = K(P)$ be as in Exercise 5. Show that $P^* = \mathcal{P}$ is the only *possible* local extremum vector in $\mathcal{C}^0[0, T]$ for K, and determine whether or not this vector is, in fact, a local extremum vector for K.

2.6. Some Remarks on the Fréchet Differential†

We have seen in Section 2.4 how to calculate the Gâteaux variations of many of the functionals appearing in the extremum problems described in Section 1.3. The results of Section 2.4, and in particular equations (2.4.2), (2.4.8), and (2.4.17), show that in all these special cases the variations involved are *linear functionals of their second arguments.* (See Section 1.6 for the notion of linearity.) For example, it is easy to check that the variation of the cost functional C given by (2.4.2) satisfies the linearity relation

$$\delta C(P; a_1 \, \Delta P_1 + a_2 \, \Delta P_2) = a_1 \, \delta C(P; \Delta P_1) + a_2 \, \delta C(P; \Delta P_2)$$

for any numbers a_1 and a_2 and for any vectors ΔP_1 and ΔP_2 in the vector space $\mathfrak{X} = \mathcal{C}^0[0, T]$. Similarly, the other variations calculated in Section 2.4 are linear functionals of their second arguments. In general, however, the variation of a given functional J need *not* satisfy this linearity condition, as is shown by the example given in Exercise 6 of Section 2.3. On the other hand, there are certain situations where it is necessary to restrict consideration to functionals whose variations are linear in the second argument. For example, the results of Sections 3.8 and 3.10 require that the variations involved must satisfy several special conditions, including this linearity condition. In such cases it is often convenient to restrict consideration to functionals J which are *Fréchet differentiable.*

We say that a functional J defined on an open subset D of a normed vector space $\mathfrak{X}$ is **Fréchet differentiable,**‡ or just *differentiable*, at a vector x in D whenever there is a *continuous linear* functional $dJ(x)$ with values $dJ(x; h)$ defined for all vectors h in $\mathfrak{X}$ and for which

$$\underset{h \to 0 \text{ in } \mathfrak{X}}{\text{limit}} \frac{J(x + h) - J(x) - dJ(x; h)}{\| h \|} = 0 \tag{2.6.1}$$

holds. Here the limit is understood in the sense of formula (1.5.1), with the functional J of that formula replaced in equation (2.6.1) with the functional

†This section can be safely skipped during a first reading. The reader who so desires should turn directly to Chapter 3.

‡After the French mathematician Maurice Fréchet (1878–), who early in the twentieth century made important contributions in the studies of normed vector spaces and abstract differential calculus.

of h given as

$$\frac{J(x+h)-J(x)-dJ(x;h)}{\|h\|}$$

for $\|h\|>0$. The continuous linear functional $dJ(x)$ is called the (Fréchet) *differential of J at x*. If J is differentiable at each vector x in D, we say that J is differentiable on D.

It follows from this definition that a functional J is differentiable at x in D whenever there is a continuous linear functional $dJ(x)$ with values $dJ(x; h)$ defined for all vectors h in $\mathfrak{X}$ such that the expression $E(x; h)$ defined by

$$J(x+h)=J(x)+dJ(x;h)+E(x;h)\|h\| \qquad \text{for any small nonzero vector } h \text{ in } \mathfrak{X} \tag{2.6.2}$$

has the limit 0 at the zero vector in $\mathfrak{X}$,

$$\operatorname*{limit}_{h\to 0 \text{ in } \mathfrak{X}} E(x;h)=0. \tag{2.6.3}$$

The linearity requirement on the differential requires that

$$dJ(x;a_1h_1+a_2h_2)=a_1dJ(x;h_1)+a_2dJ(x;h_2) \tag{2.6.4}$$

must hold for any numbers a_1, a_2 and for any vectors h_1, h_2. The continuity condition amounts to the requirement that [compare with (1.6.4)]

$$|dJ(x;h)|\leq \text{constant}\,\|h\| \qquad \text{for all vectors } h \text{ in } \mathfrak{X}. \tag{2.6.5}$$

It is easy to check that *if a functional J is differentiable at x, then the variation of J at x exists and is equal to the differential,*

$$\delta J(x;h)=dJ(x;h) \qquad \text{for all } h \text{ in } \mathfrak{X}. \tag{2.6.6}$$

Indeed, if J is differentiable at x, then the linearity relation (2.6.4) implies that $dJ(x;\epsilon h)=\epsilon\, dJ(x;h)$ for any number ϵ and any vector h, so that (2.6.2) implies that

$$\frac{J(x+\epsilon h)-J(x)}{\epsilon}=dJ(x;h)+E(x;\epsilon h)\|h\|\frac{|\epsilon|}{\epsilon}$$

for any vector h in $\mathfrak{X}$ and for all small nonzero numbers ϵ. But equation (2.6.3) implies that

$$\operatorname*{limit}_{\epsilon\to 0} E(x;\epsilon h)\frac{|\epsilon|}{\epsilon}=0$$

for any fixed vector h in $\mathfrak{X}$. The last two results imply that

$$\underset{\epsilon \to 0}{\text{limit}} \frac{J(x + \epsilon h) - J(x)}{\epsilon} = dJ(x; h)$$

whenever J is differentiable at x. It follows from this last result and equation (2.3.1) that J has a variation at x given by (2.6.6).

On the other hand, a functional J may have a variation at a vector x even if J is *not* differentiable at x. Indeed, the variation is not required to satisfy the linearity and continuity conditions (2.6.4) and (2.6.5). Hence *it is easier for a functional to have a variation than to have a differential.* (See Exercises 1 and 2 in this section and Exercises 5 and 6 of Section 2.3.) We shall generally use the variation rather than the differential in this book except in certain special situations which require the added properties of the differential.

Exercises

1. Let J be a functional defined on an open set D in a normed vector space $\mathfrak{X}$. Show that J is automatically continuous at x in D if it is differentiable at x. (Compare this result with the example in Exercise 5 of Section 2.3.)

2. Let $\mathfrak{X}$ and J be as in Exercise 4 of Section 1.5. Use Exercise 1 to show that J fails to be differentiable at each fixed vector ϕ in $\mathfrak{X}$.

3. Let $\mathfrak{Y}$ and J be as in Exercise 5 of Section 1.5. Show that J is differentiable on $\mathfrak{Y}$ and find the differential of J at the vector ϕ in $\mathfrak{Y}$. (Exercises 2 and 3 show that the differentiability of a given functional defined on a vector space may depend on the particular norm used.)

4. Show that the cost functional C defined by formulas (1.3.23) and (1.3.24) on the normed vector space $\mathfrak{C}^0[0, T]$ is differentiable with differential given by (2.4.2). Use the uniform norm $\|\cdot\|$ defined by $\|P\| = \max_{0 \leq t \leq T} |P(t)|$ for any continuous function P on $[0, T]$.

3. The Euler-Lagrange Necessary Condition for an Extremum with Constraints

In this chapter we shall introduce Euler–Lagrange multipliers and show how to use them to solve various extremum problems involving certain types of equality constraints. The meaning of the multipliers will be discussed, and several examples will be worked out in detail. We shall also discuss briefly certain extremum problems which involve *inequality* constraints.

3.1. Extremum Problems with a Single Constraint

Let $\mathfrak{X}$ be a normed vector space, let D be an open subset of $\mathfrak{X}$, and let J and K be any two functionals which are defined and have variations on D.

We consider the problem of finding extremum vectors x^* for J among all those vectors x in D which satisfy the constraint

$$K(x) = k_0, \tag{3.1.1}$$

where k_0 is some specified fixed number.

To simplify our discussion, we shall use the symbol $D[K = k_0]$ to represent the *subset of D consisting of all vectors x in D which satisfy the constraint* (3.1.1). Here k_0 may be any fixed number, and we shall always assume that there is at least one vector x in D which satisfies (3.1.1) so that the set $D[K = k_0]$ is not empty. Then the extremum problem that we shall consider is to *find local extremum vectors in* $D[K = k_0]$ *for J*. Note that the *definition* of local extremum vectors for J as given in Section 2.2 applies to the set $D[K = k_0]$ even though this set is in general *not* an *open* subset of $\mathfrak{X}$ (see the example in the next paragraph). However, the variation of J need *not* vanish at a local extremum vector x^* in $D[K = k_0]$ *if this set is not open in* $\mathfrak{X}$. Indeed, the proof of the theorem in Section 2.2 on the vanishing of the variation at x^* required that x^* be a local extremum vector for J in an *open* subset of X (why?). Hence we cannot expect to be able to use equation (2.2.4) to find extremum vectors x^* in $D[K = k_0]$ for J. It is, in fact, easy to give examples where equation (2.2.4) fails to hold for such *constrained* extremum vectors x^* in $D[K = k_0]$, as we shall now see.

Let J and K be the real-valued functions defined on $\mathfrak{R}$ by

$$J(x) = x^2, \qquad K(x) = x^2 + 2x + \tfrac{3}{4} \tag{3.1.2}$$

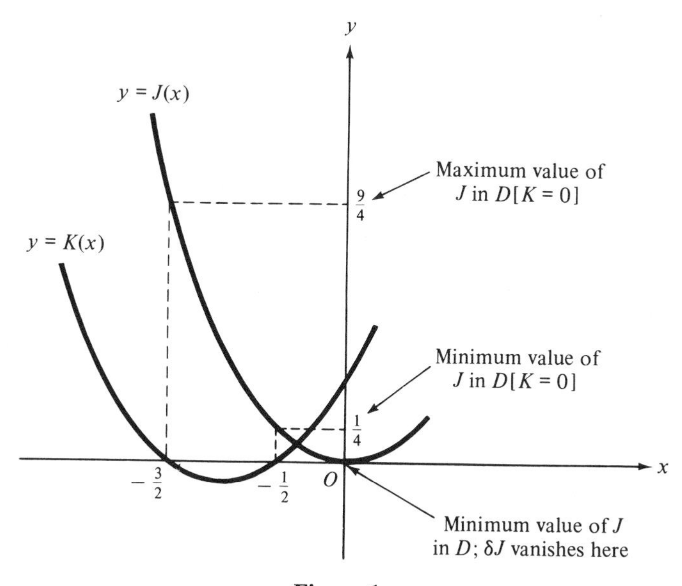

Figure 1

for any number x in $D = \mathcal{R}$. The set $D[K = 0]$ is easily seen in this case to consist of the two numbers $x = -\frac{1}{2}$ and $x = -\frac{3}{2}$ (see Figure 1), i.e.,

$$D[K = 0] = \{-\tfrac{1}{2}, -\tfrac{3}{2}\}, \tag{3.1.3}$$

and this set is not open in $D = \mathcal{R}$. Clearly $x = -\frac{1}{2}$ gives a minimum to J in $D[K = 0]$, while $x = -\frac{3}{2}$ gives a maximum to J in $D[K = 0]$. However, the variation of J fails to vanish at each point in $D[K = 0]$, and neither point furnishes a local extremum for J in $D = \mathcal{R}$.

In many extremum problems involving a constraint of the form (3.1.1) *it is not feasible to determine explicitly the set* $D[K = k_0]$ as we did in the simple example just considered [see (3.1.3)]. The method of Euler-Lagrange multipliers allows us to solve many such extremum problems without any direct consideration of the set $D[K = k_0]$. To state this multiplier method, we first need the notion of *weak continuity of variations.*

3.2. *Weak Continuity of Variations*†

If J is any functional which has a variation on an open set D contained in a normed vector space $\mathfrak{X}$, and if for some vector x in D

$$\operatorname*{limit}_{y \to x \text{ in } \mathfrak{X}} \delta J(y; \Delta x) = \delta J(x; \Delta x) \tag{3.2.1}$$

holds for every vector Δx in $\mathfrak{X}$, then we shall say that *the variation of J is* **weakly continuous at** x. Alternatively, recalling the definition of continuity of functionals [see (1.5.2)], we may say that the variation of J is weakly continuous at x in D whenever for each fixed vector Δx in $\mathfrak{X}$ the variation $\delta J(y; \Delta x)$ considered as a functional of y is continuous at $y = x$. In practice we need only show that for each fixed Δx the difference

$$\delta J(y; \Delta x) - \delta J(x; \Delta x)$$

can be made arbitrarily small for all vectors y which are sufficiently close to the vector x. We shall say that the variation of J is weakly continuous *near* x if the variation of J is weakly continuous at y for every vector y in some ball $B_\rho(x)$ centered at x. The notion of weak continuity of variations is simply a generalization to functionals of the notion of continuity of the first-order partial derivatives of real-valued functions of several real variables (see Exercise 2).

In most applications which we shall consider it will be easy to demon-

†This section can be skimmed over rather lightly during a first reading without causing any real difficulties later.

strate weak continuity for the variations involved. For example, if we consider again the cost functional C defined by formula (1.3.24), we find by (2.4.2) that

$$\delta C(Q; \Delta P) - \delta C(P; \Delta P) = 2\beta^2 \int_0^T [I_Q(t) - I_P(t)]e^{-\alpha t} \int_0^t e^{\alpha\tau} \Delta P(\tau)\, d\tau\, dt + 2 \int_0^T [Q(t) - P(t)]\, \Delta P(t)\, dt$$

for any vectors Q, P, and ΔP in the vector space $\mathcal{C}^0[0, T]$, where according to equation (1.3.23) we shall have

$$I_Q(t) - I_P(t) = e^{-\alpha t} \int_0^t e^{\alpha\tau}[Q(\tau) - P(\tau)]\, d\tau.$$

Using the last two equations, it is then easy to prove (for each fixed ΔP) that the difference

$$\delta C(Q; \Delta P) - \delta C(P; \Delta P)$$

can be made arbitrarily small by requiring $\|Q - P\|$ to be small, where we might, for example, use the uniform norm with

$$\|Q - P\| = \max_{0 \leq t \leq T} |Q(t) - P(t)|.$$

We leave the demonstration as an exercise for the reader. The conclusion in this case is that the variation $\delta C(P)$ is weakly continuous at P for each vector P in the vector space $\mathcal{C}^0[0, T]$.

Exercises

1. Let $f = f(x)$ be a real-valued function defined on some open niterval (a, b), with variation $\delta f(x)$ defined by (see Exercise 1 of Section 2.2) $\delta f(x; \Delta x) = f'(x)\, \Delta x$ for any number Δx in $\mathcal{R}$. Show that the variation of f is weakly continuous at x if and only if the function f is continuously differentiable at x in the usual sense of elementary calculus.
2. Let $F = F(x)$ be a real-valued function defined for all vectors $x = (x_1, x_2, \ldots, x_n)$ in some fixed open set D in n-dimensional Euclidean space $\mathcal{R}_n$, with variation $\delta F(x)$ defined by [see (2.2.5)] $\delta F(x; \Delta x) = \sum_{i=1}^n [\partial F(x)/\partial x_i]\, \Delta x_i$ for any vector $\Delta x = (\Delta x_1, \Delta x_2, \ldots, \Delta x_n)$ in $\mathcal{R}_n$. Show that the variation of F is weakly continuous at x if and only if the function F has continuous first-order partial derivatives $\partial F(x)/\partial x_i$ at x for $i = 1, 2, \ldots, n$.
3. Show that the functional J of Exercise 4 of Section 1.5 has a variation which is weakly continuous at each vector ϕ in $\mathcal{X}$ even though J is not itself continuous.

3.3. Statement of the Euler-Lagrange Multiplier Theorem for a Single Constraint

We are now in a position to state the multiplier theorem for an extremum problem involving a constraint of the form (3.1.1). We recall that the symbol $D[K = k_0]$ denotes the subset of D consisting of all vectors x in D which satisfy (3.1.1).

Euler-Lagrange Multiplier Theorem. *Let J and K be functionals which are defined and have variations on an open subset D of a normed vector space $\mathfrak{X}$, and let x^* be a local extremum vector in $D[K = k_0]$ for J, where k_0 is any given fixed number for which the set $D[K = k_0]$ is nonempty. Assume that both the variation of J and the variation of K are weakly continuous near x^*. Then at least one of the following two possibilities must hold:*

1. *The variation of K at x^* vanishes identically, i.e.,*

$$\delta K(x^*; \Delta x) = 0 \tag{3.3.1}$$

for every vector Δx in $\mathfrak{X}$; or

2. *The variation of J at x^* is a constant multiple of the variation of K at x^*, i.e., there is a constant λ such that*

$$\delta J(x^*; \Delta x) = \lambda\, \delta K(x^*; \Delta x) \tag{3.3.2}$$

for every vector Δx in $\mathfrak{X}$.

Before giving the proof, we note that the theorem simply guarantees that all possible local extremum vectors in $D[K = k_0]$ for J must be contained in the collection of those vectors in $D[K = k_0]$ which satisfy either (3.3.1) or (3.3.2). However, there may also be other vectors in $D[K = k_0]$ which satisfy either (3.3.1) or (3.3.2) and which are *not* local extremum vectors in $D[K = k_0]$ for J. In practice we first find all vectors x^* in D which satisfy the first condition (3.3.1) and then *retain for further consideration* only those vectors (if there are any) which also satisfy the constraint $K(x^*) = k_0$. We next find all vectors x^* which satisfy the second condition (3.3.2), where now the solutions x^* of (3.3.2) will depend in general on the value of the parameter λ. We again *retain for further consideration* only those solutions x^* which also satisfy the constraint $K(x^*) = k_0$. This requirement that x^* must satisfy the given constraint in addition to (3.3.2) will determine a fixed value (or values) for the parameter λ in terms of the given constant k_0 appearing in the constraint. Any such special value of λ for which both (3.3.2) and the constraint hold is called an **Euler–Lagrange multiplier** for the given extremum problem.

Finally, to find the desired maximum or minimum vector in $D[K = k_0]$ for J, we need only search through the collection of all those vectors x^* which have been retained for further consideration. Note that the procedure is much the same as that encountered in the use of the earlier theorem on the vanishing of the variation at an extremum vector (see Sections 2.2 and 2.5), except that rather than solve the single equation (2.2.4) for the extremum vector (as, for example, we did in Section 2.5 when solving the unconstrained extremum problem considered there for the cost functional C), we must now take into account both equations (3.3.1) and (3.3.2). Actually some authors prefer to combine these two equations into the single (more symmetrical) equation

$$\mu_0 \, \delta J(x^*; \Delta x) + \mu_1 \, \delta K(x^*; \Delta x) = 0 \tag{3.3.3}$$

for suitable constants μ_0 and μ_1. The latter equation clearly follows from our statement of the multiplier theorem given above since equation (3.3.1) corresponds to the choice $\mu_0 = 0$, $\mu_1 = 1$ in (3.3.3), while equation (3.3.2) corresponds to the choice $\mu_0 = 1$, $\mu_1 = -\lambda$ in (3.3.3).†

Leonhard Euler (1707–1789) had discovered how to solve constrained extremum problems using what amounts to equation (3.3.3) by the year 1741.‡ Joseph Lagrange (1736–1813) studied Euler's results and later formulated and publicized the multiplier theorem for the important special case in which J and K are functions of n real variables. In this special case the theorem is often referred to as the *Lagrange multiplier theorem*.§ The development of the rigorous foundations of the differential and integral calculus came later in the nineteenth century, while the development of the abstract calculus of functionals on normed vector spaces came in the twentieth century. Hence neither Euler nor Lagrange could be expected to give a proof of the multiplier theorem which would satisfy today's standards of rigor.

†According to equation (3.3.3), we may say that the variations of J and K are *linearly dependent* (as vectors in the vector space of all linear functionals on $\mathfrak{X}$) at any local extremum vector x^* in $D[K = k_0]$ for J. See John G. Hocking, *Calculus with an Introduction to Linear Algebra* (New York: Holt, Rinehart and Winston, Inc., 1970), p. 738.

‡See, for example, Chapters 5 and 6 of Leonhard Euler, *Methodus Inveniendi Lineas Curvas* (Lausanne & Geneva, 1744), where Euler uses the method of multipliers to solve constrained extemum problems involving integral functionals of the type given by our formula (2.4.9). This book was first published in 1744, but Euler had apparently completed the final manuscript in 1741. The book has been republished more recently in Euler's collected works as *Leonhardi Euleri Opera Omnia*, I, **24**.

§See Article 58 of the second part of Lagrange's *Théorie des Fonctions Analytiques* (Paris: Imp. République, 1797). This book has been republished in Lagrange's collected works as *Oeuvres de Lagrange*, **9**. Lagrange himself acknowledged Euler's earlier work on the multiplier method; see *Oeuvres de Lagrange*, **10**, p. 389.

3.4. Three Examples and Some Remarks on the Geometrical Significance of the Multiplier Theorem

Before giving the proof of the multiplier theorem (found in Section 3.5), we shall illustrate the use of the theorem by considering several simple examples. We begin by considering again the simple extremum problem already solved in Section 3.1 involving the real-valued functions J and K defined on $\mathcal{R}$ by $J(x) = x^2$ and $K(x) = x^2 + 2x + \frac{3}{4}$. For these functions the given problem is to find extremum vectors (numbers) in $D[K = 0]$ for J, where in this case $D = \mathcal{R}$. The variations of J and K at any number x are found with (2.3.4) to be given by

$$\begin{aligned} \delta J(x; \Delta x) &= 2x\, \Delta x \\ \delta K(x; \Delta x) &= 2(x+1)\, \Delta x \end{aligned} \tag{3.4.1}$$

for any number Δx. The only number x^* which satisfies $\delta K(x^*; \Delta x) = 0$ is found with (3.4.1) to be $x^* = -1$, and this number is omitted from any further consideration since it does *not* satisfy the constraint $K = 0$. Hence we need only consider further the second possibility of the Euler-Lagrange multiplier theorem; i.e., we need only consider equation (3.3.2), which with (3.4.1) becomes in the present case

$$2x^*\, \Delta x = \lambda 2(x^* + 1)\, \Delta x$$

or, equivalently,

$$2\{x^* - \lambda(x^* + 1)\}\, \Delta x = 0,$$

which must then hold for some constant λ and for *every* number Δx if x^* is a local extremum vector in $D[K = 0]$ for J. We make the special choice $\Delta x = 1$ in the last equation and conclude that any possible local extremum vector x^* in $D[K = 0]$ for J must satisfy the condition

$$x^* - \lambda(x^* + 1) = 0 \tag{3.4.2}$$

for some constant λ. Equation (3.4.2) has solutions depending on the parameter λ, given as

$$x^* = \frac{\lambda}{1 - \lambda}. \tag{3.4.3}$$

If we now insert (3.4.3) into the constraint $K(x^*) = 0$, we find the following equation for λ,

$$\lambda^2 - 2\lambda - 3 = 0,$$

with solutions $\lambda = -1$ and $\lambda = 3$. Using each of these values in turn for the Euler-Lagrange multiplier λ in (3.4.3), we find again the solutions $x^* = -\frac{1}{2}$ and $x^* = -\frac{3}{2}$, as found in Section 3.1 [see (3.1.3)]. It is a simple matter to check which value gives a maximum or minimum.

As another simple illustration of the use of the Euler-Lagrange multiplier theorem we consider the problem of finding the dimensions of the rectangle having the *smallest perimeter* among all rectangles with given *fixed area* A. We let x_1 be the length and x_2 the width of any such rectangle, so that the perimeter is $2(x_1 + x_2)$ and the area is x_1x_2. We define a *perimeter* function J and an *area* function K by

$$J(x) = 2(x_1 + x_2), \qquad K(x) = x_1x_2 \tag{3.4.4}$$

for any vector $x = (x_1, x_2)$ in two-dimensional space $\mathcal{R}_2$, and the problem is then to find a minimum point for the function J in the open set

$$D = \{x = (x_1, x_2) : x_1 > 0, x_2 > 0\} \tag{3.4.5}$$

subject to the constraint

$$K(x) = A. \tag{3.4.6}$$

Alternatively we may say that the problem is to find a minimum vector in $D[K = A]$ for J. The situation is illustrated in Figure 2.

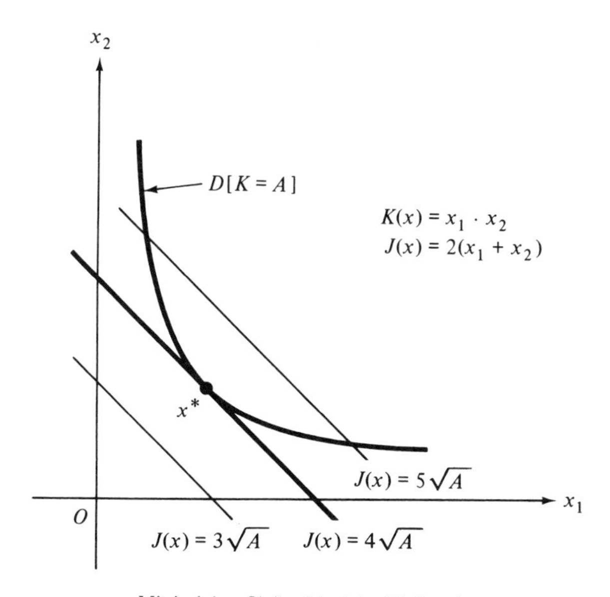

Minimizing $J(x)$ subject to $K(x) = A$

Figure 2

The variations of J and K at any point $x = (x_1, x_2)$ in D are found with (2.2.5) and (3.4.4) to be given by

$$\begin{aligned} \delta J(x; \Delta x) &= 2\Delta x_1 + 2\Delta x_2 \\ \delta K(x; \Delta x) &= x_2 \Delta x_1 + x_1 \Delta x_2 \end{aligned} \tag{3.4.7}$$

for any vector $\Delta x = (\Delta x_1, \Delta x_2)$ in $\mathcal{R}_2$. Hence if $x = (x_1, x_2)$ is any fixed vector in D, it follows from (3.4.5) and (3.4.7) that the value of the variation $\delta K(x; \Delta x)$ is *positive* in the special case $\Delta x = (x_2, x_1)$ (i.e., $\Delta x_1 = x_2$, $\Delta x_2 = x_1$), so that the variation of K does *not* vanish identically at any vector x in D. Hence the first possibility (3.3.1) of the Euler-Lagrange multiplier theorem is eliminated, and the second possibility (3.3.2) must therefore hold at any local extremum vector x^* in $D[K = A]$ for J. Therefore if $x^* = (x_1^*, x_2^*)$ is a minimum point in $D[K = A]$ for J, then

$$2\,\Delta x_1 + 2\,\Delta x_2 = \lambda x_2^*\,\Delta x_1 + \lambda x_1^*\,\Delta x_2$$

must hold, or, equivalently,

$$(2 - \lambda x_2^*)\,\Delta x_1 + (2 - \lambda x_1^*)\,\Delta x_2 = 0$$

for some constant λ and *for all numbers* Δx_1 *and* Δx_2. We make the special choices $\Delta x_1 = 2 - \lambda x_2^*$ and $\Delta x_2 = 2 - \lambda x_1^*$ in the last equation and conclude that (explain)

$$2 - \lambda x_2^* = 0, \qquad 2 - \lambda x_1^* = 0. \tag{3.4.8}$$

Hence any local extremum vector x^* in $D[K = A]$ for J must satisfy

$$x^* = (x_1^*, x_2^*) = \left(\frac{2}{\lambda}, \frac{2}{\lambda}\right) \tag{3.4.9}$$

for some constant λ. If we insert (3.4.9) back into the constraint (3.4.6) and use (3.4.4), we find for the Euler-Lagrange multiplier the value (explain)

$$\lambda = \frac{2}{\sqrt{A}}, \tag{3.4.10}$$

where we have excluded the other possible choice $\lambda = -(2/\sqrt{A})$ since it leads with (3.4.9) to a vector x^* which is not in D [see (3.4.5)]. Finally, then, the only vector x^* in D which satisfies the conditions of the Euler-Lagrange multiplier theorem is given by (3.4.9) and (3.4.10) as

$$x^* = (x_1^*, x_2^*) = (\sqrt{A}, \sqrt{A}). \tag{3.4.11}$$

We hope to conclude from (3.4.11) that the rectangle with smallest perimeter among all rectangles with given area A is a *square* with side $\sqrt{A}$. Actually we have only shown that such a square is a candidate which may give either the smallest or the largest perimeter. Indeed, the multiplier theorem which we used in obtaining (3.4.11) furnishes nothing more than a certain necessary condition [see (3.3.3)] which must be satisfied by any possible minimum and/or maximum vector, and in fact this condition may also be satisfied by certain other nonextremum vectors (see, for example, the last paragraph of Section 2.2). But as we shall see, now that we have the candidate x^* of (3.4.11) obtained from the multiplier theorem, it is an easy matter to show *directly* that x^* is, in fact, a *minimum* vector in $D[K = A]$ for J. (This fact may be quite apparent to the reader on geometrical grounds; see Figure 2.) To this end we use (3.4.4) to find

$$J(x^* + \Delta x) - J(x^*) = 2(\Delta x_1 + \Delta x_2),$$

and then if $\Delta x = (\Delta x_1, \Delta x_2)$ is any vector in $\mathcal{R}_2$ for which $x^* + \Delta x$ is in $D[K = A]$, we find from (3.4.4), (3.4.6), and (3.4.11) that

$$\sqrt{A}\,(\Delta x_1 + \Delta x_2) + \Delta x_1\, \Delta x_2 = 0.$$

The last two equations now imply that

$$J(x^* + \Delta x) - J(x^*) = \frac{(\Delta x_1)^2}{\sqrt{A} + \Delta x_1}$$

for any such admissible vector Δx, and we then conclude that

$$J(x^*) \leq J(x)$$

for any vector $x = x^* + \Delta x$ in $D[K = A]$, the desired result. The reader may wish to supply a different method of solution for this problem without using the multiplier theorem.†

The preceding example illustrates the geometric content of the Euler-Lagrange multiplier theorem, namely, that *the level curves of the two functions J and K intersect tangentially at any local extremum vector x^* in $D[K = k_0]$ for J*, as indicated in Figure 3. Indeed, it seems reasonably clear from Figure 2 in the preceding example that if x^* is the minimum vector in $D[K = A]$ for J, then the particular level curve of J which contains x^* must intersect the constraint curve $D[K = A]$ tangentially. The reader should be able to convince himself of this fact geometrically by sketching a suitable figure for the alternative situation which would obtain if these level curves were to intersect *nontangentially* at x^*. We shall see in Section 3.5 that the essential content

†See Robert Osserman, *Two-Dimensional Calculus* (New York: Harcourt Brace Jovanovich, 1968), pp. 136–137, where four different methods of solution are given.

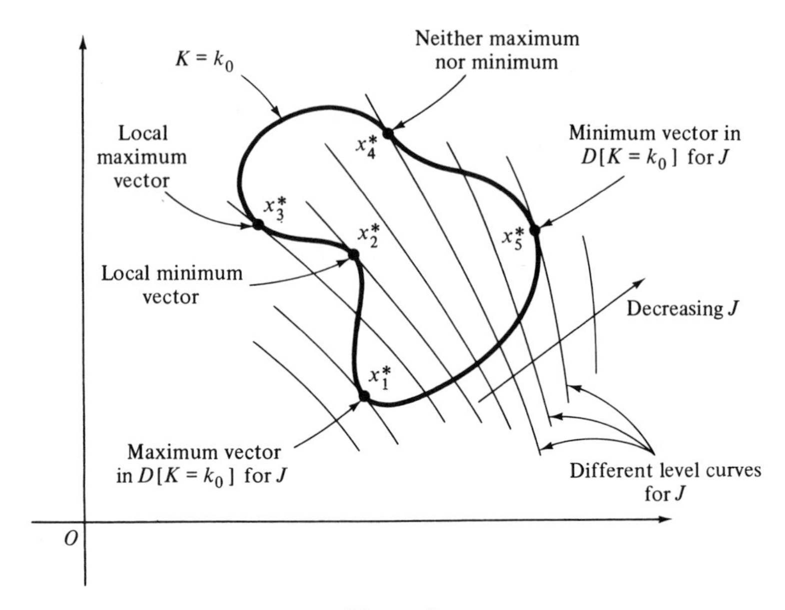

Figure 3

of this geometrical argument can be made to provide the basis for a proof of the Euler-Lagrange multiplier theorem.

The geometrical situation becomes more difficult to visualize if D is a subset not of $\mathcal{R}_2$ or of $\mathcal{R}_3$ but of some larger normed vector space $\mathfrak{X}$ such as the vector space $\mathcal{C}^0[a, b]$ consisting of all continuous functions defined on some fixed interval $[a, b]$. It is, however, customary in any case (for any normed vector space $\mathfrak{X}$) to say that the "level surfaces" $D[J = c]$ and $D[K = k]$ are *tangential* at a common point x^* if the variations of J and K at x^* are linearly dependent in the sense that equation (3.3.3) holds for some constants μ_0 and μ_1 [see the first footnote that accompanies the discussion following (3.3.3)]. This usage of the term *tangential* agrees with the common usage in the case $\mathfrak{X} = \mathcal{R}_3$ since in the latter case the *tangent plane* of any level surface such as $D[K = k_0]$ at a point x^* consists of all vectors x of the form $x = x^* + h$ for all vectors h in $\mathcal{R}_3$ which satisfy the *orthogonality condition*†

$$\nabla K(x^*) \cdot h = 0. \tag{3.4.12}$$

†Here $\nabla K(x^*) = (\partial K(x^*)/\partial x_1, \partial K(x^*)/\partial x_2, \partial K(x^*)/\partial x_3)$ is the *gradient vector* of K at x^*, and the expression $\nabla K(x^*) \cdot h$ denotes the *dot product* or *inner product* of the vectors $\nabla K(x^*)$ and h. We recall that the inner product of two vectors $y = (y_1, y_2, y_3)$ and $z = (z_1, z_2, z_3)$ in $\mathcal{R}_3$ is defined as $y \cdot z = \sum_{i=1}^{3} y_i z_i$ and satisfies the relation $y \cdot z = \|y\| \|z\| \cos \theta$, where the norm here is the Euclidean norm given by (1.4.4) and where θ is the angle between the vectors y and z. Hence two nonzero vectors are *perpendicular* or *orthogonal* if and only if their inner product vanishes. Thus the condition (3.4.12) requires that the vector h must be orthogonal to the gradient vector $\nabla K(x^*)$.

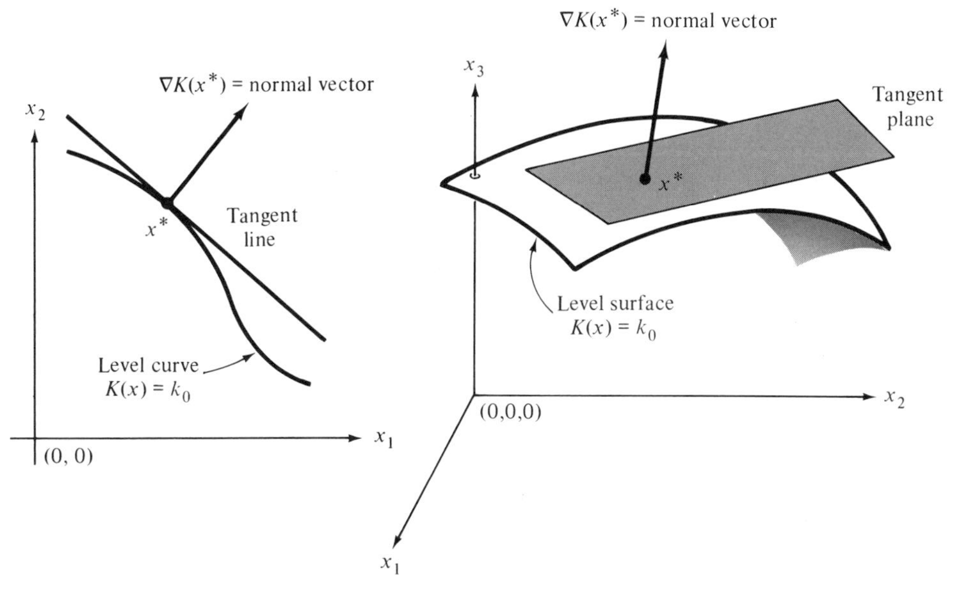

Geometric interpretation of the gradient vector

Figure 4

It is known in this case† that the gradient vector $\nabla K(x^*)$ is a *normal* vector which is perpendicular to the level surface $D[K = k_0]$ at x^* (as shown in Figure 4) unless the gradient vector vanishes identically at x^*. Hence the condition (3.4.12), or, alternatively, the condition [see (2.2.5)]

$$\delta K(x^*; h) = 0, \tag{3.4.13}$$

simply guarantees that the vector $x^* + h$ will lie in the tangent plane of the given surface at x^*. It follows from these remarks that the condition (3.3.2) of the Euler-Lagrange multiplier theorem guarantees in this case ($\mathfrak{X} = \mathcal{R}_3$) that *the tangent planes of the two surfaces* $D[J = J(x^*)]$ *and* $D[K = k_0]$ *coincide at* x^* (i.e., the two surfaces are tangential there) if x^* is a local extremum vector in $D[K = k_0]$ for J and if (3.3.1) fails to hold. In the general case where D is an open subset of any given normed vector space $\mathfrak{X}$ and $D[K = k_0]$ is any "level surface" in D, it is customary to say (by way of definition) that *the tangent plane to the surface* $D[K = k_0]$ *at a vector* x *on this surface consists of the vectors of the form* $x + h$ *for all vectors* h *in* $\mathfrak{X}$ *which satisfy the* (orthogonality) *condition* [compare with (3.4.13)]

$$\delta K(x; h) = 0.‡$$

†See Richard E. Williamson, Richard H. Crowell, and Hale F. Trotter, *Calculus of Vector Functions* (Englewood Cliffs, N.J.: Prentice-Hall, Inc., 1968), pp. 161–162.

‡Strictly speaking, one requires that K be *differentiable* at x, and one then takes $dK(x; h) = 0$ to be the defining equation for the tangent plane. See Section 2.6 for the notion of the differential.

If we use this terminology, we may then say in every case (for any normed vector space $\mathfrak{X}$) that the condition (3.3.2) of the Euler-Lagrange multiplier theorem guarantees that the tangent planes of the two surfaces $D[J = J(x^*)]$ and $D[K = k_0]$ coincide at x^* if x^* is a local extremum vector in $D[K = k_0]$ for J and if (3.3.1) fails to hold.

We close this section with one other illustrative example. We consider the problem of minimizing the value of the functional

$$J(\phi) = \int_1^2 x\phi(x)^2\,dx$$

on the vector space $\mathcal{C}^0[1, 2]$ subject to the constraint $K(\phi) = \log 2$, where the functional K is defined by

$$K(\phi) = \int_1^2 \phi(x)\,dx$$

for any function ϕ of class $\mathcal{C}^0$ on $[1, 2]$. The variations of J and K at any vector ϕ in $\mathcal{C}^0[1, 2]$ are found to be given by

$$\delta J(\phi; \Delta\phi) = 2\int_1^2 x\phi(x)\,\Delta\phi(x)\,dx$$

$$\delta K(\phi; \Delta\phi) = \int_1^2 \Delta\phi(x)\,dx$$

for any vector $\Delta\phi$ in $\mathcal{C}^0[1, 2]$. One easily checks that the first possibility of the Euler-Lagrange multiplier theorem does *not* obtain in this case, so if ϕ^* is a minimum vector for J subject to the given constraint, then

$$2\int_1^2 x\phi^*(x)\,\Delta\phi(x)\,dx = \lambda\int_1^2 \Delta\phi(x)\,dx$$

must hold, or, equivalently,

$$\int_1^2 [2x\phi^*(x) - \lambda]\,\Delta\phi(x)\,dx = 0$$

for *all* continuous functions $\Delta\phi = \Delta\phi(x)$ on the interval $[1, 2]$. If we make the special choice

$$\Delta\phi(x) = 2x\phi^*(x) - \lambda$$

in the last equation, we can conclude that (explain)

$$\phi^*(x) = \frac{\lambda}{2x} \qquad \text{for } 1 \leq x \leq 2,$$

which must hold for any such extremum vector. If we now impose the given constraint $K(\phi^*) = \log 2$, we find that the multiplier λ must take on the value $\lambda = 2$, giving

$$\phi^*(x) = \frac{1}{x} \qquad \text{for } 1 \leq x \leq 2.$$

Finally, we shall verify that this particular vector ϕ^* actually minimizes J subject to the constraint $K(\phi) = \log 2$. Indeed, if $\phi^* + \psi$ is any admissible vector in $\mathcal{C}^0[1, 2]$ with $K(\phi^* + \psi) = \log 2$, then we find that ψ must necessarily satisfy the condition (why?)

$$\int_1^2 \psi(x)\, dx = 0. \tag{3.4.14}$$

On the other hand, a direct calculation gives

$$J(\phi^* + \psi) = J(\phi^*) + 2\int_1^2 x\phi^*(x)\psi(x)\, dx + \int_1^2 x\psi(x)^2\, dx.$$

Since $x\phi^*(x) = 1$, we find that

$$2\int_1^2 x\phi^*(x)\psi(x)\, dx = 2\int_1^2 \psi(x)\, dx = 0$$

by (3.4.14), so that

$$J(\phi^* + \psi) = J(\phi^*) + \int_1^2 x\psi(x)^2\, dx$$

holds for any admissible vector $\phi^* + \psi$. The last equation implies the desired result $J(\phi^* + \psi) \geq J(\phi^*)$ for any admissible vector $\phi = \phi^* + \psi$ in $\mathcal{C}^0[1, 2]$.

Exercises

1. Use graphical methods based on a consideration of suitable level lines to maximize the function $x_1 - x_2$ on $\mathcal{R}_2$ subject to the constraint $x_1^2 + x_2^2 = 1$.

2. Use graphical methods as in Exercise 1 to minimize the function $x_1^2 + x_2^2$ subject to the constraint $x_1 - x_2 = \sqrt{2}$.

3. Maximize the function $\exp(-(x_1^2 + 2x_2^2))$ subject to the constraint $x_1 + 2x_2 = 3$.

4. Minimize the function $x_1^2 + x_2^2$ on $\mathcal{R}_2$ subject to the constraint $x_1^2 - (x_2 - 1)^3 = 0$. Solve this problem both geometrically and by the Euler-Lagrange multiplier theorem.

5. Maximize the function $x_1 + x_2$ on $\mathcal{R}_2$ subject to the constraint $x_1^2 + x_2^2 = 0$.

6. Minimize the function $x_1^2 + x_2^2 + 2x_3^2 - x_1 - x_2x_3$ on $\mathcal{R}_3$ subject to the constraint $x_1 + x_2 + x_3 = 35$.

7. Let D be the open subset of $\mathcal{R}_3$ consisting of all vectors $x = (x_1, x_2, x_3)$ for which $x_1 > 0$, $x_2 > 0$, and $x_3 > 0$. Minimize the function $x_1^2 + x_2^2 + x_3^2 + x_1x_2x_3$ on D subject to the constraint $x_1x_2x_3 = 8$.

8. Find the maximum value of the functional $J(\phi) = \int_0^1 \phi(x)\,dx$ on the vector space $\mathcal{C}^0[0, 1]$ subject to the fixed constraint $\int_0^1 [\phi(x)^2 + x\phi(x)]\,dx = 3\frac{11}{12}$. Verify directly that $J(\phi^*)$ is a maximum. *Hint:* If ϕ^* is the candidate for the maximum vector, you should be able to verify directly that $J(\phi^* + \psi) \leqslant J(\phi^*)$ for any *admissible* vector $\phi^* + \psi$.

9. Find the minimum value of the functional $J(\phi) = \int_0^1 x\phi(x)\,dx$ on the vector space $\mathcal{C}^0[0, 1]$ subject to the constraint $\int_0^1 \phi(x)^2\,dx = \frac{1}{12}$. Verify directly that $J(\phi^*)$ is a minimum.

3.5. Proof of the Euler-Lagrange Multiplier Theorem†

We come finally to the proof of the Euler-Lagrange multiplier theorem for a single constraint as stated in Section 3.3. It is sufficient to prove that the second possibility (3.3.2) must hold in the event that the first possibility (3.3.1) fails to hold. Hence we shall assume that equation (3.3.1) fails to hold in general, so that we may choose a fixed vector $\Delta\bar{x}$ in $\mathfrak{X}$ such that

$$\delta K(x^*; \Delta\bar{x}) \neq 0, \tag{3.5.1}$$

and we shall then prove in this case that equation (3.3.2) must necessarily hold for every vector Δx in $\mathfrak{X}$.

Actually (3.3.2) will follow directly from (3.5.1) and from the relation

$$\det\begin{pmatrix} \delta J(x^*; \Delta x) & \delta J(x^*; \Delta y) \\ \delta K(x^*; \Delta x) & \delta K(x^*; \Delta y) \end{pmatrix} = 0, \tag{3.5.2}$$

which we shall show must hold for every pair of vectors Δx and Δy in $\mathfrak{X}$. Indeed, *if* (3.5.2) holds for all Δx and for all Δy in $\mathfrak{X}$, we may take $\Delta y = \Delta\bar{x}$ for any fixed $\Delta\bar{x}$ as in (3.5.1), and we then find upon expanding the determinant of (3.5.2) and dividing by $\delta K(x^*; \Delta\bar{x})$ that

$$\delta J(x^*; \Delta x) = \lambda\, \delta K(x^*; \Delta x) \tag{3.5.3}$$

†Some readers may wish to bypass this section during a first reading and turn directly now to Section 3.6.

for all Δx in $\mathfrak{X}$, where here we have set [see (3.5.1)]

$$\lambda = \frac{\delta J(x^*; \Delta \bar{x})}{\delta K(x^*; \Delta \bar{x})}.$$

Since (3.5.3) is exactly the desired result (3.3.2), it follows then that we need only prove (3.5.2), to which proof we now turn. [Note that (3.5.2) will hold trivially if J and K are functions defined on $\mathfrak{X} = \mathfrak{R}$ as in the first example considered in Section 3.4; see formulas (3.1.2) and (3.4.1).]

Since (3.5.2) clearly holds if either Δx or Δy is the zero vector in $\mathfrak{X}$ (why?), we need only consider arbitrary *nonzero* vectors. Hence let Δx and Δy now denote any such fixed nonzero vectors in $\mathfrak{X}$. For all sufficiently small numbers α and β the vector

$$x^* + \alpha\,\Delta x + \beta\,\Delta y$$

will be in the open set D (why?), so we may consider the functional values

$$\begin{aligned} &J(x^* + \alpha\,\Delta x + \beta\,\Delta y) \\ &K(x^* + \alpha\,\Delta x + \beta\,\Delta y) \end{aligned} \tag{3.5.4}$$

as functions of the numbers α and β defined in some open disc

$$U = \{(\alpha, \beta) \colon \alpha^2 + \beta^2 < \rho^2\}$$

centered at the origin in the (α, β)-plane. Without any loss we may and shall choose the radius ρ of U to be so small that both the variation of J and the variation of K are weakly continuous at each vector in the corresponding open set

$$\mathfrak{U} = \{x^* + \alpha\,\Delta x + \beta\,\Delta y \colon \text{all } (\alpha, \beta) \text{ in } U\} \tag{3.5.5}$$

in D. We shall use the expressions $\mathcal{J}(\alpha, \beta)$ and $\mathcal{K}(\alpha, \beta)$ to denote the functional values given by (3.5.4) when considered as functions of α and β; i.e.,

$$\begin{aligned} \mathcal{J}(\alpha, \beta) &= J(x^* + \alpha\,\Delta x + \beta\,\Delta y) \\ \mathcal{K}(\alpha, \beta) &= K(x^* + \alpha\,\Delta x + \beta\,\Delta y) \end{aligned} \tag{3.5.6}$$

for any point (α, β) contained in the stated disc U centered at the origin in the (α, β)-plane. Finally we consider the mapping

$$\begin{aligned} j &= \mathcal{J}(\alpha, \beta) \\ k &= \mathcal{K}(\alpha, \beta), \end{aligned} \tag{3.5.7}$$

which maps the disc U of the (α, β)-plane into the (j, k)-plane. The origin in the (α, β)-plane maps onto the point $(j, k) = (J(x^*), k_0)$ since x^* is an ex-

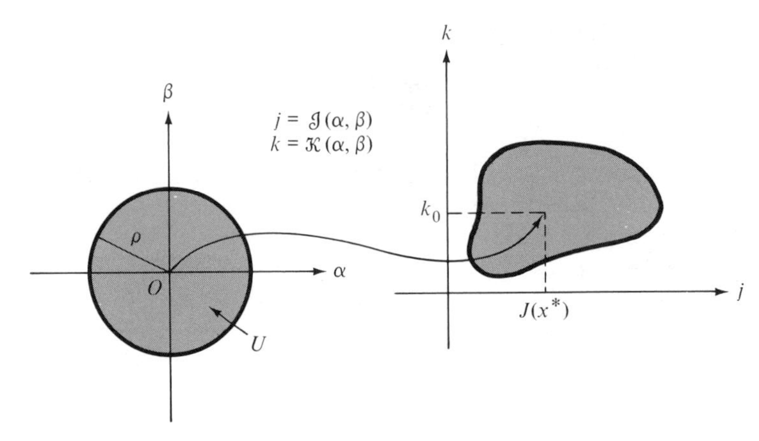

Figure 5

tremum vector in $D[K = k_0]$ for J (see Figure 5). We shall see below that the functions $\mathcal{J}$ and $\mathcal{K}$ are continuous on U even though the continuity of the functionals J and K on $\mathcal{U}$ may be open to question. (See, for example, Exercise 3 of Section 3.2.)

We wish to consider the possibility of solving (inverting) the equations of (3.5.7) near the origin $(\alpha, \beta) = (0, 0)$ to obtain α and β as functions of j and k near the point $(j, k) = (J(x^*), k_0)$. *This is possible* according to the inverse function theorem of calculus† *if the functions* $\mathcal{J}(\alpha, \beta)$ *and* $\mathcal{K}(\alpha, \beta)$ *defined by* (3.5.6) *are continuously differentiable* (in the usual sense of calculus) *near* $(\alpha, \beta) = (0, 0)$ *with nonzero Jacobian determinant*

$$\det\begin{pmatrix} \dfrac{\partial\mathcal{J}(\alpha, \beta)}{\partial\alpha} & \dfrac{\partial\mathcal{J}(\alpha, \beta)}{\partial\beta} \\ \dfrac{\partial\mathcal{K}(\alpha, \beta)}{\partial\alpha} & \dfrac{\partial\mathcal{K}(\alpha, \beta)}{\partial\beta} \end{pmatrix} \neq 0 \tag{3.5.8}$$

at $(\alpha, \beta) = (0, 0)$.

To check whether or not the inverse function theorem will apply, we must compute the partial derivatives appearing in the Jacobian matrix

$$\begin{pmatrix} \dfrac{\partial\mathcal{J}(\alpha, \beta)}{\partial\alpha} & \dfrac{\partial\mathcal{J}(\alpha, \beta)}{\partial\beta} \\ \dfrac{\partial\mathcal{K}(\alpha, \beta)}{\partial\alpha} & \dfrac{\partial\mathcal{K}(\alpha, \beta)}{\partial\beta} \end{pmatrix},$$

where, for example,

$$\frac{\partial\mathcal{J}(\alpha, \beta)}{\partial\alpha} = \operatorname*{limit}_{\epsilon\to 0}\frac{\mathcal{J}(\alpha+\epsilon, \beta) - \mathcal{J}(\alpha,\beta)}{\epsilon} \tag{3.5.9}$$

if the limit exists. In the present case we find with (3.5.6) that

†Williamson et al., *Op. cit.*, p. 201.

$$\frac{\mathcal{J}(\alpha + \epsilon, \beta) - \mathcal{J}(\alpha, \beta)}{\epsilon}$$
$$= \frac{J(x^* + \alpha\,\Delta x + \beta\,\Delta y + \epsilon\,\Delta x) - J(x^* + \alpha\,\Delta x + \beta\,\Delta y)}{\epsilon},$$

and it then follows from (2.3.4) that the limit appearing in (3.5.9) exists and is given by the variation of J as

$$\frac{\partial \mathcal{J}(\alpha, \beta)}{\partial \alpha} = \delta J(x^* + \alpha\,\Delta x + \beta\,\Delta y; \Delta x).$$

The other partial derivatives appearing in the Jacobian matrix are similarly computed, giving together

$$\begin{pmatrix} \dfrac{\partial \mathcal{J}(\alpha, \beta)}{\partial \alpha} & \dfrac{\partial \mathcal{J}(\alpha, \beta)}{\partial \beta} \\ \dfrac{\partial \mathcal{K}(\alpha, \beta)}{\partial \alpha} & \dfrac{\partial \mathcal{K}(\alpha, \beta)}{\partial \beta} \end{pmatrix}$$
$$= \begin{pmatrix} \delta J(x^* + \alpha\,\Delta x + \beta\,\Delta y; \Delta x) & \delta J(x^* + \alpha\,\Delta x + \beta\,\Delta y; \Delta y) \\ \delta K(x^* + \alpha\,\Delta x + \beta\,\Delta y; \Delta x) & \delta K(x^* + \alpha\,\Delta x + \beta\,\Delta y; \Delta y) \end{pmatrix}. \tag{3.5.10}$$

It follows from (3.5.10) and the fact that both the variation of J and the variation of K are weakly continuous in $\mathcal{U}$ [see (3.5.5)] that the functions $\mathcal{J}$ and $\mathcal{K}$ defined by (3.5.6) are continuously differentiable in the usual sense of calculus, everywhere in U. Moreover, according to (3.5.10), the Jacobian determinant of the mapping (3.5.7) evaluated at the origin $(\alpha, \beta) = (0, 0)$ is given as

$$\det \begin{pmatrix} \dfrac{\partial \mathcal{J}}{\partial \alpha} & \dfrac{\partial \mathcal{J}}{\partial \beta} \\ \dfrac{\partial \mathcal{K}}{\partial \alpha} & \dfrac{\partial \mathcal{K}}{\partial \beta} \end{pmatrix}\Bigg|_{\alpha=\beta=0} = \det \begin{pmatrix} \delta J(x^*; \Delta x) & \delta J(x^*; \Delta y) \\ \delta K(x^*; \Delta x) & \delta K(x^*; \Delta y) \end{pmatrix}.$$

Hence (3.5.8) holds at $(\alpha, \beta) = (0, 0)$ if and only if

$$\det \begin{pmatrix} \delta J(x^*; \Delta x) & \delta J(x^*; \Delta y) \\ \delta K(x^*; \Delta x) & \delta K(x^*; \Delta y) \end{pmatrix} \neq 0 \tag{3.5.11}$$

holds, so that we have finally by the inverse function theorem [see the paragraph containing (3.5.8)] the following result: *Equation* (3.5.7) *can be solved or inverted near the origin to give* α *and* β *as functions of* j *and* k *if* (3.5.11) *holds.*

But now we shall observe that (3.5.11), if valid, leads to a contradiction [from which we shall conclude that the desired result (3.5.2) holds]. Indeed,

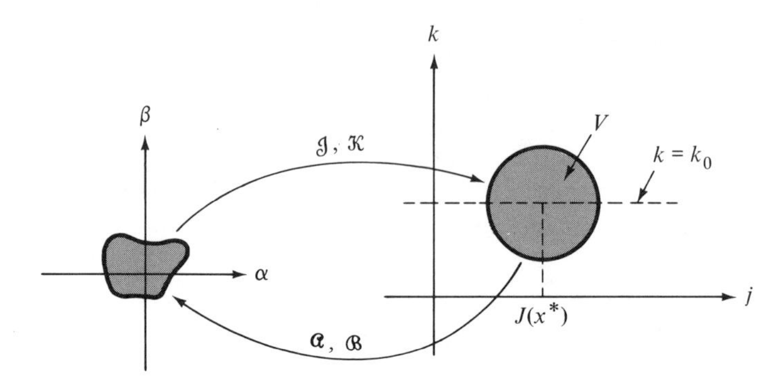

Figure 6

suppose that (3.5.11) were to hold, and let

$$\begin{aligned} \alpha &= \mathcal{A}(j, k) \\ \beta &= \mathcal{B}(j, k) \end{aligned} \tag{3.5.12}$$

be the resulting inverse map for (3.5.7) given by the inverse function theorem. Here the functions $\mathcal{A}$ and $\mathcal{B}$ are guaranteed by the inverse function theorem to be defined and continuous for all points (j, k) in some disc V centered at the point $(J(x^*), k_0)$ in the (j, k)-plane (see Figure 6) and to satisfy the *inverse relations*

$$\begin{aligned} \mathcal{J}(\mathcal{A}(j, k), \mathcal{B}(j, k)) &= j \\ \mathcal{K}(\mathcal{A}(j, k), \mathcal{B}(j, k)) &= k \end{aligned} \tag{3.5.13}$$

for all (j, k) in V. Moreover, $\mathcal{A}(j, k) = \mathcal{B}(j, k) = 0$ holds at the point $(j, k) = (J(x^*), k_0)$.

Let now (j^*, k_0) be any point in V whose k-coordinate is k_0, and consider the corresponding point (α^*, β^*) in the (α, β)-plane given as

$$\begin{aligned} \alpha^* &= \mathcal{A}(j^*, k_0) \\ \beta^* &= \mathcal{B}(j^*, k_0). \end{aligned} \tag{3.5.14}$$

Then by (3.5.6), (3.5.14), and (3.5.13) we find that

$$\begin{aligned} J(x^* + \alpha^* \Delta x + \beta^* \Delta y) &= \mathcal{J}(\alpha^*, \beta^*) = \mathcal{J}(\mathcal{A}(j^*, k_0), \mathcal{B}(j^*, k_0)) = j^* \\ K(x^* + \alpha^* \Delta x + \beta^* \Delta y) &= \mathcal{K}(\alpha^*, \beta^*) = \mathcal{K}(\mathcal{A}(j^*, k_0), \mathcal{B}(j^*, k_0)) = k_0. \end{aligned}$$

Hence by choosing the number j^* sufficiently near the value $J(x^*)$ with either $j^* > J(x^*)$ or $j^* < J(x^*)$, we can get corresponding points $x^* + \alpha^* \Delta x + \beta^* \Delta y$ in $D[K = k_0]$ arbitrarily near the point x^*, with the value $J(x^* + \alpha^* \Delta x + \beta^* \Delta y)$ either greater than or less than $J(x^*)$. *This contradicts the fact that x^* is a local extremum vector in $D[K = k_0]$ for J and therefore proves that* (3.5.11) *cannot hold.* Since Δx and Δy were arbitrary (nonzero) vectors in $\mathcal{X}$, this proves that (3.5.2) holds and *completes the proof of the Euler-Lagrange multiplier theorem.*

It should perhaps be emphasized that the essential result which is demonstrated by this proof of the multiplier theorem is the validity of equation (3.5.2) whenever x^* is a local extremum vector in $D[K = k_0]$ for J. The Euler-Lagrange multiplier theorem itself follows directly from (3.5.2). We mention also that (3.5.2) is equivalent to the single equation (3.3.3) with

$$\mu_0 = \delta K(x^*; \Delta y), \qquad \mu_1 = -\delta J(x^*; \Delta y),$$

as the reader can verify.

Exercises

1. Let J and K be functionals which are defined and have variations on an open subset D of a normed vector space $\mathfrak{X}$, and let x^* be a local extremum vector in $D[K \leq k_0]$ for J, where $D[K \leq k_0]$ is the subset of D which consists of all vectors x in D which satisfy the following *inequality* constraint:

$$K(x) \leq k_0. \tag{3.5.15}$$

   Assume that both the variation of J and the variation of K are weakly continuous near x^* and that $K(x^*) = k_0$. Show that at least one of the two possibilities (3.3.1) and (3.3.2) must necessarily hold at x^*. *Hint:* The same proof which was given for the case of the equality constraint (3.1.1) may also suffice for the present case of the inequality constraint (3.5.15). (The combined results of this exercise and Exercise 2 are contained in a broader result concerning inequality constraints which was proved by Fritz John in 1948. We shall discuss John's result briefly in Section 3.10.)

2. Let J and K be functionals which are defined on an open subset D of a normed vector space $\mathfrak{X}$, and let x^* be a local extremum vector in $D[K \leq k_0]$ for J. Assume that J has a variation at x^*, while K is continuous at x^* with $K(x^*) < k_0$. Is it possible in this case to conclude that (3.3.2) must hold? Explain. (The inequality constraint is said to be *inactive* in this case.)

3.6. The Euler-Lagrange Multiplier Theorem for Many Constraints

The result of the theorem which is proved in Section 3.5 can easily be extended to handle extremum problems which involve any finite number of constraints such as $K(x) = k$. Let $K_1, K_2, \ldots K_m$ be any collection of functionals which are defined and have variations on an open subset D of a normed vector space $\mathfrak{X}$, and let $D[K_i = k_i \text{ for } i = 1, 2, \ldots, m]$ denote the *subset of D which consists of all vectors x in D which simultaneously satisfy all the following constraints:*

$$K_1(x) = k_1, K_2(x) = k_2, \ldots, K_m(x) = k_m. \tag{3.6.1}$$

Here $k_1, k_2, \ldots, k_m$ may be any given numbers, and we assume always that there is at least one vector in D which satisfies all the constraints of (3.6.1) so that the set $D[K_i = k_i$ for $i = 1, 2, \ldots, m]$ is not empty.

In this case we have the following multiplier theorem.

Theorem. *Let $J, K_1, K_2, \ldots, K_m$ be functionals which are defined and have variations on an open subset D of a normed vector space $\mathfrak{X}$, and let x^* be a local extremum vector in $D[K_i = k_i$ for $i = 1, 2, \ldots, m]$ for J, where $k_1, k_2, \ldots, k_m$ are any given fixed numbers for which the set $D[K_i = k_i$ for $i = 1, 2, \ldots, m]$ is nonempty. Assume that the variation of J and the variation of each K_i (for $i = 1, 2, \ldots, m$) are weakly continuous near x^*. Then at least one of the following two possibilities must hold:*

1. *The following determinant vanishes identically,*

$$\det \begin{vmatrix} \delta K_1(x^*; \Delta x_1) & \delta K_1(x^*; \Delta x_2) & \cdots & \delta K_1(x^*; \Delta x_m) \\ \delta K_2(x^*; \Delta x_1) & \delta K_2(x^*; \Delta x_2) & \cdots & \delta K_2(x^*; \Delta x_m) \\ \vdots & \vdots & & \vdots \\ \delta K_m(x^*; \Delta x_1) & \delta K_m(x^*; \Delta x_2) & \cdots & \delta K_m(x^*; \Delta x_m) \end{vmatrix} = 0 \tag{3.6.2}$$

for all vectors $\Delta x_1, \Delta x_2, \ldots, \Delta x_m$ in $\mathfrak{X}$; or

2. *The variation of J at x^* is a linear combination of the variations of $K_1, K_2, \ldots, K_m$ at x^*, i.e., there are constants $\lambda_1, \lambda_2, \ldots, \lambda_m$ such that*

$$\begin{aligned} \delta J(x^*; \Delta x) &= \lambda_1\, \delta K_1(x^*; \Delta x) + \lambda_2\, \delta K_2(x^*; \Delta x) + \cdots + \lambda_m\, \delta K_m(x^*; \Delta x) \\ &= \sum_{i=1}^{m} \lambda_i\, \delta K_i(x^*; \Delta x) \end{aligned} \tag{3.6.3}$$

holds for every vector Δx in $\mathfrak{X}$.

Proof.† The stated result will follow directly from the relation [compare with (3.5.2)]

$$\det \begin{vmatrix} \delta J(x^*; \Delta x) & \delta J(x^*; \Delta x_1) & \delta J(x^*; \Delta x_2) & \cdots & \delta J(x^*; \Delta x_m) \\ \delta K_1(x^*; \Delta x) & \delta K_1(x^*; \Delta x_1) & \delta K_1(x^*; \Delta x_2) & \cdots & \delta K_1(x^*; \Delta x_m) \\ \delta K_2(x^*; \Delta x) & \delta K_2(x^*; \Delta x_1) & \delta K_2(x^*; \Delta x_2) & \cdots & \delta K_2(x^*; \Delta x_m) \\ \vdots & \vdots & \vdots & & \vdots \\ \delta K_m(x^*; \Delta x) & \delta K_m(x^*; \Delta x_1) & \delta K_m(x^*; \Delta x_2) & \cdots & \delta K_m(x^*; \Delta x_m) \end{vmatrix} = 0, \tag{3.6.4}$$

which can be shown to hold for all vectors $\Delta x, \Delta x_1, \Delta x_2, \ldots, \Delta x_m$ in $\mathfrak{X}$ whenever x^* is a local extremum vector in $D[K_i = k_i$ for $i = 1, 2, \ldots, m]$

†This proof can be safely skipped during a first reading. The reader who so desires should turn directly to Section 3.7.

for J. Indeed, if we assume the validity of (3.6.4) for the moment and expand the determinant there by cofactors about the first column,† we find that [compare with (3.3.3)]

$$\mu_0\, \delta J(x^*; \Delta x) + \sum_{i=1}^{m} \mu_i\, \delta K_i(x^*; \Delta x) = 0 \qquad \text{for every vector } \Delta x \text{ in } \mathfrak{X}, \tag{3.6.5}$$

where

$$\mu_0 = \det \begin{vmatrix} \delta K_1(x^*; \Delta x_1) & \delta K_1(x^*; \Delta x_2) & \cdots & \delta K_1(x^*; \Delta x_m) \\ \delta K_2(x^*; \Delta x_1) & \delta K_2(x^*; \Delta x_2) & \cdots & \delta K_2(x^*; \Delta x_m) \\ \vdots & \vdots & & \vdots \\ \delta K_m(x^*; \Delta x_1) & \delta K_m(x^*; \Delta x_2) & \cdots & \delta K_m(x^*; \Delta x_m) \end{vmatrix}$$

and

$$\mu_i = -\det \begin{vmatrix} \delta K_1(x^*; \Delta x_1) & \delta K_1(x^*; \Delta x_2) & \cdots & \delta K_1(x^*; \Delta x_m) \\ \delta K_2(x^*; \Delta x_1) & \delta K_2(x^*; \Delta x_2) & \cdots & \delta K_2(x^*; \Delta x_m) \\ \vdots & \vdots & & \vdots \\ \delta K_{i-1}(x^*; \Delta x_1) & \delta K_{i-1}(x^*; \Delta x_2) & \cdots & \delta K_{i-1}(x^*; \Delta x_m) \\ \delta J(x^*; \Delta x_1) & \delta J(x^*; \Delta x_2) & \cdots & \delta J(x^*; \Delta x_m) \\ \delta K_{i+1}(x^*; \Delta x_1) & \delta K_{i+1}(x^*; \Delta x_2) & \cdots & \delta K_{i+1}(x^*; \Delta x_m) \\ \vdots & \vdots & & \vdots \\ \delta K_m(x^*; \Delta x_1) & \delta K_m(x^*; \Delta x_2) & \cdots & \delta K_m(x^*; \Delta x_m) \end{vmatrix} \begin{matrix} \\ \\ \\ \\ \longleftarrow \text{ith row} \\ \\ \\ \\ \end{matrix}$$

for $i = 1, 2, \ldots, m$ and for any vectors $\Delta x_1, \Delta x_2, \ldots, \Delta x_m$ in $\mathfrak{X}$. Hence if we assume the validity of (3.6.4), it is clear that *the second possibility* (3.6.3) *must hold if the first possibility* (3.6.2) *fails*. Indeed, if (3.6.2) fails to hold, then there are vectors $\Delta x_1, \Delta x_2, \ldots, \Delta x_m$ in $\mathfrak{X}$ for which $\mu_0 \neq 0$. If we use such a collection of vectors $\Delta x_1, \Delta x_2, \ldots, \Delta x_m$ in (3.6.4) and (3.6.5), we can then divide (3.6.5) by μ_0 and obtain (3.6.3) with $\lambda_i = -\mu_i/\mu_0$ (explain). Hence we need only prove the result (3.6.4).

To prove (3.6.4), we apply the inverse function theorem‡ to the map [compare with (3.5.7)]

$$\begin{aligned} j &= \mathcal{J}(\alpha, \beta_1, \beta_2, \cdots, \beta_m) \\ k_1 &= \mathcal{K}_1(\alpha, \beta_1, \beta_2, \cdots, \beta_m) \\ k_2 &= \mathcal{K}_2(\alpha, \beta_1, \beta_2, \cdots, \beta_m) \\ &\;\vdots \\ k_m &= \mathcal{K}_m(\alpha, \beta_1, \beta_2, \cdots, \beta_m), \end{aligned} \tag{3.6.6}$$

†See Williamson et al., *op. cit.*, p. 44.
‡Ibid., p. 201.

which maps $(\alpha, \beta_1, \beta_2, \ldots, \beta_m)$-space into $(j, k_1, k_2, \ldots, k_m)$-space, where the functions $\mathcal{J}, \mathcal{K}_1, \mathcal{K}_2, \ldots, \mathcal{K}_m$ are defined for any fixed collection of nonzero vectors $\Delta x_1, \Delta x_2, \ldots, \Delta x_m$ by [compare with (3.5.6)]

$$\mathcal{J}(\alpha, \beta_1, \beta_2, \cdots, \beta_m) = J(x^* + \alpha\,\Delta x + \beta_1\,\Delta x_1 + \beta_2\,\Delta x_2 + \cdots + \beta_m\,\Delta x_m)$$

$$\mathcal{K}_1(\alpha, \beta_1, \beta_2, \cdots, \beta_m) = K_1(x^* + \alpha\,\Delta x + \beta_1\,\Delta x_1 + \beta_2\,\Delta x_2 + \cdots + \beta_m\,\Delta x_m)$$

$$\mathcal{K}_2(\alpha, \beta_1, \beta_2, \cdots, \beta_m) = K_2(x^* + \alpha\,\Delta x + \beta_1\,\Delta x_1 + \beta_2\,\Delta x_2 + \cdots + \beta_m\,\Delta x_m)$$

$$\vdots$$

$$\mathcal{K}_m(\alpha, \beta_1, \beta_2, \cdots, \beta_m) = K_m(x^* + \alpha\,\Delta x + \beta_1\,\Delta x_1 + \beta_2\,\Delta x_2 + \cdots + \beta_m\,\Delta x_m).$$

If we now argue just as in the case of a single constraint (see Section 3.5), we can show that the map (3.6.6) can be inverted near the origin in $(\alpha, \beta_1, \beta_2, \ldots, \beta_m)$-space so as to give $\alpha, \beta_1, \beta_2, \ldots, \beta_m$ as functions of $j, k_1, k_2, \ldots, k_m$ *if* (3.6.4) *fails to hold* for some vectors $\Delta x, \Delta x_1, \Delta x_2, \ldots, \Delta x_m$. This, however, would lead to a contradiction of the fact that x^* is a local extremum vector in $D[K_i = k_i \text{ for } i = 1, 2, \ldots, m]$ for J. The proof is similar to the proof already given in Section 3.5 for the case of a single constraint, and we leave the details as an exercise for the reader.

It may be useful to note that if the first possibility (3.6.2) holds, then μ_0 vanishes in (3.6.5) and in this case (3.6.5) shows that the variations of the constraint functionals are related to each other at x^* by the relation

$$\sum_{i=1}^{m} \mu_i\, \delta K_i(x^*; \Delta x) = 0, \tag{3.6.7}$$

which must hold for all vectors Δx in $\mathfrak{X}$. In this case the (variations of the) constraints are *not* independent at x^*. We leave the geometrical interpretation of (3.6.7) to the reader. (See the discussion in Section 3.4.)

3.7. An Optimum Consumption Policy with Terminal Savings Constraint During a Period of Inflation

We consider a problem in investment planning for a person who has a certain known annual income (e.g., salary) from some source and who in addition has some accumulated savings which he has invested and which earn him a known annual return (e.g., dividends or interest). We assume that the savings can be easily liquidated and converted into consumable goods, so that the person's *total available annual resources for consumption* consist of his current annual income, his previous savings, and his current annual return on those savings which were invested.

The person wishes to have accumulated a certain specified level of savings at the end of T years, and in the meantime he wishes to plan his savings program so as to maximize the satisfaction he may derive from the consumption of the remaining part of his available annual resources which is not invested. The problem then is to decide how much of the total available annual resources should be consumed and how much should be reinvested annually so as to maximize the satisfaction received from consumption while taking into account the terminal constraint which has been specified for the savings level at the end of T years.

We shall let $S = S(t)$ denote the savings which are accumulated and invested at time t. The savings level S may be affected by three factors, as indicated schematically in Figure 7. First, the known annual **income** I may be used to increase the savings level; second, the current annual **return** R earned on the savings may be reinvested to increase the savings level; and, third, the current annual **consumption** C acts to decrease the savings.

For simplicity we shall assume that the quantities involved may change continuously throughout each year. For example, the savings may be on deposit in a savings bank which compounds interest at a continuous rate, and the individual's annual consumption rate C is allowed to change from day to day. In this case it is natural to relate the variables S, I, R, and C through the differential equation (see Figure 7)

$$\dot{S} = I + R - C, \tag{3.7.1}$$

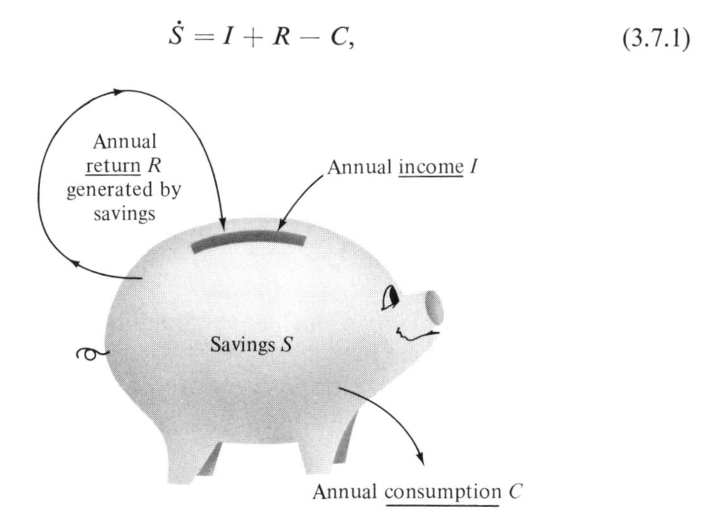

Figure 7

which simply states that *the instantaneous rate of change of savings* ($\dot{S} = dS/dt$) *is given by the difference of the total rate of income* ($I + R$) *and the total rate of expenditure* (C). We assume that the initial level of savings accumulated is known, say,

$$S(t) = S_0 \qquad \text{at } t = 0 \tag{3.7.2}$$

for a given nonnegative constant S_0.

Again for simplicity we shall assume that the return R earned on the savings is proportional to the savings level itself, so that

$$R = \alpha S \tag{3.7.3}$$

for a given positive constant α; i.e., we assume that the savings earn interest compounded continuously at the fixed rate of $100\alpha\,\%$.

If we insert (3.7.3) into (3.7.1), we find the equation

$$\dot{S} - \alpha S = I - C,$$

which may be integrated† with (3.7.2) to give the savings at time t as

$$S(t) = e^{\alpha t}S_0 + e^{\alpha t}\int_0^t e^{-\alpha\tau}[I(\tau) - C(\tau)]\,d\tau. \tag{3.7.4}$$

Here we assume that the income function $I = I(t)$ is known (at least over some fixed time period $0 \leq t \leq T$), and the optimization problem will involve making a suitable choice for the unknown consumption function $C = C(t)$.

Finally, we assume that it is required to have a specified level of savings at the end of the time period $0 \leq t \leq T$, say,

$$S(t) = S_T \qquad \text{at } t = T, \tag{3.7.5}$$

where S_T is a given nonnegative constant. Evaluating (3.7.4) at $t = T$ and using (3.7.5), we then find the requirement

$$\int_0^T e^{-\alpha t}C(t)\,dt = S_0 - e^{-\alpha T}S_T + \int_0^T e^{-\alpha t}I(t)\,dt, \tag{3.7.6}$$

which must be satisfied by any admissible consumption rate C, where again we emphasize that the income function $I = I(t)$ is assumed known so that *the right-hand side of* (3.7.6) *represents a specified constant*. If we define a functional K on the vector space $\mathcal{C}^0[0, T]$ by

$$K(C) = \int_0^T e^{-\alpha t}C(t)\,dt \tag{3.7.7}$$

†See George B. Thomas, Jr., *Calculus and Analytic Geometry* (Reading, Mass.: Addison-Wesley Publishing Company, Inc., 1968), Chapter 18.

for any function $C = C(t)$ of class $\mathcal{C}^0$ on the interval $0 \leq t \leq T$, then the constraint (3.7.6) can be given in the form

$$K(C) = S_0 - e^{-\alpha T} S_T + \int_0^T e^{-\alpha t} I(t)\, dt. \tag{3.7.8}$$

Roughly speaking, then, the optimization problem we shall consider is to maximize the satisfaction derived from consumption subject to the constraint (3.7.8). Of course we need some suitable measure of the *satisfaction* derived from any given consumption rate $C = C(t)$ over the time interval $[0, T]$. Such a measure of satisfaction might take the form of an integral such as

$$\int_0^T F(t, C(t))\, dt, \tag{3.7.9}$$

where $F = F(t, C)$ would be some suitable given function of t and C.

Now if the integral of (3.7.9) is to be a *reasonable* measure of the satisfaction derived from the consumption rate C, then the specified function F appearing there must satisfy certain natural conditions. For example, the function $F = F(t, C)$ should be an *increasing function* of its second argument C so that the satisfaction as measured by (3.7.9) will increase whenever consumption increases. In this case *the maximization of* (3.7.9) *will place a premium on the selection of a large consumption rate.*

In fact, for definiteness and simplicity we shall consider in detail only the case where F is defined by

$$F(t, C) = e^{-\beta t} \log(1 + C) \tag{3.7.10}$$

for any number $t \geq 0$ and for any number $C > 0$ (we consider only positive consumption rates). Since the logarithm function

$$\log(1 + C)$$

increases with increasing C, the choice (3.7.10) in (3.7.9) will indeed place a *preference on larger consumption rates* if the integral of (3.7.9) is to be maximized. The quantity β appearing in (3.7.10) is a *discount rate* which allows for the fact that a unit of income which might be used today in consumption may later have a different true value. We shall assume that $\beta > 0$, so that the term $e^{-\beta t}$ will decay (decrease) with increasing time and may be taken to represent the effects of inflation.

Finally, then, using (3.7.10) and (3.7.9), we define a *satisfaction functional* $\mathcal{S}$ by

$$\mathcal{S}(C) = \int_0^T e^{-\beta t} \log[1 + C(t)]\, dt \tag{3.7.11}$$

for any suitable consumption function $C = C(t)$. Specifically, we shall take the domain of $\mathcal{S}$ to be that subset D of the vector space $\mathcal{C}^0[0, T]$ which consists of all continuous functions $C = C(t)$ on $[0, T]$ satisfying the condition

$$C(t) > 0. \tag{3.7.12}$$

Here we shall use the uniform norm on $\mathcal{C}^0[0, T]$, with

$$||C|| = \max_{0 \leq t \leq T} |C(t)|$$

for any function C of class $\mathcal{C}^0$ on $[0, T]$. Note that the subset D of $\mathcal{C}^0[0, T]$ characterized by (3.7.12) is an open set in $\mathcal{C}^0[0, T]$.

We shall now seek to maximize the functional $\mathcal{S}$ on D subject to the constraint (3.7.8). If we define a constant k_0 by

$$k_0 = S_0 - e^{-\alpha T} S_T + \int_0^T e^{-\alpha t} I(t)\, dt, \tag{3.7.13}$$

then in the notation of Section 3.1 the problem is to find a maximum vector C^* in the set $D[K = k_0]$ for the functional $\mathcal{S}$ [see the definitions (3.7.7) and (3.7.11)]. We shall use the Euler-Lagrange multiplier theorem of Section 3.3 to search for such an extremum vector.

To this end we must calculate the variations of the functionals K and $\mathcal{S}$. If we use (3.7.7), we find that

$$\frac{K(C + \epsilon\, \Delta C) - K(C)}{\epsilon} = K(\Delta C),$$

from which follows directly the result

$$\delta K(C; \Delta C) = K(\Delta C) = \int_0^T e^{-\alpha t}\, \Delta C(t)\, dt \tag{3.7.14}$$

for any vector $\Delta C = \Delta C(t)$ in $\mathcal{C}^0[0, T]$ and for any fixed function $C = C(t)$ of class $\mathcal{C}^0$ on $[0, T]$. Similarly, if we write [see (3.7.11) and (3.7.10)]

$$\mathcal{S}(C) = \int_0^T F(t, C(t))\, dt \tag{3.7.15}$$

with $F = F(t, C)$ defined by (3.7.10), we find, using the same methods as were used in going from (2.4.9) to (2.4.17) (the reader should be able to fill in the details)

$$\delta\mathcal{S}(C; \Delta C) = \int_0^T \frac{\partial}{\partial C} F(t, C(t))\, \Delta C(t)\, dt \tag{3.7.16}$$

for any vector ΔC in $\mathcal{C}^0[0, T]$ and for any fixed vector C in the domain D of

$\mathcal{S}$. In the present case we may use (3.7.10) to find

$$\frac{\partial}{\partial C} F(t, C) = \frac{e^{-\beta t}}{1 + C}, \tag{3.7.17}$$

so that (3.7.16) gives

$$\delta \mathcal{S}(C; \Delta C) = \int_0^T \frac{e^{-\beta t}}{1 + C(t)} \Delta C(t)\, dt \tag{3.7.18}$$

for any vector ΔC in $\mathcal{C}^0[0, T]$ and for any C in D.

It is easy to check with (3.7.14) and (3.7.18) that both the variation of K and the variation of $\mathcal{S}$ are weakly continuous at each vector C in D (see Section 3.2). Moreover, the variation of K clearly does *not* vanish identically for all vectors ΔC in $C^0[0, T]$ [see (3.3.1)]; in fact, by (3.7.14)

$$\delta K(C; \Delta C) = \int_0^T e^{-\alpha t}\, dt > 0 \qquad \text{for } \Delta C \equiv 1.$$

Hence if C^* is a local extremum vector in $D[K = k_0]$ for $\mathcal{S}$, it then follows by the Euler-Lagrange multiplier theorem that there must be a constant λ such that [see (3.3.2)]

$$\delta \mathcal{S}(C^*; \Delta C) = \lambda\, \delta K(C^*; \Delta C)$$

for *all* vectors ΔC in $\mathcal{C}^0[0, T]$, or, using (3.7.14) and (3.7.16), we find the necessary condition

$$\int_0^T \left[\frac{\partial}{\partial C} F(t, C^*(t)) - \lambda e^{-\alpha t}\right] \Delta C(t)\, dt = 0, \tag{3.7.19}$$

which must hold for *every* function ΔC of class $\mathcal{C}^0$ on $[0\ T]$. In particular, if we take ΔC to be the continuous function defined by

$$\Delta C(t) = \frac{\partial}{\partial C} F(t, C^*(t)) - \lambda e^{-\alpha t}$$

for $0 \leq t \leq T$, where $C^*(t)$ is the fixed extremum function and λ is the constant appearing in (3.7.19), we then find that

$$\int_0^T \left[\frac{\partial}{\partial C} F(t, C^*(t)) - \lambda e^{-\alpha t}\right]^2 dt = 0,$$

from which we conclude that

$$\frac{\partial}{\partial C} F(t, C^*(t)) - \lambda e^{-\alpha t} = 0 \tag{3.7.20}$$

for all t in the interval $[0, T]$. On the other hand, the condition (3.7.20) also implies (3.7.19), and so the requirements (3.7.19) (for all ΔC) and (3.7.20) (for all t) are actually equivalent. If we use (3.7.17) in (3.7.20), we find in the present case the necessary condition

$$\frac{e^{-\beta t}}{1 + C^*(t)} = \lambda e^{-\alpha t} \qquad (3.7.21)$$

for all $0 \leq t \leq T$.

Condition (3.7.21) is similar to the conditions (3.4.2) and (3.4.8) which appeared in the simple examples of Section 3.4, and just as in those examples (3.7.21) can be solved to give solution vectors depending on the parameter λ. In this case we find from (3.7.21) that [compare with (3.4.3) and (3.4.9)]

$$C^*(t) = -1 + \frac{1}{\lambda} e^{(\alpha - \beta)t} \qquad (3.7.22)$$

for $0 \leq t \leq T$. If we now insert (3.7.22) back into the constraint (3.7.6), we find the following equation for λ:

$$\frac{1}{\lambda} = \left(S_0 - e^{-\alpha T} S_T + \int_0^T e^{-\alpha t} I(t)\, dt + \frac{1 - e^{-\alpha T}}{\alpha}\right)\left(\frac{\beta}{1 - e^{-\beta T}}\right). \qquad (3.7.23)$$

We may now insert the value of λ obtained from (3.7.23) back into (3.7.22) to find a unique candidate for the desired extremum vector C^* in $D[K = k_0]$ for $\mathcal{S}$.

Hence relations (3.7.22) and (3.7.23) define the only function C^* of class $\mathcal{C}^0$ on $[0, T]$ which satisfies the necessary condition

$$\delta \mathcal{S}(C^*; \Delta C) = \lambda\, \delta K(C^*; \Delta C)$$

of the Euler-Lagrange multiplier theorem. However, we have not yet proved that the present optimization problem actually has a solution. We must still check whether or not our extremum vector C^* actually furnishes a maximum value in $D[K = k_0]$ for the functional $\mathcal{S}$. In fact, it may not! The vector C^* may not even be in D since $C^* = C^*(t)$ may take on negative values and violate (3.7.12). In this case the individual has been overly optimistic in his requests as compared with his resources (he may have hoped for an excessively high final savings level S_T), and the stated optimization problem has no solution. We shall discuss this situation further in Section 5.2.

For the moment we shall assume that the function C^* defined by (3.7.22) and (3.7.23) satisfies the condition (3.7.12). For example, if the interest rate α exceeds the inflation rate β, then C^* will satisfy (3.7.12) if

$$S_0 + \int_0^T e^{-\alpha t} I(t)\, dt + \frac{1 - e^{-\alpha T}}{\alpha} \geq e^{-\alpha T} S_T + \frac{1 - e^{-\beta T}}{\beta} \qquad (3.7.24)$$

holds (see Exercise 2 at the end of Section 5.2). In this case the individual's resources as measured by S_0 and $I(t)$ are adequate to permit the attaining of his goals, and C^* will be in the set $D[K = k_0]$. Moreover, we can easily show in this case that C^* is, in fact, the desired maximum vector in $D[K = k_0]$ for the satisfaction functional $\mathcal{S}$. For this purpose we use (3.7.11) and (3.7.22) to find

$$\begin{aligned}\mathcal{S}(C^* + \Delta C) - \mathcal{S}(C^*) &= \int_0^T e^{-\beta t} \log\left[1 + \frac{\Delta C(t)}{1 + C^*(t)}\right] dt \\ &= \int_0^T e^{-\beta t} \log\left[1 + \lambda e^{(\beta-\alpha)t}\, \Delta C(t)\right] dt \qquad (3.7.25)\end{aligned}$$

for any function ΔC of class $\mathcal{C}^0$ on $[0, T]$ such that $C^* + \Delta C$ is in $D[K = k_0]$. Here $C^* + \Delta C$ must satisfy the condition (3.7.12) along with the constraint (3.7.6). Since C^* already satisfies (3.7.6) because of the special choice of λ given by (3.7.23), it follows that ΔC must satisfy the following constraint (explain):

$$\int_0^T e^{-\alpha t}\, \Delta C(t)\, dt = 0. \qquad (3.7.26)$$

Now a special case of Taylor's theorem with the remainder in integral form gives for any function f of class $\mathcal{C}^2$

$$f(x) = f(0) + f'(0)x + \int_0^x (x - t) f''(t)\, dt, \qquad (3.7.27)$$

which can easily be verified by integrating the last term of (3.7.27) by parts. If we take $f(x) = \log(1 + x)$ in (3.7.27), we find that

$$\log(1 + x) = x - \int_0^x \frac{x - t}{(1 + t)^2}\, dt$$

for any number x with $1 + x > 0$, from which we easily find that (explain)

$$\log(1 + x) \le x,$$

which is valid for any number $x > -1$, as illustrated in Figure 8. In particular, the last inequality implies that

$$\log[1 + \lambda e^{(\beta-\alpha)t}\, \Delta C(t)] \le \lambda e^{(\beta-\alpha)t}\, \Delta C(t)$$

for all $0 \le t \le T$ and for all admissible functions $\Delta C(t)$, and this inequality can be multiplied by $e^{-\beta t}$ and integrated to give

$$\int_0^T e^{-\beta t} \log[1 + \lambda e^{(\beta-\alpha)t}\, \Delta C(t)]\, dt \le \lambda \int_0^T e^{-\alpha t}\, \Delta C(t)\, dt.$$

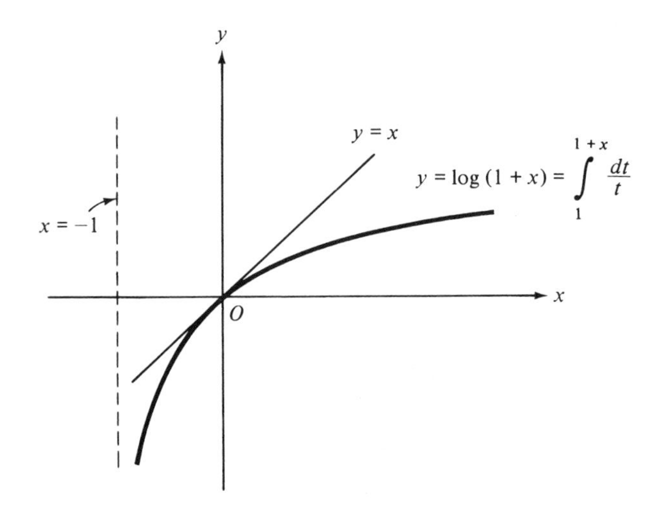

Figure 8

Finally, then, we may use the last result with (3.7.25) to find the inequality

$$\mathcal{S}(C^* + \Delta C) - \mathcal{S}(C^*) \leq \lambda \int_0^T e^{-\alpha t}\,\Delta C(t)\,dt$$

for any admissible function ΔC, and then (3.7.26) implies that

$$\mathcal{S}(C^* + \Delta C) - \mathcal{S}(C^*) \leq 0 \tag{3.7.28}$$

for any vector ΔC such that $C^* + \Delta C$ is in $D[K = k_0]$. This proves that *the satisfaction functional* $\mathcal{S}$ *of* (3.7.11) *has a maximum in* $D[K = k_0]$ *at the vector* C^* *defined by the formulas* (3.7.22) *and* (3.7.23), as expected.

We shall discuss this optimization problem further in Section 5.2. For the moment we only observe that according to (3.7.22) *the optimum consumption policy depends critically on the relative sizes of the investment return rate* α *and the discount* (inflation) *rate* β. For example, suppose that the Euler-Lagrange multiplier λ of (3.7.23) is positive. [This will be the case if (3.7.24) holds, although it may also occur without (3.7.24).] Then if inflation dominates investment return with $\beta > \alpha$, it follows that total satisfaction from consumption will be maximized while still meeting the terminal savings constraint (3.7.5) through a policy of consuming more and investing less during the early years of the period $0 \leq t \leq T$ and then consuming less and investing more in later years so as to achieve the constraint (3.7.5). On the other hand, if investment return dominates inflation with $\alpha > \beta$, then one should invest more and consume less during the early years and then consume more in later years. Finally, if investment return just equals inflation with $\alpha = \beta$, then (3.7.22) shows that one should consume at a constant rate during the given time period. For example, if *both* the inflation rate and the investment return rate are .05 (i.e., 5% per year), if an individual has an annual income of $10,000 per year with no initial savings, and if the individual wishes to accumulate $15,000 over a 5-year period while maximizing the satisfaction from consumption as measured by (3.7.11), then the individual should consume at a constant rate of about $7358 per year (see Exercise 2). If the remaining $2642 per year (from the annual income) is invested and earns 5% per year (compounded continuously), then the goal of $15,000 will be achieved while maximizing (3.5.11).

We mention finally that if instead of (3.7.10) we use some other suitable function $F = F(t, C)$ in the functional $\mathcal{S}$ of (3.7.15), we shall then be led to equations (3.7.20) and (3.7.6) for the determination of the optimum consumption rate C^*. Of course, it may *not* be possible in such a case to find C^* in such a simple closed form as that given by (3.7.22). Rather it may be necessary to compute approximate values for $C^* = C^*(t)$ numerically at certain chosen specified times $t = t_1, t_2, \ldots, t_N$. This could be done with an appropriate algorithm using an electronic computer and would ordinarily be acceptable in practice. We might also mention that the above analysis can easily be broadened to include the effects of other factors, such as taxes.

Exercises

1. Prove that the variation of $\mathcal{S}$ given by (3.7.18) is weakly continuous at each vector C in $\mathcal{C}^0[0, T]$ which satisfies (3.7.12).
2. Verify that the optimum consumption rate in the example given before the last paragraph above is about $7358 per year, when $\alpha = \beta = .05$, $I = 10{,}000$, $S_0 = 0$, $S_T = 15{,}000$, and $T = 5$.

3.8. The Meaning of the Euler-Lagrange Multipliers

We found for the value of the Euler-Lagrange multiplier λ in the preceding problem on consumption policy that [see (3.7.23) and (3.7.13)]

$$\lambda = \frac{1}{k + [(1 - e^{-\alpha T})/\alpha]} \frac{1 - e^{-\beta T}}{\beta}, \tag{3.8.1}$$

where $k = k_0$ is the value of the specified constraint

$$K(C) = k \tag{3.8.2}$$

which was imposed on all competing consumption policies $C = C(t)$, with k

given as [see (3.7.8) and (3.7.13)]

$$k = S_0 - e^{-\alpha T} S_T + \int_0^T e^{-\alpha t} I(t)\, dt. \tag{3.8.3}$$

If we use (3.8.1), we may rewrite the optimum consumption rate C^* of (3.7.22) in terms of the constraint value k as

$$C^*(t) = -1 + \left(k + \frac{1 - e^{-\alpha T}}{\alpha}\right)\left(\frac{\beta}{1 - e^{-\beta T}}\right) e^{(\alpha - \beta)t} \tag{3.8.4}$$

for $0 \leq t \leq T$, and the resulting maximum value of the satisfaction functional $\mathcal{S}$ subject to (3.8.2) is found from (3.7.11) and (3.8.4) by integration to be

$$\mathcal{S}(C^*) = \frac{T(\beta - \alpha)e^{-\beta T}}{\beta} + \left(\frac{1 - e^{-\beta T}}{\beta}\right)\left[\frac{\alpha - \beta}{\beta} + \log \frac{\beta}{1 - e^{-\beta T}} + \log\left(k + \frac{1 - e^{-\alpha T}}{\alpha}\right)\right]. \tag{3.8.5}$$

Hence the maximum value of $\mathcal{S}$ in $D[K = k]$ may be considered to be a function of the particular value of k used in the constraint (3.8.2), and it is often useful in applications to know how changes in k may affect the extreme value $\mathcal{S}(C^*)$. For this purpose we compute the derivative of $\mathcal{S}(C^*)$ with respect to k from (3.8.5) and find that

$$\frac{\partial \mathcal{S}(C^*)}{\partial k} = \frac{1}{k + [(1 - e^{-\alpha T})/\alpha]} \frac{1 - e^{-\beta T}}{\beta},$$

which with (3.8.1) implies in this case that

$$\frac{\partial \mathcal{S}(C^*)}{\partial k} = \lambda. \tag{3.8.6}$$

Hence, at least in this case, *the Euler-Lagrange multiplier* λ *given by* (3.7.23) *also gives the rate of change of the extreme value* $\mathcal{S}(C^*)$ *with respect to the constraint value* k. This is useful information in practice and makes it doubly important to study the multiplier λ of (3.7.23) [which is needed in any case in (3.7.22)]. For example, suppose we find from (3.7.23) that λ is positive. Then (3.8.6) tells us that the maximum satisfaction from consumption will *increase* if the constraint value k of (3.8.3) is increased (why?). Of course, this is to be expected in the particular optimization problem under consideration since according to (3.8.3) k gives a measure of the total resources known initially to be available for consumption during the entire given time period $0 \leq t \leq T$. Hence if the income $I = I(t)$ or the initial savings level S_0 were to be raised or if the required terminal savings level S_T were to be reduced, then the maximum satisfaction $\mathcal{S}(C^*)$ would naturally increase.

The result (3.8.6)—that the Euler-Lagrange multiplier gives the rate of change of the extreme functional value with respect to the constraint—is generally true for such optimization problems with any finite number of constraints such as those considered in Section 3.6. Hence let J, K_1, $K_2, \ldots,$ K_m be any given functionals which are defined on an open set D contained in a normed vector space $\mathfrak{X}$. *We shall assume for simplicity that all these functionals are differentiable on D.* (See Section 2.6.) In fact, the *linearity* and *continuity* of the differentials with respect to their second arguments will play an important part in the proof below. We assume also that the differentials of all the functionals concerned are weakly continuous at all vectors x considered in D, where weak continuity for differentials is defined exactly as for variations by replacing δJ with dJ in equation (3.2.1). As usual we let $D[K_i = k_i$ for $i = 1, 2, \ldots, m]$ be the subset of D consisting of all vectors x in D which satisfy all the constraints of (3.6.1), where $k_1, k_2, \ldots, k_m$ may in general be any given numbers. We shall, however, assume that the set $D[K_i = k_i$ for $i = 1, 2, \ldots, m]$ is nonempty for all choices of $k_1, k_2, \ldots, k_m$ considered.

Suppose now that J has a local maximum (or minimum) in $D[K_i = k_i$ for $i = 1, 2, \ldots, m]$ at some vector x^*. If the determinant (3.6.2) is *not* identically zero, then the Euler-Lagrange multiplier theorem imples that there are numbers $\lambda_1, \lambda_2, \ldots, \lambda_m$ such that [see (3.6.3), where, however, we have replaced the variations with differentials; recall (2.6.6)]

$$dJ(x^*; \Delta x) = \sum_{j=1}^{m} \lambda_j \, dK_j(x^*; \Delta x) \tag{3.8.7}$$

for every vector Δx in $\mathfrak{X}$.

Any such extremum vector x^* will in general depend on the values $k_1,$ $k_2, \ldots, k_m$ used in the constraints of (3.6.1), much as the extremum vector C^* of (3.8.4) depends on $k = k_1$. To indicate this dependency of x^* on $k_1,$ $k_2, \ldots, k_m$, we shall write

$$x^* = x^*(k_1, k_2, \ldots, k_m), \tag{3.8.8}$$

where we assume that the optimization problem considered for x^* has a solution given by (3.8.8) for all choices of $k_1, k_2, \ldots, k_m$ considered. The local extreme value of the functional J in $D[K_i = k_i$ for $i = 1, 2, \ldots, m]$ corresponding to the extremum vector x^* of (3.8.8) then becomes

$$J(x^*(k_1, k_2, \ldots, k_m)), \tag{3.8.9}$$

which also depends on the values $k_1, k_2, \ldots, k_m$ used in the constraints, much as the extreme value $\mathcal{S}(C^*)$ of (3.8.5) depends on $k = k_1$.

Suppose now that the real-valued function of $k_1, k_2, \ldots, k_m$ defined by (3.8.9) has continuous first-order partial derivatives with respect to the variables $k_1, k_2, \ldots, k_m$, denoted as

$$\frac{\partial}{\partial k_i} J(x^*(k_1, k_2, \cdots, k_m)) \qquad \text{for } i = 1, 2, \cdots, m.$$

In this case it turns out that these partial derivatives are related to the corresponding Euler-Lagrange multipliers appearing in (3.8.7) by the simple equation

$$\frac{\partial}{\partial k_i} J(x^*(k_1, k_2, \cdots, k_m)) = \lambda_i \qquad \text{for } i = 1, 2, \cdots, m, \tag{3.8.10}$$

which is the general form of the special result (3.8.6) which was shown to hold above for the satisfaction functional $\mathcal{S}$.

Hence *the multiplier λ_i gives the rate of change of the extreme value J of* (3.8.9) *with respect to the* ith *constraint k_i when all other constraint values k_j* (for all $j \neq i$) *are held fixed.* Alternatively, we may say that λ_i measures the sensitivity of $J(x^*(k_1, k_2, \ldots, k_m))$ to changes in the ith constraint value k_i. It sometimes happens in optimization problems in economics that the functional J represents a profit or a cost which is to be maximized or minimized, while the functionals $K_1, K_2, \ldots, K_m$ represent the quantities of m different resources which are to be used in producing the profit or cost. Then, according to (3.8.10), the multiplier λ_i tells us how much the maximum profit or the minimum cost will be changed if the available quantity of the ith resource is changed by one unit. Hence λ_i can be interpreted in this case as a *value* or *price* per unit of resource i. In such cases the Euler-Lagrange multipliers are sometimes called **shadow values** or **shadow prices**. In the preceding problem on the optimization of consumption it is seen from (3.8.3) and (3.8.6) that the Euler-Lagrange multiplier λ furnishes in that case a measure of the *price* placed on the terminal savings level S_T since λ measures the amount of increase that may be obtained in the satisfaction functional by a unit *decrease* in the terminal savings level S_T (assuming that the resources S_0 and I are fixed).

We shall now indicate briefly a proof† of relation (3.8.10) in the special case that the functionals are defined and differentiable on an open set D contained in *n-dimensional Euclidean space* $\mathcal{X} = \mathcal{R}_n$ equipped with any suitable norm [as, for example, the norm given by (1.4.4)]. The proof in the more general case of an arbitrary normed vector space $\mathcal{X}$ follows *exactly* the same lines, except in the more general case the partial derivatives $\partial x^*/\partial k_i$ and the remainder R appearing below in equation (3.8.14) are functions with *vector values* in $\mathcal{X}$.

Hence suppose that $J, K_1, K_2, \ldots, K_m$ are differentiable functions defined on an open set D in n-space $\mathcal{R}_n$, and let J have a local maximum or local minimum in $D[K_i = k_i \text{ for } i = 1, 2, \ldots, m]$ at some point $x^* = (x_1^*, x_2^*, \ldots, x_n^*)$ in $\mathcal{R}_n$. As in (3.8.8) we write $x^* = x^*(k_1, k_2, \ldots, k_m)$ to indicate the dependency of x^* on the constraint values $k_1, k_2, \ldots, k_m$, or, more briefly,

†The rest of this section can safely be skipped during a first reading. In this case the reader should turn directly to Section 3.9.

$$x^* = x^*(k),$$

where k may be any suitable m-tuple of numbers,

$$k = (k_1, k_2, \ldots, k_m).$$

In this case each component of x^* depends on k, and so we may also write

$$x_j^* = x_j^*(k) \qquad \text{for } j = 1, 2, \ldots, n, \tag{3.8.11}$$

where $x^* = (x_1^*, x_2^*, \ldots, x_n^*)$.

We assume that the real-valued functions given by (3.8.11) are continuously differentiable (in the usual sense of calculus) at all m-tuples k considered, in which case for all suitable (small) m-tuples $\Delta k = (\Delta k_1, \Delta k_2, \ldots, \Delta k_m)$ [compare with (2.3.6) and (2.6.2)]

$$x_j^*(k + \Delta k) = x_j^*(k) + \sum_{i=1}^{m} \frac{\partial x_j^*(k)}{\partial k_i} \Delta k_i + R_j(k; \Delta k) \| \Delta k \| \qquad \text{for } j = 1, 2, \ldots, n \tag{3.8.12}$$

will hold for suitable functions $R_j(k; \Delta k)$, which satisfy

$$\operatorname*{limit}_{\Delta k \to 0 \text{ in } \mathcal{R}_m} R_j(k; \Delta k) = 0 \qquad \text{for } j = 1, 2, \ldots, n. \tag{3.8.13}$$

We may write (3.8.12) more briefly in vector form as

$$x^*(k + \Delta k) = x^*(k) + \sum_{i=1}^{m} \frac{\partial x^*(k)}{\partial k_i} \Delta k_i + R(k; \Delta k) \| \Delta k \|, \tag{3.8.14}$$

where now x^*, $\partial x^*/\partial k_i$, and R are all vectors with n components, as, for example,

$$\frac{\partial x^*(k)}{\partial k_i} = \left(\frac{\partial x_1^*(k)}{\partial k_i}, \frac{\partial x_2^*(k)}{\partial k_i}, \ldots, \frac{\partial x_n^*(k)}{\partial k_i} \right).$$

Now, (3.8.14) holds for all small m-tuples $\Delta k = (\Delta k_1, \Delta k_2, \ldots, \Delta k_m)$. In particular, if we take all components of Δk to be zero except for the ith component Δk_i, we find that

$$x^*(k + \Delta k_i e_i) = x^*(k) + \frac{\partial x^*(k)}{\partial k_i} \Delta k_i + R(k; \Delta k_i e_i) | \Delta k_i | \tag{3.8.15}$$

for all small nonzero numbers Δk_i, where e_i here denotes the ith unit vector

$$e_i = (0, \ldots, 0, 1, 0, \ldots, 0)$$

in $\mathcal{R}_m$ with all components zero except for the ith component which equals 1.

Hence we find from (3.8.15) that

$$J(x^*(k + \Delta k_i e_i)) = J\left(x^*(k) + \frac{\partial x^*(k)}{\partial k_i}\Delta k_i + R(k; \Delta k_i e_i)\,|\Delta k_i|\right),$$

and then using (2.6.2) with $x = x^*(k)$ and

$$h = \Delta x = \frac{\partial x^*(k)}{\partial k_i}\Delta k_i + R(k; \Delta k_i e_i)\,|\Delta k_i|, \tag{3.8.16}$$

we have (explain)

$$\begin{aligned} &J(x^*(k + \Delta k_i e_i)) \\ &\quad = J(x^*(k)) + dJ\left(x^*(k); \frac{\partial x^*(k)}{\partial k_i}\Delta k_i + R\,|\Delta k_i|\right) + E(x^*(k); \Delta x)\,\|\Delta x\|. \end{aligned}$$

If we use the linearity of the differential as a function of its second argument, from the last equation we can derive

$$\begin{aligned} \frac{J(x^*(k + \Delta k_i e_i)) - J(x^*(k))}{\Delta k_i} &= dJ\left(x^*(k); \frac{\partial x^*(k)}{\partial k_i}\right) \\ &\quad + \frac{|\Delta k_i|}{\Delta k_i}\, dJ(x^*(k); R(k; \Delta k_i\, e_i)) \\ &\quad + \frac{\|\Delta x\|}{\Delta k_i}\, E(x^*(k); \Delta x), \end{aligned} \tag{3.8.17}$$

and then (3.8.13), (3.8.16), (3.8.17), (2.6.3), and the continuity of the differential as a function of its second argument imply that [we also use the result $dJ\,(x^*; 0) = 0$]

$$\operatorname*{limit}_{\Delta k_i \to 0} \frac{J(x^*(k + \Delta k_i e_i)) - J(x^*(k))}{\Delta k_i} = dJ\left(x^*(k); \frac{\partial x^*(k)}{\partial k_i}\right). \tag{3.8.18}$$

But the partial derivative $\partial J(x^*(k))/\partial k_i$ is by definition given as the limit

$$\operatorname*{limit}_{\Delta k_i \to 0} \frac{J(x^*(k + \Delta k_i e_i)) - J(x^*(k))}{\Delta k_i} = \frac{\partial J(x^*(k))}{\partial k_i},$$

so that (3.8.18) implies that

$$\frac{\partial J(x^*(k))}{\partial k_i} = dJ\left(x^*; \frac{\partial x^*}{\partial k_i}\right).$$

If we now take $\Delta x = \partial x^*/\partial k_i$ in (3.8.7), we find with this last result that

$$\frac{\partial J(x^*(k))}{\partial k_i} = \sum_{j=1}^{m} \lambda_j dK_j\left(x^*(k); \frac{\partial x^*(k)}{\partial k_i}\right),$$

from which the desired result (3.8.10) *will follow provided that*

$$dK_j\left(x^*(k);\ \frac{\partial x^*(k)}{\partial k_i}\right) = \begin{cases} 1 & \text{if } j = i \\ 0 & \text{if } j \neq i \end{cases} \tag{3.8.19}$$

holds (explain). We can easily check that (3.8.19) does, in fact, hold by computing [compare with (3.8.17)]

$$\frac{K_j(x^*(k + \Delta k_i e_i)) - K_j(x^*(k))}{\Delta k_i}$$
$$= dK_j\left(x^*(k);\ \frac{\partial x^*(k)}{\partial k_i}\right) + \text{terms which vanish as } \Delta k_i \to 0. \tag{3.8.20}$$

Since $x^*(k + \Delta k_i e_i)$ is by definition in the set $D[K_j = k_j$ for all $j \neq i$, $K_i = k_i + \Delta k_i]$, we have

$$K_j(x^*(k + \Delta k_i e_i)) = \begin{cases} k_i + \Delta k_i & \text{if } j = i \\ k_j & \text{if } j \neq i \end{cases}$$

and, similarly,

$$K_j(x^*(k)) = k_j \qquad \text{for all } j.$$

Inserting these results into (3.8.20) and letting Δk_i tend toward zero, we obtain (3.8.19), and the desired result follows.

Exercises

1. Let $J(x) = x_1^2 + x_2^2 + 2x_3^2 - x_1 - x_2 x_3$ and $K(x) = x_1 + x_2 + x_3$ for any vector $x = (x_1, x_2, x_3)$ in $\mathcal{R}_3$. Find $\partial J(x^*)/\partial k$ evaluated at $k = 35$ if x^* is a minimum vector in $\mathcal{R}_3$ for J subject to the constraint $K(x) = k$. (See Exercise 6 of Section 3.4.)
2. Let D be the open subset of $\mathcal{R}_3$ consisting of all vectors $x = (x_1, x_2, x_3)$ for which $x_1 > 0$, $x_2 > 0$, and $x_3 > 0$, and let $J(x) = x_1^2 + x_2^2 + x_3^2 + x_1 x_2 x_3$ and $K(x) = x_1 x_2 x_3$ for any vector x in D. Find $\partial J(x^*)/\partial k$ evaluated at $k = 8$ if x^* is a maximum vector in D for J subject to the constraint $K(x) = k$. (See Exercise 7 of Section 3.4.)

3.9. Chaplygin's Problem, or a Modern Version of Queen Dido's Problem

We consider Chaplygin's problem described in Section 1.3 for the area functional $A = A(\alpha)$ defined by formula (1.3.11) on the normed vector space $\mathcal{C}^0[0, T]$ consisting of all functions (steering controls) $\alpha = \alpha(t)$ of class $\mathcal{C}^0$ on the fixed interval $[0, T]$, equipped with the uniform norm. Briefly, the problem is to find that particular steering control $\alpha^*(t)$ that will allow an airplane to encircle a *maximum area* in time T while flying at constant natural speed v_0 (relative to the surrounding air) and while a constant wind is blowing. We refer the reader to Section 1.3 for a more complete description of the problem.

Derivation of the Extremum Equation. We seek an extremum vector α^* in $\mathcal{C}^0[0, T]$ that will furnish a maximum value for the area functional A of (1.3.11) subject to the constraints of (1.3.9) and (1.3.10). If we define functionals K_1, K_2, and K_3 on $\mathcal{C}^0[0, T]$ by

$$K_1(\alpha) = \int_0^T \cos \alpha(t)\, dt \tag{3.9.1}$$

$$K_2(\alpha) = \int_0^T \sin \alpha(t)\, dt \tag{3.9.2}$$

and

$$K_3(\alpha) = \alpha(0) \tag{3.9.3}$$

for any continuous function $\alpha = \alpha(t)$ on $[0, T]$, then the constraints (1.3.9) and (1.3.10) can be given as

$$K_1(\alpha) = -\frac{w_0}{v_0} T \tag{3.9.4}$$

$$K_2(\alpha) = 0 \tag{3.9.5}$$

and

$$K_3(\alpha) = \alpha_0. \tag{3.9.6}$$

The optimization problem is then to find an extremum vector α^* in $D[K_1 = -(w_0/v_0)T, K_2 = 0, K_3 = \alpha_0]$ for the functional A, where here we take the open set D to be the entire normed vector space $\mathcal{C}^0[0, T]$.

To proceed, we need to calculate the variations of the functionals involved. We use (3.9.1) to calculate

$$K_1(\alpha + \epsilon\, \Delta\alpha) = \int_0^T \cos\, [\alpha(t) + \epsilon\, \Delta\alpha(t)]\, dt$$

for any number ϵ and any pair of vectors α and $\Delta\alpha$ in $\mathcal{C}^0[0, T]$, and this result implies, upon differentiation, that [compare with (2.4.5) and (2.4.6)]

$$\frac{d}{d\epsilon} K_1(\alpha + \epsilon\, \Delta\alpha) = \int_0^T \frac{\partial}{\partial\epsilon} \cos\, [\alpha(t) + \epsilon\, \Delta\alpha(t)]\, dt.$$

We may use the chain rule of differential calculus to find for each fixed t that (explain)

$$\frac{\partial}{\partial \epsilon} \cos [\alpha(t) + \epsilon \, \Delta\alpha(t)] = -\sin[\alpha(t) + \epsilon \, \Delta\alpha(t)] \, \Delta\alpha(t),$$

and the last two equations with (2.3.4) then give

$$\delta K_1(\alpha; \Delta\alpha) = \frac{d}{d\epsilon} K_1(\alpha + \epsilon \, \Delta\alpha)\Big|_{\epsilon=0} = -\int_0^T \sin \alpha(t) \, \Delta\alpha(t) \, dt \tag{3.9.7}$$

for any function $\Delta\alpha$ of class $\mathcal{C}^0$ on $[0, T]$. Similarly, from (3.9.2) and (2.3.4) we find that

$$\delta K_2(\alpha; \Delta\alpha) = \int_0^T \cos \alpha(t) \, \Delta\alpha(t) \, dt \tag{3.9.8}$$

for any $\Delta\alpha$ in $\mathcal{C}^0[0, T]$. Finally, the variation of K_3 is found easily from (3.9.3) and (2.3.4) to be

$$\delta K_3(\alpha; \Delta\alpha) = \Delta\alpha(0), \tag{3.9.9}$$

again for any $\Delta\alpha$ in $\mathcal{C}^0[0, T]$. The variation of the area functional A is given by (2.4.8).

Now it can be easily checked that the variations of all the functionals A, K_1, K_2, and K_3 are weakly continuous everywhere on $\mathcal{C}^0[0, T]$. Moreover, it can be shown that the determinant [see (3.6.2)]

$$\det \begin{vmatrix} \delta K_1(\alpha; \Delta\alpha_1) & \delta K_1(\alpha; \Delta\alpha_2) & \delta K_1(\alpha; \Delta\alpha_3) \\ \delta K_2(\alpha; \Delta\alpha_1) & \delta K_2(\alpha; \Delta\alpha_2) & \delta K_2(\alpha; \Delta\alpha_3) \\ \delta K_3(\alpha; \Delta\alpha_1) & \delta K_3(\alpha; \Delta\alpha_2) & \delta K_3(\alpha; \Delta\alpha_3) \end{vmatrix} \tag{3.9.10}$$

does *not* vanish identically for all $\Delta\alpha_1$, $\Delta\alpha_2$, and $\Delta\alpha_3$ in $\mathcal{C}^0[0, T]$, at least provided that the function $\alpha = \alpha(t)$ is continuously differentiable with $d\alpha(t)/dt \neq 0$ for $0 \leq t \leq T$. (See Exercise 2 for an indication of a method of proof for this assertion.) Hence if α^* is any local extremum vector in $D[K_1 = -Tw_0/v_0, K_2 = 0, K_3 = \alpha_0]$ for A satisfying $d\alpha^*/dt \neq 0$, it follows from the Euler-Lagrange multiplier theorem that there are constants λ_1, λ_2, and λ_3 such that [see (3.6.3)]

$$\delta A(\alpha^*; \Delta\alpha) = \lambda_1 \, \delta K_1(\alpha^*; \Delta\alpha) + \lambda_2 \, \delta K_2(\alpha^*; \Delta\alpha) + \lambda_3 \, \delta K_3(\alpha^*; \Delta\alpha) \tag{3.9.11}$$

for *all* vectors $\Delta\alpha$ in $\mathcal{C}^0[0, T]$. If we use (2.4.8), (3.9.7), (3.9.8), and (3.9.9) in

(3.9.11), we find the necessary condition

$$\frac{v_0^2}{2}\int_0^T \Delta\alpha(t)\int_0^t \cos\,[\alpha^*(t)-\alpha^*(\tau)]\,d\tau\,dt - \frac{v_0^2}{2}\int_0^T\int_0^t \Big\{\cos\,[\alpha^*(t)-\alpha^*(\tau)] + \frac{w_0}{v_0}\cos\alpha^*(\tau)\Big\}\,\Delta\alpha(\tau)\,d\tau\,dt = \int_0^T \Big\{-\sin\alpha^*(t)\Big[\lambda_1 + \frac{v_0 y_0}{2}\Big] + \cos\alpha^*(t)\Big[\lambda_2 - \frac{v_0}{2}(w_0 t + x_0)\Big]\Big\}\,\Delta\alpha(t)\,dt + \lambda_3\,\Delta\alpha(0), \tag{3.9.12}$$

which must hold for all functions $\Delta\alpha = \Delta\alpha(t)$ of class $\mathcal{C}^0$ on $[0, T]$. If we interchange the orders of the repeated integrations in the second double integral on the left-hand side of (3.9.12) [see (2.5.17)], we find from (3.9.12) after some integration and rearrangement that

$$\int_0^T \Big\{\frac{v_0^2}{2}\int_0^t \cos\,[\alpha^*(t)-\alpha^*(\tau)]\,d\tau - \frac{v_0 w_0}{2}(T-t)\cos\alpha^*(t) - \frac{v_0^2}{2}\int_t^T \cos\,[\alpha^*(t)-\alpha^*(\tau)]\,d\tau + \Big(\lambda_1 + \frac{v_0 y_0}{2}\Big)\sin\alpha^*(t) + \Big(-\lambda_2 + \frac{v_0 x_0}{2} + \frac{v_0 w_0}{2}\,t\Big)\cos\alpha^*(t)\Big\}\,\Delta\alpha(t)\,dt = \lambda_3\,\Delta\alpha(0). \tag{3.9.13}$$

On the other hand, a well-known result from integral calculus implies that

$$\int_t^T \cos\,[\alpha^*(t)-\alpha^*(\tau)]\,d\tau = \int_0^T \cos\,[\alpha^*(t)-\alpha^*(\tau)]\,d\tau - \int_0^t \cos\,[\alpha^*(t)-\alpha^*(\tau)]\,d\tau,$$

which along with the identity

$$\cos\,[\alpha^*(t)-\alpha^*(\tau)] = \cos\alpha^*(t)\cos\alpha^*(\tau) + \sin\alpha^*(t)\sin\alpha^*(\tau) \tag{3.9.14}$$

and the constraints (3.9.4) and (3.9.5) allows us to rewrite (3.9.13) finally as

$$\int_0^T \Big\{v_0^2\int_0^t \cos\,[\alpha^*(t)-\alpha^*(\tau)]\,d\tau + \Big(\lambda_1 + \frac{v_0 y_0}{2}\Big)\sin\alpha^*(t) + \Big(-\lambda_2 + \frac{v_0 x_0}{2} + v_0 w_0 t\Big)\cos\alpha^*(t)\Big\}\,\Delta\alpha(t)\,dt = \lambda_3\,\Delta\alpha(0), \tag{3.9.15}$$

which must now hold for *all continuous functions* $\Delta\alpha = \Delta\alpha(t)$ on $[0, T]$.

Equation (3.9.15), which involves the arbitrary vector $\Delta\alpha$, is a direct consequence of the Euler-Lagrange multiplier theorem. We now wish to eliminate the arbitrary vector $\Delta\alpha$ from consideration so as to get a simpler necessary condition involving only the desired extremum vector α^*. This would be easy to do *if* we knew that the right-hand side of equation (3.9.15) were zero for all $\Delta\alpha$, that is, if we knew that

$$\lambda_3 = 0 \tag{3.9.16}$$

held. Indeed, then we could argue as in Section 3.7 in going from (3.7.19) to (3.7.20) and find in this case that

$$v_0^2 \int_0^t \cos[\alpha^*(t) - \alpha^*(\tau)]\,d\tau + \left(\lambda_1 + \frac{v_0 y_0}{2}\right)\sin\alpha^*(t) + \left(-\lambda_2 + \frac{v_0 x_0}{2} + v_0 w_0 t\right)\cos\alpha^*(t) = 0, \tag{3.9.17}$$

which would then be required to hold for all t in the interval $[0, T]$. Equation (3.9.17) would then be the desired simpler equation involving only the unknown extremum vector α^*, and hopefully it could be solved to give α^* just as equations (2.5.3) and (3.7.20) were solved, respectively, to give the extremum vectors P^* and C^*.

Unfortunately we have no justification at the moment to assume that (3.9.16) holds, and it is therefore not yet clear whether or not equation (3.9.17) must necessarily follow from (3.9.15). But, in fact, it must! Indeed, since (3.9.15) must hold for *all* continuous functions $\Delta\alpha$ on $[0, T]$, it must certainly hold in particular for all such continuous functions $\Delta\alpha$ which vanish at the left end point of the given interval $[0, T]$. But for any such function $\Delta\alpha$ satisfying

$$\Delta\alpha(0) = 0, \tag{3.9.18}$$

the necessary condition (3.9.15) becomes

$$\int_0^T \left\{v_0^2 \int_0^t \cos[\alpha^*(t) - \alpha^*(\tau)]d\tau + \left(\lambda_1 + \frac{v_0 y_0}{2}\right)\sin\alpha^*(t) + \left(-\lambda_2 + \frac{v_0 x_0}{2} + v_0 w_0 t\right)\cos\alpha^*(t)\right\}\Delta\alpha(t)\,dt = 0, \tag{3.9.19}$$

which must then necessarily hold for all continuous functions $\Delta\alpha$ on $[0, T]$ satisfying (3.9.18). *We can now use a basic lemma of the calculus of variations* (see Section A4 of the Appendix) *to conclude from the last result that* (3.9.17) *must also necessarily hold.* We refer the reader to Section A4 of the Appendix, where a proof of this assertion is given.

Hence, using the lemma of Section A4 of the Appendix, we can, in fact, conclude that *any local extremum vector* α^* *in* $D[K_1 = -(w_0/v_0)T, K_2 = 0, K_3 = \alpha_0]$ *for A must satisfy the necessary equation* (3.9.17) *for all t in the interval* $[0, T]$. It then follows also that *the condition* (3.9.16) *must hold* since (3.9.17) implies that the left-hand side of (3.9.15) vanishes for every $\Delta\alpha$, and hence the right-hand side must also vanish for *every* continuous function $\Delta\alpha$

on $[0, T]$. In particular, if we take $\Delta\alpha$ to be the constant function $\Delta\alpha(t) \equiv 1$, we then obtain the result (3.9.16) (explain). Note that (3.9.16) along with (3.8.10) and (3.9.3) leads us to predict that

$$\frac{\partial}{\partial \alpha_0} A(\alpha^*) = 0, \tag{3.9.20}$$

where $A(\alpha^*)$ is the maximum area encircled by the airplane and α_0 is the initial steering angle. We shall see below that although the choice of α_0 *does affect* the optimum flight path, it does so in such a way as *not* to affect the resulting maximum area encircled [see (3.9.43) below]. Hence (3.9.20) will indeed be true.

The Extremum Path. We shall now attempt to solve equation (3.9.17) for the optimum steering control $\alpha^* = \alpha^*(t)$.† First, however, we shall show that the Euler-Lagrange multipliers λ_1 and λ_2 appearing in (3.9.17) can both be given in terms of a single parameter μ. In fact, if we first choose $t = 0$ in (3.9.17) and then $t = T$, we find with (3.9.14) and the constraints (3.9.4), (3.9.5), and (3.9.6) the two conditions

$$\begin{gathered}\left(\lambda_1 + \frac{v_0 y_0}{2}\right)\sin\alpha_0 + \left(-\lambda_2 + \frac{v_0 x_0}{2}\right)\cos\alpha_0 = 0\\ \left(\lambda_1 + \frac{v_0 y_0}{2}\right)\sin\alpha^*(T) + \left(-\lambda_2 + \frac{v_0 x_0}{2}\right)\cos\alpha^*(T) = 0,\end{gathered} \tag{3.9.21}$$

which must be satisfied by the multipliers λ_1 and λ_2. If the final steering angle $\alpha^*(T)$ satisfies

$$\alpha^*(T) = \alpha_0 + 2\pi n \tag{3.9.22}$$

for some integer n, so that

$$\tan\alpha^*(T) = \tan\alpha_0,$$

then the system of equations (3.9.21) is *singular*‡ and has infinitely many solutions§ for λ_1 and λ_2 in the form

$$\begin{aligned}\lambda_1 &= -\frac{v_0 y_0}{2} - \mu v_0^2 \cos\alpha_0\\ \lambda_2 &= \frac{v_0 x_0}{2} - \mu v_0^2 \sin\alpha_0\end{aligned} \tag{3.9.23}$$

†The rest of this section may be skimmed over rather lightly without causing any later difficulties.

‡See John G. Hocking, *Calculus with an Introduction to Linear Algebra* (New York: Holt, Rinehart and Winston, Inc., 1970), p. 721.

§*Ibid.*, p. 745.

for any arbitrary constant μ. On the other hand, if (3.9.22) does *not* hold, then the system (3.9.21) is *nonsingular*,† and the unique solution is given by (3.9.23) with $\mu = 0$. Hence in every case (3.9.23) includes all possible solutions of (3.9.21). We shall see later how to specify the unknown constant μ. For the moment we simply use (3.9.23) to rewrite the necessary condition (3.9.17) in the form

$$\int_0^t \cos [\alpha^*(t) - \alpha^*(\tau)]\, d\tau = \mu \sin [\alpha^*(t) - \alpha_0] - et \cos \alpha^*(t), \tag{3.9.24}$$

where we have used the trigonometric identity

$$\sin (\alpha^* - \alpha_0) = \sin \alpha^* \cos \alpha_0 - \sin \alpha_0 \cos \alpha^*, \tag{3.9.25}$$

and where we have defined the number e as the known ratio

$$e = \frac{w_0}{v_0}. \tag{3.9.26}$$

As in Section 1.3 we assume that $0 \leq e < 1$.

We can now use equation (3.9.24) along with equations (1.3.7) with $\alpha = \alpha^*$ to show that *the extremum curve flown by the airplane must be an ellipse* in the (x, y)-plane. In fact, if we integrate both sides of the identity (3.9.14) with respect to τ for $0 \leq \tau \leq t$ and then use equations (1.3.7) with $\alpha = \alpha^*$, we find directly that

$$\int_0^t \cos [\alpha^*(t) - \alpha^*(\tau)]\, d\tau$$
$$= \left(\frac{X(t) - x_0}{v_0} - et\right) \cos \alpha^*(t) + \left(\frac{Y(t) - y_0}{v_0}\right) \sin \alpha^*(t),$$

where e is again given by (3.9.26). Then (3.9.24), (3.9.25), and the last result give

$$(X(t) - x_0 + \mu v_0 \sin \alpha_0) \cos \alpha^*(t) + (Y(t) - y_0 - \mu v_0 \cos \alpha_0) \sin \alpha^*(t) = 0,$$

which must hold for points $(x, y) = (X(t),\ Y(t))$ along any extremum curve. This relation gives in turn,

$$\tan \alpha^*(t) = \frac{X(t) - x_0 + \mu v_0 \sin \alpha_0}{-Y(t) + y_0 + \mu v_0 \cos \alpha_0}$$

from which we find that (see Figure 9)

$$\begin{aligned} X(t) - x_0 + \mu v_0 \sin \alpha_0 &= R(t) \sin \alpha^*(t) \\ Y(t) - y_0 - \mu v_0 \cos \alpha_0 &= -R(t) \cos \alpha^*(t), \end{aligned} \tag{3.9.27}$$

†*Ibid.*, p. 721.

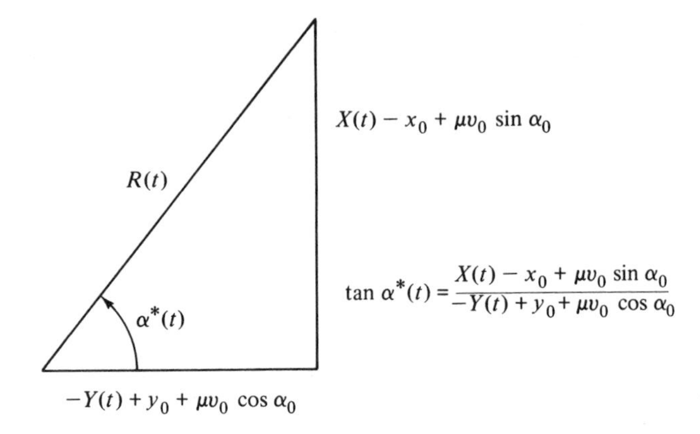

Figure 9

with

$$R(t)^2 = [X(t) - x_0 + \mu v_0 \sin \alpha_0]^2 + [Y(t) - y_0 - \mu v_0 \cos \alpha_0]^2. \tag{3.9.28}$$

If we differentiate both sides of (3.9.28) with respect to t and use (3.9.27), we find that

$$R'(t) = \sin \alpha^*(t) X'(t) - \cos \alpha^*(t) Y'(t),$$

which with (1.3.6) and (3.9.26) implies finally that

$$R'(t) = eY'(t).$$

This equation can be integrated to give

$$R(t) = eY(t) + c$$

for some constant of integration c, and then this last result with (3.9.28) implies that

$$[X(t) - x_0 + \mu v_0 \sin \alpha_0]^2 + [Y(t) - y_0 - \mu v_0 \cos \alpha_0]^2 = [eY(t) + c]^2 \tag{3.9.29}$$

for $(x, y) = (X(t), Y(t))$ along any extremum curve. It follows from (3.9.29) and the initial condition (1.3.8) that the constant c must take on the value (explain)

$$c = \mu v_0 - e y_0,$$

and then (3.9.29) can be rewritten as

$$(x - x_0 + \mu v_0 \sin \alpha_0)^2 + (y - y_0 - \mu v_0 \cos \alpha_0)^2 = [e(y - y_0) + \mu v_0]^2 \tag{3.9.30}$$

for $(x, y) = (X(t), Y(t))$.

Equation (3.9.30) is the equation of an ellipse with eccentricity e and major axis perpendicular to the x-axis, as shown in Figure 10. In fact, (3.9.30) can be written in the familiar form

$$\frac{(x - x_1)^2}{a^2} + \frac{(y - y_1)^2}{b^2} = 1, \tag{3.9.31}$$

with

$$\begin{aligned} x_1 &= x_0 - \mu v_0 \sin \alpha_0 \\ y_1 &= y_0 + \mu v_0 \frac{e + \cos \alpha_0}{1 - e^2} \end{aligned} \tag{3.9.32}$$

and

$$\begin{aligned} a &= \mu v_0 \frac{1 + e \cos \alpha_0}{\sqrt{1 - e^2}} \\ b &= \mu v_0 \frac{1 + e \cos \alpha_0}{1 - e^2}. \end{aligned} \tag{3.9.33}$$

The center of the ellipse is at the point (x_1, y_1) with coordinates given by (3.9.32), as shown in Figure 10, while the area enclosed by any such ellipse is

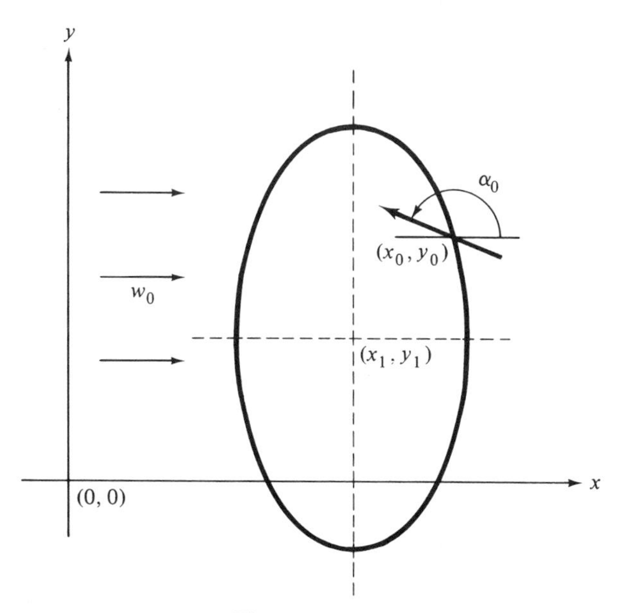

Figure 10

given by the well-known formula (which the reader should be able to derive)

$$\text{area} = \pi ab.$$

If we use (3.9.33) in the latter formula, we find in the present case that

$$\text{area} = \pi\mu^2 v_0^2 \frac{(1 + e\cos\alpha_0)^2}{(1 - e^2)^{3/2}}. \tag{3.9.34}$$

Determining the Multiplier. We must finally show how to specify the unknown constant μ. The key to this specification lies in equation (3.9.31). Indeed, (3.9.31) implies that *there is a well-defined tangent line at each point* $(x, y) = (X(t), Y(t))$ *on any possible extremum curve* in the (x, y)-plane, and this implies with (1.3.4) that (explain)

$$X'(T) = X'(0), \qquad Y'(T) = Y'(0).$$

The latter result along with (1.3.6) and (1.3.10) implies in turn for any possible optimum steering control angle $\alpha = \alpha^*(t)$ the conditions

$$\cos\alpha^*(T) = \cos\alpha_0$$
$$\sin\alpha^*(T) = \sin\alpha_0,$$

which imply that *the optimum steering angle* α^* *must automatically satisfy condition* (3.9.22) *for some integer n*. Hence condition (3.9.22) is a *natural boundary condition* similar to the condition (2.5.16) which appeared in the solution of the optimization problem considered in Section 2.5. The integer n appearing in (3.9.22) is determined by the number of times the airplane traverses the ellipse (3.9.31) during the given time period $0 \le t \le T$.

We can now use the natural boundary condition (3.9.22) to determine the constant μ appearing in (3.9.32) and (3.9.33), and at the same time we shall obtain the optimum steering control $\alpha^*(t)$. If we differentiate the integral equation (3.9.24) twice with respect to time t, we easily find the differential equation [with $\dot{\alpha}^* = (d/dt)\alpha^*$, $\ddot{\alpha}^* = (d^2/dt^2)\alpha^*$]

$$\ddot{\alpha}^*(1 + e\cos\alpha^*) + 2(\dot{\alpha}^*)^2 e\sin\alpha^* = 0, \tag{3.9.35}$$

which must be satisfied by the optimum steering control angle $\alpha = \alpha^*(t)$. [The calculation involved here in obtaining (3.9.35) from (3.9.24) is similar to the calculation used in Section 2.5 in obtaining (2.5.7) from (2.5.3).] We may rewrite (3.9.35) as

$$\frac{\ddot{\alpha}^*}{\dot{\alpha}^*} = \frac{-2\dot{\alpha}^* e\sin\alpha^*}{1 + e\cos\alpha^*}$$

or

$$\frac{d}{dt}\log\dot{\alpha}^* = 2\frac{d}{dt}\log(1 + e\cos\alpha^*).$$

The last equation can be integrated to give

$$\dot{\alpha}^* = \dot{\alpha}_0 \left(\frac{1 + e \cos \alpha^*}{1 + e \cos \alpha_0}\right)^2, \tag{3.9.36}$$

where we have used the initial condition (3.9.6), and where the constant of integration $\dot{\alpha}_0$ is given as

$$\dot{\alpha}_0 = \dot{\alpha}^*(t) \qquad \text{at } t = 0. \tag{3.9.37}$$

The differential equation (3.9.36) for $\alpha^* = \alpha^*(t)$ is *separable* and can be integrated to give†

$$\int_{\alpha_0}^{\alpha^*} \frac{d\alpha}{(1 + e \cos \alpha)^2} = \frac{\dot{\alpha}_0 t}{(1 + e \cos \alpha_0)^2}, \tag{3.9.38}$$

where we again used the initial condition (3.9.6). The constant $\dot{\alpha}_0$ of (3.9.37) can be related to the previous constant μ by differentiating (3.9.24) once and setting $t = 0$ in the resulting equation; we find that

$$\dot{\alpha}_0 = \frac{1 + e \cos \alpha_0}{\mu}, \tag{3.9.39}$$

where we have used (3.9.6) again along with (3.9.37). Hence (3.9.38) and the last result give‡

$$\int_{\alpha_0}^{\alpha^*} \frac{d\alpha}{(1 + e \cos \alpha)^2} = \frac{t}{\mu(1 + e \cos \alpha_0)}, \tag{3.9.40}$$

where the integral on the left-hand side can be evaluated using the formula§

$$\int \frac{d\alpha}{(1 + e \cos \alpha)^2} = \frac{1}{1 - e^2}\left[\frac{-e \sin \alpha}{1 + e \cos \alpha} + \frac{2}{\sqrt{1 - e^2}} \tan^{-1}\sqrt{\frac{1 - e}{1 + e}} \tan\frac{\alpha}{2}\right] \tag{3.9.41}$$

†See Thomas, *op. cit.*, p. 693.

‡It is interesting to note that this same equation (3.9.40) occurs also in celestial mechanics in the derivation of *Kepler's equation* for the motion of a planet around the sun, except in that case the constant μ is related to the *angular momentum* of the planet, and the present steering angle α corresponds to the *true anomaly* of the (elliptical) planetary motion. See Chapter 6 of J. M. A. Danby, *Fundamentals of Celestial Mechanics* (New York: The Macmillan Company, 1964). The methods used in celestial mechanics to obtain numerical solutions of Kepler's equation can be used to solve (3.9.40).

§*Standard Mathematical Tables*, 18th ed. (Cleveland: The Chemical Rubber Co., 1970), p. 426.

which is valid for $e^2 < 1$. But now finally we can calculate μ from (3.9.40) and the natural boundary condition (3.9.22). If we set $t = T$ in (3.9.40) and use (3.9.22) along with (3.9.41), we find for $\mu = \mu_n$ the value (the reader should carry through this calculation)

$$\mu_n = \frac{(1 - e^2)^{3/2}T}{2\pi n(1 + e \cos \alpha_0)} \tag{3.9.42}$$

for any nonzero integer n as appears in the natural boundary condition (3.9.22). For notational convenience we shall always take n to be a *positive* integer

$$n = 1, 2, 3, \ldots,$$

and then we shall write

$$\mu = \mu_{-n} = -\mu_n$$

whenever we wish to consider negative integers in (3.9.42). We let $\alpha_n^* = \alpha_n^*(t)$ and $\alpha_{-n}^* = \alpha_{-n}^*(t)$ be the steering controls obtained, respectively, using $\mu = \mu_n$ and $\mu = \mu_{-n}$ in (3.9.40). [These functions α_n^* and α_{-n}^* are uniquely determined for each positive integer n since the left-hand side of (3.9.40) is an increasing function of α^*.]

Hence we have for each positive integer n *two* possible values for the parameter μ, namely $\mu = \mu_n$ and $\mu = \mu_{-n}$, with corresponding optimum steering controls α_n^* and α_{-n}^*. From (3.9.42) we find that

$$\mu_n^2 = \mu_{-n}^2,$$

and since the area of the ellipse (3.9.31) as given by (3.9.34) depends only on μ^2, it follows that *the steering controls α_n^* and α_{-n}^* give two distinct ellipses having equal areas.* These two ellipses have the same shape but different centers, given, respectively, by using $\mu = \mu_n$ and $\mu = \mu_{-n}$ in (3.9.32). Moreover, the resulting ellipses are traversed in different directions since α_n^* increases with increasing time, while α_{-n}^* decreases [see (3.9.36), (3.9.39), and (3.9.42)]. The situation is illustrated in Figure 11.

The choice $n = 1$ gives the two optimum steering controls α_1^* and α_{-1}^* leading to corresponding extremum ellipses with *maximum area* [see (3.9.34) and (3.9.42)], while any other choice of $n(n = 2, 3, \ldots)$ leads to *smaller ellipses*, which, according to (3.9.22), must be traversed n times. For example, the steering control α_{-2}^* requires the airplane to travel *twice* in a clockwise direction around the smaller ellipse obtained by using

$$\mu = \mu_{-2} = -\frac{(1 - e^2)^{3/2}T}{4\pi(1 + e \cos \alpha_0)}$$

in (3.9.31), (3.9.32), and (3.9.33). The *total area* encircled by the airplane in

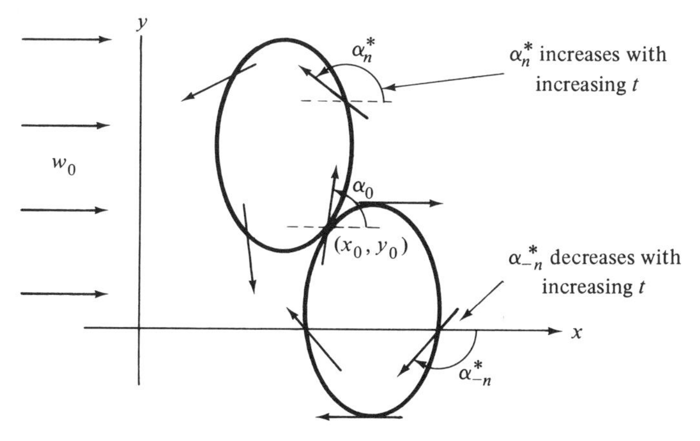

Figure 11

time T during n traversals of the ellipse corresponding to either of the steering controls α_n^* or α_{-n}^* is found by taking $\mu^2 = \mu_n^2 = \mu_{-n}^2$ in (3.9.34) and multiplying the result by n to account for the n traversals. We find in this way the value

$$A(\alpha_n^*) = A(\alpha_{-n}^*) = \frac{(1 - e^2)^{3/2}(v_0 T)^2}{4\pi n} \tag{3.9.43}$$

for the area functional A of (1.3.11) evaluated at either of the local extremum vectors $\alpha = \alpha_n^*$ or $\alpha = \alpha_{-n}^*$. The choice $n = 1$ leads to extremum vectors α_1^* and α_{-1}^*, giving *maximum area* among all admissible functions $\alpha = \alpha(t)$ in the vector space $\mathcal{C}^0[0, T]$ subject to the constraints (3.9.4), (3.9.5), and (3.9.6). In fact, α_1^* is the unique maximum vector among all such admissible functions α for which $\dot{\alpha} > 0$, while α_{-1}^* is the maximum vector among all admissible α for which $\dot{\alpha} < 0$. All other choices of $n (n = 2, 3, \ldots)$ lead to *local extremum vectors* α_n^* and α_{-n}^*, which give maximum area among all *nearby* admissible functions α for which, respectively, either $\|\alpha - \alpha_n^*\|$ or $\|\alpha - \alpha_{-n}^*\|$ is sufficiently small. These assertions can be proved by direct (but lengthy) calculations starting with (1.3.11), similar to the analogous calculations given in Sections 2.5 and 3.7 [see (2.5.14) and (3.7.28)].

We note finally that the previously anticipated result (3.9.20) is realized by (3.9.43) since the latter equation indeed shows that the initial steering angle α_0 does not affect the maximum area encircled. In fact, α_0 in no way affects the lengths of the major and minor axes of the extremum ellipses, as is shown by (3.9.33) and (3.9.42), which give for these lengths

$$a_n = \frac{(1 - e^2)v_0 T}{2\pi n}$$

$$b_n = \frac{\sqrt{1 - e^2}\, v_0 T}{2\pi n} = \frac{a_n}{\sqrt{1 - e^2}}.$$

The initial steering angle α_0 does, however, affect the location of the centers of these extremum ellipses as given by (3.9.32) and (3.9.42).

Exercises

1. Show that the variation of K_1 given by (3.9.7) is weakly continuous everywhere on $\mathcal{C}^0[0, T]$.

2. Show that the determinant given by (3.9.10) does *not* vanish identically for all $\Delta\alpha_1, \Delta\alpha_2, \Delta\alpha_3$ if the function α satisfies the conditions $d\alpha(t)/dt \neq 0$ (for all $0 \leq t \leq T$) and $\alpha(T) = \alpha(0) + 2\pi n$ (for some nonzero integer n). *Hints:* Take $\Delta\alpha_1(t) = [\sin \alpha(t)]\, d\alpha(t)/dt$ and $\Delta\alpha_2(t) = [\cos \alpha(t)]\, d\alpha(t)/dt$ and show in this case that the value of the given determinant becomes *determinant* $= -\pi^2 n^2 \Delta\alpha_3(0) + \pi n \dot{\alpha}(0) \int_0^T \cos[\alpha(t) - \alpha(0)]\, \Delta\alpha_3(t)\, dt$, where $\Delta\alpha_3$ may be any continuous function on $[0, T]$. You should be able to choose a particular function $\Delta\alpha_3$ for which this determinant will be nonzero. For example, take $\Delta\alpha_3(t) = 1 - kt$ for $0 \leq t \leq k^{-1}$, and $\Delta\alpha_3(t) = 0$ for $k^{-1} \leq t \leq T$, for some suitably large positive number k.

3. Describe the optimum paths in the (x, y)-plane in the case of no wind, $w_0 = 0$.

4. Find the optimum steering angles $\alpha = \alpha_1^*(t)$ and $\alpha = \alpha_{-1}^*(t)$ in the case of no wind ($w_0 = 0$), and calculate the resulting positions of the airplane for $0 \leq t \leq T$ using (1.3.7).

5. A person starts at the point $(x_0, y_0) = (0, 0)$ and walks with *fixed speed* v_0 along a path γ given parametrically in terms of *time* t as

$$\gamma : \begin{cases} x = X(t) \\ y = Y(t) \qquad \text{for} \quad 0 \leq t \leq T. \end{cases}$$

Let $\alpha = \alpha(t)$ be the person's *steering angle*, with [see (1.3.6)] $X'(t) = v_0 \cos \alpha(t)$, $Y'(t) = v_0 \sin \alpha(t)$, and $X(t) = v_0 \int_0^t \cos \alpha(s)\, ds$, $Y(t) = v_0 \int_0^t \sin \alpha(s)\, ds$. If γ terminates somewhere on the x-axis at time $t = T$, then the *area* enclosed between γ and the x-axis is given as [see (1.3.5)] *area* $= \frac{1}{2} \int_0^T [X(t)Y'(t) - Y(t)X'(t)]\, dt = (v_0^2/2) \int_0^T \left[\sin \alpha(t) \int_0^t \cos \alpha(s) ds - \cos \alpha(t) \int_0^t \sin \alpha(s) ds\right] dt$, while the *total length* traversed in time T at the constant speed v_0 is given as $v_0 T =$ length of γ. Show that *a suitable circular arc maximizes the area enclosed with the x-axis* among all such paths γ which terminate on the positive x-axis and which have given fixed total length l (i.e., $v_0 T = l$).

3.10. *The John Multiplier Theorem*

We have seen in previous sections how the Euler-Lagrange multiplier theorem of Section 3.6 can be used to find extremum vectors for a functional

J on a constraint set of the type $D[K_i = k_i \text{ for } i = 1, 2, \ldots, m]$, where D is a given open set in a normed vector space $\mathfrak{X}$ and where all the functionals involved are assumed to have variations which are weakly continuous on D. In certain situations, however, one may wish to find an extremum vector for a functional J in a constraint set of the type $D[K_i \leq k_i \text{ for } i = 1, 2, \ldots, m]$, where the latter set consists of all vectors x in D which satisfy the following *inequality* constraints:

$$K_i(x) \leq k_i \qquad \text{for all } i = 1, 2, \ldots, m.$$

For example, in the simplest case $m = 1$ we may wish to minimize J among all vectors x which satisfy the single inequality constraint

$$K_1(x) \leq k_1,$$

where k_1 is a fixed specified constant.

Suppose in this case that the vector x^* is a local minimum vector in $D[K_1 \leq k_1]$ for J, i.e., $K_1(x^*) \leq k_1$, and there is some ball $B_\rho(x^*)$ in $\mathfrak{X}$ centered at x^* such that $J(x^*) \leq J(x)$ for all vectors x which simultaneously are in $B_\rho(x^*)$ and satisfy $K_1(x) \leq k_1$. We consider in this case the following two possibilities:

$$K_1(x^*) = k_1 \tag{3.10.1}$$

and

$$K_1(x^*) < k_1. \tag{3.10.2}$$

If (3.10.1) holds for such a local minimum vector x^* in $D[K_1 \leq k_1]$ for J, it is then clear that x^* is also a local minimum vector in $D[K_1 = k_1]$ for J (why?), and we can then appeal to the Euler-Lagrange multiplier theorem to find that [see (3.3.3) and (3.6.5)]

$$\mu_0\, \delta J(x^*; \Delta x) + \mu_1\, \delta K_1(x^*; \Delta x) = 0 \qquad \text{for all vectors } \Delta x \text{ in } \mathfrak{X} \tag{3.10.3}$$

for suitable multipliers μ_0 and μ_1 *which do not both vanish*. In fact, we need only consider the two possibilities [see the discussion which follows equation (3.3.3)]

$$\mu_0 = 0, \qquad \mu_1 = 1$$

and

$$\mu_0 = 1, \qquad \mu_1 = -\lambda,$$

where λ is the Euler-Lagrange multiplier which appears in equation (3.3.2). Hence we are guaranteed that both of the multipliers μ_0 and μ_1 which appear in (3.10.3) may be taken to be *nonnegative* provided that $\lambda \leq 0$. If we assume that the functionals J and K_1 are differentiable, then we can use the methods of Section 3.8 to find [see (3.8.9) and (3.8.10)]

$$\lambda = \frac{\partial}{\partial k_1} J(x^*(k_1)),$$

so that λ will satisfy $\lambda \leq 0$ provided that

$$\frac{\partial}{\partial k_1} J(x^*(k_1)) \leq 0. \tag{3.10.4}$$

But, clearly, $J(x^*(k_1))$ will *increase* (or in any case it will not decrease) if k_1 decreases since x^* is a local *minimum* vector in $D[K_1 \leq k_1]$ for J (explain). This shows that (3.10.4) holds, and we can therefore conclude that both of the multipliers μ_0 and μ_1 which appear in (3.10.3) may be taken to be *non-negative* if x^* is a local minimum vector in $D[K_1 \leq k_1]$ for J and if (3.10.1) holds.

Suppose now that the other case (3.10.2) holds. Then by continuity† $K_1(x) < k_1$ will hold for all vectors x in some ball $B_\rho(x^*)$ centered at x^*, and it follows that x^* is a local minimum vector in D for J (why?). Hence in this case the inequality constraint is *inactive*, and we can appeal to the theorem of Section 2.2 to find that [see equation (2.2.4)]

$$\delta J(x^*; \Delta x) = 0 \qquad \text{for all vectors } \Delta x \text{ in } \mathfrak{X}.$$

Hence this case can be included also in (3.10.3) if we take $\mu_0 = 1$ and $\mu_1 = 0$ in (3.10.3). Note also in this case that we may take both μ_0 and μ_1 to be nonnegative.

We can summarize our results as follows. If J and K_1 are differentiable functionals on an open subset D of a normed vector space $\mathfrak{X}$ and if x^* is a local *minimum* vector in $D[K_1 \leq k_1]$ for J, then *there are nonnegative constants μ_0 and μ_1, which do not both vanish, such that* (3.10.3) *holds*. Moreover, our discussion above shows that

$$[K_1(x^*) - k_1]\mu_1 = 0 \tag{3.10.5}$$

also holds, which simply guarantees that *at most one* of the two inequalities $K_1(x^*) \leq k_1$ and $\mu_1 \geq 0$ may reduce to *strict* inequality (i.e., at most one of these two inequalities may show *slack*). Condition (3.10.5) is sometimes called the *complementary slackness* condition.

The results just described for the case of a single inequality constraint constitute a partial result of the *John multiplier theorem*, which was proved by Fritz John in 1948.‡ In the case of several inequality constraints the results are as follows.

†We assume that the functional K_1 is continuous at x^*; this will automatically be the case if K_1 is differentiable at x^*. (See, however, Exercise 3 of Section 3.2.)

‡Fritz John, "Extremum Problems with Inequalities as Subsidiary Conditions," in *Studies and Essays, Courant Anniversary Volume* (New York: John Wiley & Sons, Inc. (Interscience Division), 1948), pp. 187–204. John considered the case in which $\mathfrak{X}$ is a finite dimensional vector space.

John Multiplier Theorem. *Let $J, K_1, K_2, \ldots, K_m$ be differentiable functionals on an open subset D of a normed vector space $\mathfrak{X}$, and let x^* be a local minimum vector in $D[K_i \leq k_i$ for $i = 1, 2, \ldots, m]$ for J. Assume also that the differential of J and the differential of each K_i (for $i = 1, 2, \ldots, m$) are weakly continuous*† *near x^*. Then there are nonnegative constants $\mu_0, \mu_1, \mu_2, \ldots, \mu_m$, which do not all vanish, such that*

$$\mu_0 \, dJ(x^*; \Delta x) + \sum_{i=1}^{m} \mu_i \, dK_i(x^*; \Delta x) = 0 \qquad \text{for all vectors } \Delta x \text{ in } \mathfrak{X} \tag{3.10.6}$$

and such that

$$[K_i(x^*) - k_i]\mu_i = 0 \qquad \text{for each } i = 1, 2, \ldots, m. \tag{3.10.7}$$

Condition (3.10.6) is completely analogous to equation (3.6.5) of the Euler-Lagrange multiplier theorem, whereas the nonnegativeness of the multipliers μ_i (for $i = 0, 1, 2, \ldots, m$) and the associated complementary slackness conditions (3.10.7) have no counterparts in the Euler-Lagrange multiplier theorem. These additional conditions guarantee in the present case that the ith inequality constraint will be *inactive* for the minimum vector x^* if $K_i(x^*) < k_i$; in this case $\mu_i = 0$ will hold in (3.10.6), and the ith inequality constraint will play no active role in the determination of x^*.

There is, of course, a similar result for a local *maximum* vector for J, which can be obtained from the above result by replacing J with $-J$. Also, if any of the inequality constraints are given initially in the form $K_i(x) \geq k_i$, we can replace K_i and k_i, respectively, with $-K_i$ and $-k_i$ so as to reduce the problem to that already considered above. We shall make no use of the John multiplier theorem in this book,‡ and we therefore omit its proof except for the discussion already given earlier for the case $m = 1$.

A theorem which is closely related to the John multiplier theorem is the Kuhn-Tucker theorem§ of nonlinear programming. Hadley¶ discusses this result and related topics in the area of *mathematical programming*. We mention finally that in 1967 Mangasarian and Fromovitz‖ integrated the results of the Euler-Lagrange multiplier theorem and the John multiplier theorem

†Weak continuity for differentials is defined exactly as for variations by replacing δJ with dJ in equation (3.2.1).

‡See, however, Chapter 5, where we consider certain extremum problems which involve pointwise global inequality constraints.

§H. W. Kuhn and A. W. Tucker, "Nonlinear Programming," in *Proceedings of the Second Berkeley Symposium on Mathematical Statistics and Probability* (J. Neyman, ed., Berkeley: University of California Press, 1951), pp. 481–492.

¶G. Hadley, *Nonlinear and Dynamic Programming* (Reading, Mass.: Addison-Wesley Publishing Company, Inc., 1964).

‖O. L. Mangasarian and S. Fromovitz, "The Fritz John Necessary Optimality Conditions in the Presence of Equality and Inequality Constraints," *Journal of Mathematical Analysis and Applications*, **17** (1967), 37–47.

and showed how to handle extremum problems which involve simultaneously both inequality and equality constraints in the case in which $\mathfrak{X}$ is a finite dimensional vector space such as $\mathcal{R}_n$.

Exercises

1. Minimize the function $x_1^2 + x_2^2 + 2x_3^2 - x_1 - x_2x_3$ on $\mathcal{R}_3$ subject to the inequality constraint $x_1 + x_2 + x_3 \geq 35$. (Compare with Exercise 6 in Section 3.4.)
2. Let D be the open subset of $\mathcal{R}_3$ consisting of all vectors $x = (x_1, x_2, x_3)$ for which $x_1 > 0$, $x_2 > 0$, and $x_3 > 0$. Minimize the function $x_1^2 + x_2^2 + x_3^2 + x_1x_2x_3$ on D subject to the inequality constraint $x_1x_2x_3 \geq 8$. (Compare with Exercise 7 in Section 3.4.)

4. Applications of the Euler-Lagrange Multiplier Theorem in the Calculus of Variations

In this chapter we shall show how to use the Euler-Lagrange multiplier theorem to solve problems involving fixed and variable end points in the calculus of variations. Such problems arise in physics, engineering, and geometry. We shall consider several examples in detail, including several problems on the minimum transit time of a boat, several brachistochrone problems, and some problems on geodesic curves. We shall also consider some of the variational principles of physics, notably the principle of least action and Fermat's principle of least time.

4.1. Problems with Fixed End Points

We consider the problem of maximizing or minimizing the value of a functional $J = J(Y)$ defined by

$$J(Y) = \int_{x_0}^{x_1} F(x, Y(x), Y'(x))\,dx \tag{4.1.1}$$

in terms of a given known function F as in (2.4.9). Here $Y = Y(x)$ is required to be a function of class $\mathcal{C}^1$ on the fixed interval $[x_0, x_1]$, and we shall suppose that the values of Y are specified at the end points as

$$Y(x_0) = y_0, \qquad Y(x_1) = y_1 \tag{4.1.2}$$

for given constants y_0 and y_1.

The brachistochrone functional $T = T(Y)$ of (1.3.3) which gives the time of descent of a bead sliding down a wire joining two points is of the form (4.1.1), with the function F given by (2.4.10). In this case the constants y_0 and y_1 are the y-coordinates of the two specified end points of the wire.

If we define functionals K_0 and K_1 by

$$K_0(Y) = Y(x_0), \qquad K_1(Y) = Y(x_1) \tag{4.1.3}$$

for any function $Y = Y(x)$ in the vector space $\mathcal{C}^1[x_0, x_1]$, then the problem considered is to find extremum vectors in $D[K_0 = y_0, K_1 = y_1]$ for the functional J of (4.1.1). Here we take the open set D to be the entire normed vector space $\mathcal{C}^1[x_0, x_1]$ with any suitable norm as, for example, the norm defined by (2.4.12).

We easily find from (4.1.3) and the definition of the variation [see (2.3.4)] that

$$\begin{aligned} \delta K_0(Y; \Delta Y) &= \Delta Y(x_0) \\ \delta K_1(Y; \Delta Y) &= \Delta Y(x_1) \end{aligned} \tag{4.1.4}$$

for any function ΔY in $\mathcal{C}^1[x_0, x_1]$, while, according to (2.4.17), the variation of J is given by

$$\begin{aligned} \delta J(Y; \Delta Y) = \int_{x_0}^{x_1} [&F_Y(x, Y(x), Y'(x))\,\Delta Y(x) \\ &+ F_{Y'}(x, Y(x), Y'(x))\,\Delta Y'(x)]\,dx \end{aligned} \tag{4.1.5}$$

for any continuously differentiable function $\Delta Y = \Delta Y(x)$ on $[x_0, x_1]$. Here $Y = Y(x)$ need *not* satisfy the constraints (4.1.2), but rather $Y(x)$ may be any suitable function in $\mathcal{C}^1[x_0, x_1]$. The expressions $F_Y(x, Y(x), Y'(x))$ and $F_{Y'}(x, Y(X), Y'(x))$ are defined by (2.4.16).

One easily checks that all the hypotheses of the Euler-Lagrange multiplier theorem of Section 3.6 are satisfied, at least provided that the given function F is "nice." Hence *at least one of the two possibilities concluded in that theorem must hold* for any local extremum vector Y in $D[K_0 = y_0, K_1 = y_1]$. We can eliminate the first possibility by observing with (4.1.4) that the determinant [see (3.6.2)]

$$\det\begin{vmatrix} \delta K_0(Y;\Delta Y_0) & \delta K_0(Y;\Delta Y_1) \\ \delta K_1(Y;\Delta Y_0) & \delta K_1(Y;\Delta Y_1) \end{vmatrix} = \Delta Y_0(x_0)\Delta Y_1(x_1) - \Delta Y_0(x_1)\Delta Y_1(x_0)$$

does *not* vanish identically for all functions ΔY_0 and ΔY_1 in $\mathfrak{C}^1[x_0, x_1]$. In fact, if we take

$$\Delta Y_0(x) = 1 \qquad \text{for all } x$$

and

$$\Delta Y_1(x) = \frac{x - x_0}{x_1 - x_0} \qquad \text{for any } x,$$

then the corresponding value of the determinant is seen to be 1. This eliminates the first possibility, and hence *the second possibility of the multiplier theorem must obtain*, so that if $Y = Y(x)$ is a local extremum vector in $D[K_0 = y_0, K_1 = y_1]$ for J, there will be constants λ_0 and λ_1 such that [see (3.6.3)]

$$\delta J(Y;\Delta Y) = \lambda_0\, \delta K_0(Y;\Delta Y) + \lambda_1\, \delta K_1(Y;\Delta Y)$$

holds for *all* vectors ΔY in $\mathfrak{C}^1[x_0, x_1]$. If we use (4.1.4) and (4.1.5), we may rewrite this condition as

$$\int_{x_0}^{x_1} [F_Y(x, Y(x), Y'(x))\,\Delta Y(x) + F_{Y'}(x, Y(x), Y'(x))\,\Delta Y'(x)]\,dx$$
$$= \lambda_0\,\Delta Y(x_0) + \lambda_1\,\Delta Y(x_1), \tag{4.1.6}$$

which must necessarily hold for all continuously differentiable functions $\Delta Y = \Delta Y(x)$.

As usual we wish somehow to eliminate from consideration the arbitrary function ΔY which appears in (4.1.6) so as to obtain a simpler equation which will involve only the extremum vector and which may be solved to give $Y = Y(x)$. However, equation (4.1.6) differs in one respect from the analogous equations (3.7.19) and (3.9.15) which occurred in the optimization problems of Sections 3.7 and 3.9. Indeed, equation (4.1.6) involves not only the arbitrary function $\Delta Y = \Delta Y(x)$ but *also its derivative* $\Delta Y'(x) = (d/dx)\,\Delta Y(x)$. If we can somehow first eliminate the derivative $\Delta Y'$ from (4.1.6), it may then be possible to eliminate ΔY altogether by using the same approach as used in Section 3.9 based on the lemma of Section A4 of the Appendix.

In fact, Joseph Lagrange showed in 1755 (at the age of 19) how to eliminate the derivative $\Delta Y'$ from (4.1.6) *provided that the function of x given by the expression*

$$F_{Y'}(x,\ Y(x),\ Y'(x)) \tag{4.1.7}$$

is itself continuously differentiable with respect to x. Indeed, in this case the product rule of differentiation gives

$$\begin{aligned}\frac{d}{dx}[F_{Y'}(x, Y(x), Y'(x))\,\Delta Y(x)] &= F_{Y'}(x, Y(x), Y'(x))\,\Delta Y'(x)\\ &\quad + \left[\frac{d}{dx}F_{Y'}(x, Y(x), Y'(x))\right]\Delta Y(x),\end{aligned}$$

and this equation can be integrated to give

$$\begin{aligned}\int_{x_0}^{x_1}\frac{d}{dx}[F_{Y'}(x, Y(x), Y'(x))\,\Delta Y(x)]\,dx &= \int_{x_0}^{x_1}F_{Y'}(x, Y(x), Y'(x))\,\Delta Y'(x)\,dx\\ &\quad + \int_{x_0}^{x_1}\left[\frac{d}{dx}F_{Y'}(x, Y(x), Y'(x))\right]\Delta Y(x)\,dx.\end{aligned}$$

But the fundamental theorem of calculus implies that

$$\begin{aligned}&\int_{x_0}^{x_1}\frac{d}{dx}[F_{Y'}(x, Y(x), Y'(x))\,\Delta Y(x)]\,dx\\ &\qquad = F_{Y'}(x_1, Y(x_1), Y'(x_1))\Delta Y(x_1) - F_{Y'}(x_0, Y(x_0), Y'(x_0))\,\Delta Y(x_0),\end{aligned}$$

so that we find that (explain)

$$\begin{aligned}\int_{x_0}^{x_1}F_{Y'}(x, Y(x), Y'(x))\,\Delta Y'(x)\,dx &= F_{Y'}(x_1, Y(x_1), Y'(x_1))\,\Delta Y(x_1)\\ - F_{Y'}(x_0, Y(x_0), Y'(x_0))\,\Delta Y(x_0) &- \int_{x_0}^{x_1}\left[\frac{d}{dx}F_{Y'}(x, Y(x), Y'(x))\right]\Delta Y(x)\,dx,\end{aligned} \tag{4.1.8}$$

which is, in fact, just a special case of the general formula for integration by parts. We now use this result to eliminate the term involving $\Delta Y'$ in (4.1.6) so as to find

$$\begin{aligned}&\int_{x_0}^{x_1}\left[F_Y(x, Y(x), Y'(x)) - \frac{d}{dx}F_{Y'}(x, Y(x), Y'(x))\right]\Delta Y(x)\,dx\\ &\quad = [\lambda_0 + F_{Y'}(x_0, y_0, Y'(x_0))]\,\Delta Y(x_0) + [\lambda_1 - F_{Y'}(x_1, y_1, Y'(x_1))]\,\Delta Y(x_1),\end{aligned} \tag{4.1.9}$$

which must then hold for all vectors ΔY in the vector space $\mathcal{C}^1[x_0, x_1]$. We have used in (4.1.9) the constraints (4.1.2), which are satisfied by any extremum vector Y in $D[K_0 = y_0, K_1 = y_1]$.

Since equation (4.1.9) involves only ΔY and not $\Delta Y'$, it can be handled similarly to equation (3.9.15) in Chaplygin's problem. In fact, if we first consider *only* those functions ΔY which *vanish at the end points* $x = x_0$ and $x = x_1$, we find from (4.1.9) the condition [compare with (3.9.19)]

$$\int_{x_0}^{x_1} \left[F_Y(x, Y(x), Y'(x)) - \frac{d}{dx} F_{Y'}(x, Y(x), Y'(x)) \right] \Delta Y(x)\, dx = 0, \tag{4.1.10}$$

which must hold for all functions ΔY *of class* $\mathfrak{C}^1$ *on* $[x_0, x_1]$ *which satisfy the additional requirements* $\Delta Y(x_0) = 0$ *and* $\Delta Y(x_1) = 0$. Since the function

$$F_Y(x, Y(x), Y'(x)) - \frac{d}{dx} F_{Y'}(x, Y(x), Y'(x))$$

in square brackets in (4.1.10) will itself be a continuous function of x [provided that the function of (4.1.7) is continuously differentiable], it now follows from (4.1.10) and the lemma of Section A4 of the Appendix (with $n = 1$) that *the extremum function* $Y(x)$ *must satisfy* (explain)

$$F_Y(x, Y(x), Y'(x)) - \frac{d}{dx} F_{Y'}(x, Y(x), Y'(x)) = 0 \tag{4.1.11}$$

for all x *in the interval* $[x_0, x_1]$. This is the desired simplified equation obtained from eliminating the arbitrary function ΔY and its derivative $\Delta Y'$ from (4.1.6).

The Swiss mathematician Leonhard Euler† had obtained equation (4.1.11) by 1741, using an ingenious but awkward procedure in which he approximated the integral in (4.1.1) by a sum and the extremum function $Y(x)$ by its ordinates at certain points and then varied these ordinates one at a time. Lagrange studied Euler's results while in his teens, and he wrote a letter to Euler in 1755 in which he obtained the same equation (4.1.11) by a method closely related to that used here. Euler recognized the generality and simplicity of the method of Lagrange, and he encouraged Lagrange in his work. Lagrange was elected into the Berlin Academy of Sciences the following year as a foreign member under the sponsorship of Euler. Euler later introduced

†Euler, who studied under John Bernoulli, is both the most important mathematician and the most important theoretical physicist of the eighteenth century. He made fundamental contributions in differential calculus, integral calculus, number theory, algebra, geometry, analysis, differential equations, finite differences, the calculus of variations, the lunar theory, hydrodynamics, elasticity, and analytical mechanics. He also integrated, unified, systematized, and extended large areas which previously contained only scattered and vague partial results. Euler's work is noted for its clarity and order.

the name *calculus of variations* to describe Lagrange's method. We call equation (4.1.11) the **Euler-Lagrange equation** after these two men.†

In 1879 Paul Du Bois-Reymond, a professor at the University of Tübingen in Germany, pointed out that Lagrange's derivation of (4.1.11) was in a sense incomplete since Lagrange gave no justification for the added assumption that the function given by (4.1.7) must be continuously differentiable. Indeed, if $Y(x)$ is any local extremum function in $D[K_0 = y_0, K_1 = y_1]$ for J with $D = \mathcal{C}^1[x_0, x_1]$, it is then conceivable that the derivative $Y'(x)$ may itself *not* be differentiable, and if it is not, then the function given by (4.1.7) will in general not be differentiable. [Consider, for example, the situation with $F(x, y, z) = z^2$, giving with (2.4.16) the result $F_{Y'}(x, Y(x), Y'(x)) = 2Y'(x)$.] But if the function of (4.1.7) is not differentiable, then Lagrange's derivation breaks down. Du Bois-Reymond showed, in fact, that *the function given by* (4.1.7) *will automatically be continuously differentiable for the extremum function* $Y(x)$ *as a consequence of the necessary condition* (4.1.6). Hence Du Bois-Reymond made Lagrange's derivation complete. We give Du Bois-Reymond's proof of this result in Section A5 of the Appendix.

In practice the Euler-Lagrange equation (4.1.11) can often be solved for the extremum function $Y(x)$ subject to the constraints of (4.1.2). It happens that the Euler-Lagrange multipliers λ_0 and λ_1 do not appear in (4.1.11) and (4.1.2), and they need not be considered further. It may be of interest, however, to note that the values of the multipliers in this case are given as

$$\begin{aligned} \lambda_0 &= -F_{Y'}(x_0, y_0, Y'(x_0)) \\ \lambda_1 &= F_{Y'}(x_1, y_1, Y'(x_1)), \end{aligned} \tag{4.1.12}$$

as follows by using (4.1.11) in (4.1.9) and then taking first

$$\Delta Y(x) = \frac{x_1 - x}{x_1 - x_0}$$

in (4.1.9) and next

$$\Delta Y(x) = \frac{x - x_0}{x_1 - x_0}.$$

†Lagrange was first drawn to mathematics at the age of 17 when he happened upon an essay by Edmund Halley (Newton's friend) that excited his interest. A year later he was an accomplished mathematician and became professor of mathematics at the artillery school in Turin, Italy. It is interesting to note that Lagrange's father had earlier obtained considerable money and social position through inheritance and marriage, only to lose the money in unsuccessful speculation, so that there was none left to pass on to Lagrange. Lagrange looked back later and judged himself fortunate on this account. "If I had inherited a fortune in money I should probably not have cast my lot with mathematics." See E. T. Bell, *Men of Mathematics* (New York: Simon and Schuster, 1937, 1965), p. 153, and Rouse Ball, *A Short Account of the History of Mathematics* (New York: Dover Publications, Inc., 1960), p. 402.

We may use (4.1.12) and (3.8.10) to find the rates of change of the extreme value $J(Y)$ with respect to the parameters y_0 and y_1; we find in this way that (explain)

$$\frac{\partial J(Y)}{\partial y_0} = -F_{Y'}(x_0, y_0, Y'(x_0))$$
$$\frac{\partial J(Y)}{\partial y_1} = F_{Y'}(x_1, y_1, Y'(x_1)),$$

where $Y = Y(x)$ is any fixed local extremum vector in $D[K_0 = y_0, K_1 = y_1]$ for J.

Shortest Distance Between Two Points. As an illustration of the above procedure we shall consider briefly the simple problem of finding a curve that minimizes the distance between two given points $P_0 = (x_0, y_0)$ and $P_1 = (x_1, y_1)$ in the plane. We consider curves which can be represented parametrically in terms of x as $y = Y(x)$ for $x_0 \leq x \leq x_1$ and for some suitable function $Y(x)$ which will depend on the particular curve considered. We let J denote the length of any such curve, given as

$$J(Y) = \int_{x_0}^{x_1} \sqrt{1 + Y'(x)^2}\, dx.$$

We have written $J(Y)$ here to indicate that the length of the curve is a functional which depends on the particular function $Y = Y(x)$ which appears here. This length functional $J(Y)$ can be written in the form of (4.1.1) with $F(x, y, z) = \sqrt{1 + z^2}$. In this case $\partial F/\partial y = 0$ and $\partial F/\partial z = z/\sqrt{1 + z^2}$, and so the Euler-Lagrange equation (4.1.11) becomes [see (2.4.16)]

$$\frac{d}{dx}\left[\frac{Y'(x)}{\sqrt{1 + Y'(x)^2}}\right] = 0.$$

This equation can be integrated to give

$$\frac{Y'(x)}{\sqrt{1 + Y'(x)^2}} = \text{constant},$$

which in turn implies that $Y'(x) = A$ for some suitable constant A. Finally, then, we find for the extremum function the result $Y(x) = Ax + B$, where the two constants of integration A and B can be determined from the two conditions $Y(x_0) = y_0$ and $Y(x_1) = y_1$. As expected, we find that the shortest curve joining P_0 and P_1 is the straight-line segment connecting these two points.

The Euler-Lagrange equation (4.1.11) can be simplified in certain special cases. For example, suppose that the given function $F = F(x, y, z)$ does *not depend on the first argument* x, as happens, for example, in the case of the above length functional and also in the case of John Bernoulli's brachistochrone functional T of (1.3.3), where we have

$$F(y, z) = \sqrt{\frac{1 + z^2}{2g(y_0 - y)}}, \tag{4.1.13}$$

which is independent of x. In any such case with

$$F = F(y, z) \tag{4.1.14}$$

independent of x, we may differentiate the expression

$$F(Y(x), Y'(x)) - Y'(x)F_{Y'}(Y(x), Y'(x))$$

with respect to x using the chain rule of differentiation along with (2.4.16) and (4.1.14) to find for any smooth function $Y(x)$ that

$$\begin{aligned} \frac{d}{dx}&[F(Y(x), Y'(x)) - Y'(x)F_{Y'}(Y(x), Y'(x))] \\ &= F_Y(Y(x), Y'(x))\, Y'(x) + F_{Y'}(Y(x), Y'(x))\, Y''(x) \\ &- Y''(x)F_{Y'}(Y(x), Y'(x)) - Y'(x)\frac{d}{dx}F_{Y'}(Y(x), Y'(x)) \\ &= Y'(x)\left[F_Y(Y(x), Y'(x)) - \frac{d}{dx}F_{Y'}(Y(x), Y'(x))\right]. \end{aligned}$$

Hence if $Y(x)$ is any solution of the Euler-Lagrange equation (4.1.11), we find in this case that (explain)

$$\frac{d}{dx}[F(Y(x), Y'(x)) - Y'(x)F_{Y'}(Y(x), Y'(x))] = 0,$$

which may be integrated to give

$$F(Y(x), Y'(x)) - Y'(x)F_{Y'}(Y(x), Y'(x)) = C \tag{4.1.15}$$

for some constant of integration C. Hence in the special case (4.1.14) we may replace the Euler-Lagrange equation (4.1.11) with the simpler equation (4.1.15). The latter equation is a *first*-order differential equation for $Y(x)$, while (4.1.11) is in general a more difficult *second*-order differential equation for $Y(x)$.†

†In certain physical applications in which the independent variable x is *time* and the integrand function F is the *Lagrangian* function, as, for example, in the motion of a single particle in a conservative force field, it can be shown that the expression on the left-hand side of equation (4.1.15) equals the (negative of the) total energy of motion. In such cases (4.1.15) simply requires that the total energy remain constant during the motion. See Section 4.8, where we shall discuss this notion of conservation of energy for physical systems in motion further.

Another special case occurs whenever the given function $F = F(x, y, z)$ does *not depend on the second argument* y. This case occurs for the transit time functional $T = T(Y)$ of (1.3.17) since in that case we have

$$F(x, z) = \frac{\sqrt{1 - e(x)^2 + z^2} - e(x)z}{v_0[1 - e(x)^2]}, \tag{4.1.16}$$

which is independent of y. Here $e(x)$ is the known function of x given by (1.3.15) and v_0 is a constant. In any such case with

$$F = F(x, z) \tag{4.1.17}$$

independent of y, we have $\partial F/\partial y = 0$, and it then follows from (2.4.16) and (4.1.17) that the Euler-Lagrange equation (4.1.11) reduces in this case to

$$\frac{d}{dx} F_{Y'}(x, Y'(x)) = 0.$$

This equation can be integrated to give

$$F_{Y'}(x, Y'(x)) = C \tag{4.1.18}$$

for some constant of integration C. Therefore in the special case (4.1.17) we may replace the Euler-Lagrange equation (4.1.11) with the simpler equation (4.1.18).

Minimum Transit Time of a Boat. As an example of the latter situation, we consider the problem of finding the minimum transit time that can be achieved by varying the path of a boat crossing a river of width l from a given initial point P_0 to a given terminal point P_1 as described in Section 1.3. We assume (as in Section 1.3) that the boat travels with constant natural speed v_0 and that the river has no cross currents. We take $P_0 = (x_0, y_0) = (0, 0)$ and $P_1 = (x_1, y_1) = (l, y_1)$, as shown in Figure 1, and we then seek a curve γ connecting P_0 and P_1 given as

$$\gamma\colon\quad y = Y(x), \qquad 0 \leq x \leq l \tag{4.1.19}$$

along which the boat can travel from P_0 to P_1 in minimum time. The transit time is given by the functional T of (1.3.17), which is of the form (4.1.1), with F given by (4.1.16). Hence the minimum transit time is achieved along a curve γ which satisfies (4.1.18). In this case we find from (4.1.16) and (2.4.16) that (the reader should carry through the calculation)

$$F_{Y'}(x, Y'(x)) = \frac{Y'(x) - e(x)\sqrt{1 - e(x)^2 + Y'(x)^2}}{v_0[1 - e(x)^2]\sqrt{1 - e(x)^2 + Y'(x)^2}},$$

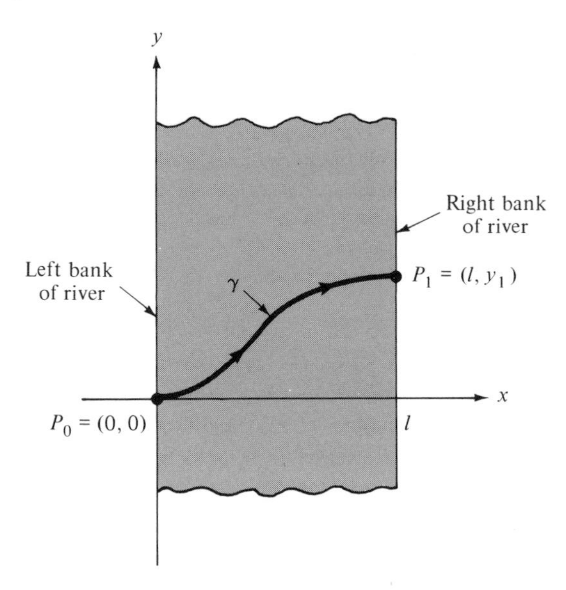

Figure 1

and then (4.1.18) implies that

$$\frac{Y'(x) - e(x)\sqrt{1 - e(x)^2 + Y'(x)^2}}{v_0[1 - e(x)^2]\sqrt{1 - e(x)^2 + Y'(x)^2}} = C,$$

which can be simplified after a short calculation to give

$$Y'(x)^2 = \frac{\{e(x) + A[1 - e(x)^2]\}^2}{1 - 2Ae(x) - A^2[1 - e(x)^2]} \tag{4.1.20}$$

for some constant $A = v_0 C$.

Hence in this case the Euler-Lagrange equation furnishes the differential equation (4.1.20), which we hope to solve for the extremum function $Y(x)$. If we take the square root of both sides of (4.1.20) and integrate the result, we find that

$$Y(x) = \int_0^x \frac{e(\xi) + A[1 - e(\xi)^2]}{\sqrt{1 - 2Ae(\xi) - A^2[1 - e(\xi)^2]}}\, d\xi \tag{4.1.21}$$

for $0 \leq x \leq l$, where we have imposed the constraint $Y(0) = 0$. [We have also assumed that $Y(x) \geq 0$ here in taking the *positive* square root; see, however, Exercise 2.] The remaining constraint $Y(l) = y_1$ can be used to determine the constant A in terms of the given data l, y_1, and $e(x)$ through the relation (explain)

$$y_1 = \int_0^l \frac{e(x) + A[1 - e(x)^2]}{\sqrt{1 - 2Ae(x) - A^2[1 - e(x)^2]}}\, dx. \tag{4.1.22}$$

Unfortunately equation (4.1.22) cannot in general be solved in closed form for A; rather this equation must in general be solved numerically for an approximate value of A. Finally, we can use (4.1.20), (1.3.16), and (1.3.17) to find the minimum transit time $T = T_{\text{minimum}}$ and the optimum steering control angle $\alpha = \alpha(x)$ that will guide the boat along the minimizing path γ; we find that

$$\cos\alpha(x) = \frac{\sqrt{1 - 2Ae(x) - A^2[1 - e(x)^2]}}{1 - Ae(x)} \tag{4.1.23}$$

for $0 \le x \le l$ and that

$$T_{\text{minimum}} = \frac{1}{v_0}\int_0^l \frac{1 - Ae(x)}{\sqrt{1 - 2Ae(x) - A^2[1 - e(x)^2]}}\,dx, \tag{4.1.24}$$

where the constant A is determined as before by (4.1.22). The reader may wish to check whether or not these results are in agreement with the "obvious solution" in the case of no river current, $e(x) = 0$. Finally, it is interesting to note that any increase in both the boat speed v_0 and the river current speed $w(x)$ which leaves *unchanged* the ratio $w(x)/v_0 = e(x)$ will *not* affect the extremum path γ and the optimum steering angle α given by (4.1.19), (4.1.21), (4.1.22), and (4.1.23), although it will give a shorter minimum transit time (4.1.24).

Exercises

1. A boat is to cross a uniformly flowing stream 300 feet wide from a given initial point to a terminal point located $300/\sqrt{3}$ feet downstream on the opposite bank, as shown in Figure 2. The boat will travel at the constant natural speed of 88 feet per minute (which is 1 mile per hour), and the stream current has a constant uniform speed of $88/\sqrt{3}$ feet per minute. Find the minimum transit time of the boat, the minimizing path, and the optimum steering angle α.

2. The boat of Exercise 1 is to recross the stream under the same conditions as before except this time the terminal point is located $300/\sqrt{3}$ feet *upstream* on the opposite bank, as shown in Figure 3. (Such a trip might correspond to a *return* trip.) Find the path of minimum transit time and the optimum steering angle, and compare the minimum transit time with that of the downstream trip of Exercise 1.

3. Let $P_0 = (x_0, y_0)$ and $P_1 = (x_1, y_1)$ be two given points in the plane with $x_0 < x_1$ and $y_0 > 0$, $y_1 > 0$, and let γ be any curve which connects P_0 and P_1 given as $\gamma\colon y = Y(x)$, $x_0 \le x \le x_1$. Such a curve γ is sought that will yield the *least surface area* for the resulting surface of revolution obtained by rotating γ about the x-axis as shown in Figure 4. The area A of such a surface of revolution can be calculated using the techniques of differential and integral calculus and is found to be $A(Y) = 2\pi \int_{x_0}^{x_1} Y(x)\sqrt{1 + Y'(x)^2}\,dx$. Show that if $Y(x)$ is a continuously

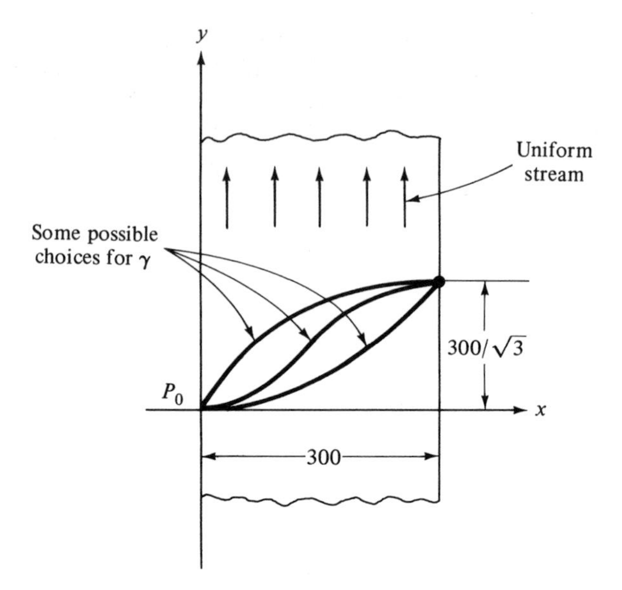

Figure 2

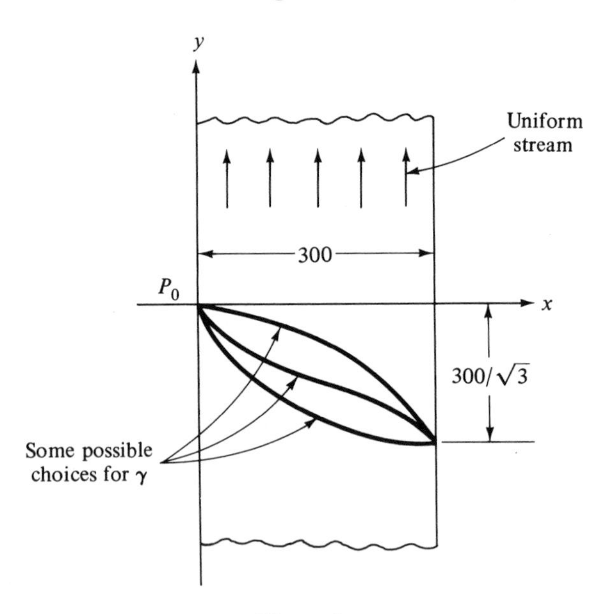

Figure 3

differentiable function which minimizes this area functional subject to the constraints $Y(x_0) = y_0$ and $Y(x_1) = y_1$, then $Y(x)$ must be a *catenary* given as $Y(x) = a \cosh [(x - b)/a]$ for suitable constants a and b. *Hint:* You may be able to use one of the simpler forms of the Euler-Lagrange equation since F has the special form $F(y, z) = 2\pi y\sqrt{1 + z^2}$. The resulting surface of revolution is called

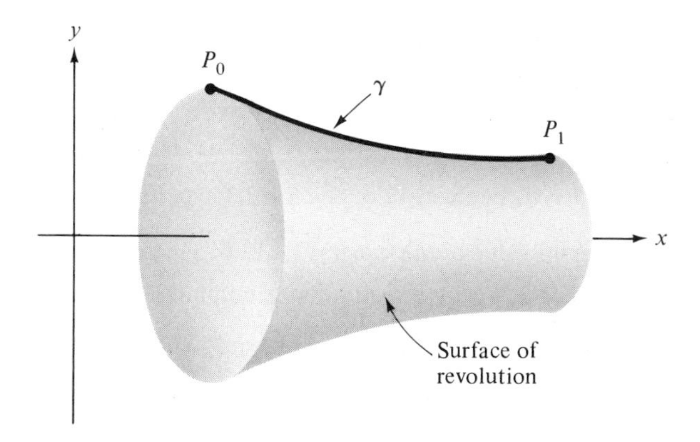

Figure 4

a *catenoid* and does indeed furnish the smallest possible area provided that the locations of the points P_0 and P_1 satisfy certain restrictions.†

4. (Queen Dido's problem) Let $P_0 = (x_0, 0)$ and $P_1 = (x_1, 0)$ be two fixed points on the x-axis with $x_0 < x_1$, and let l be any given fixed length satisfying $x_1 - x_0 < l < (\pi/2)(x_1 - x_0)$. Let γ be any curve of length l connecting P_0 and P_1 given as $\gamma: y = Y(x)$, $x_0 \le x \le x_1$, with $Y(x) \ge 0$, as shown in Figure 5. Show that a suitable *circular arc* encloses the *greatest area with the x-axis* among all such curves of length l. *Hints:* The length of any such curve is given by the integral $\int_{x_0}^{x_1} \sqrt{1 + Y'(x)^2}\, dx$, while the area enclosed with the x-axis is given as $\int_{x_0}^{x_1} Y(x)\, dx$. You should be able to use the Euler-Lagrange multiplier theorem with *three* constraints to show that the extremum curve γ must satisfy a differential equation of the form $Y'(x)^2 = (x - a)^2/[c^2 - (x - a)^2]$, which can be integrated to give $(x - a)^2 + (y - b)^2 = c^2$ for $y = Y(x)$ and for suitable constants a, b, and c. Explain how to determine the constants.

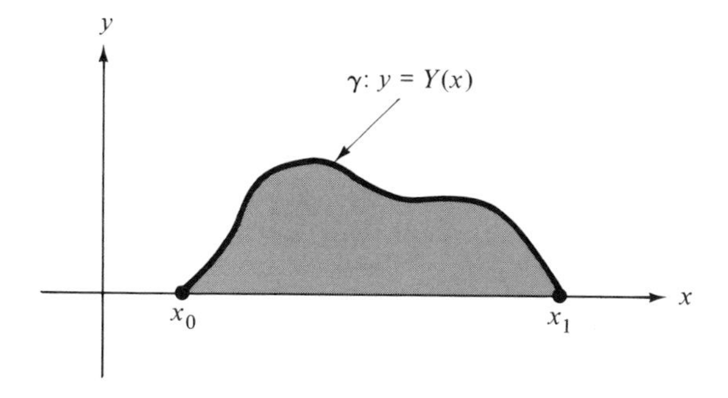

Figure 5

†Gilbert Ames Bliss, *Calculus of Variations* (Chicago: The Open Court Publishing Company, 1925), pp. 85–119. See also Hans Sagan, *Introduction to the Calculus of Variations* (New York: McGraw-Hill Book Company, 1969), pp. 62–65.

5. Prove the inequality $\int_0^1 Y(x)^2\,dx \leq (4/\pi^2)\int_0^1 Y'(x)^2\,dx$ for all continuously differentiable functions $Y(x)$ which satisfy the conditions $Y(0) = 0$ and $Y(1) = 1$. Find such a function for which the inequality reduces to *equality*. *Hint:* Minimize the functional $J = \left[\int_0^1 Y'(x)^2\,dx \Big/ \int_0^1 Y(x)^2\,dx\right]$ subject to the conditions $Y(0) = 0$, $Y(1) = 1$. Exercise 12 of Section 2.3 may be helpful.

6. Find the minimum value of the integral $\int_1^2 x^2 Y'(x)^2\,dx$ subject to the conditions $Y(1) = 1$, $Y(2) = 2$. Verify directly that a minimum is achieved.

7. Find the minimum value of the integral $\int_0^\pi [Y'(x)^2 + 2Y(x)\sin x]\,dx$ subject to the conditions $Y(0) = 0$, $Y(\pi) = 0$. Verify directly that a minimum is achieved.

8. Find the minimum value of the integral $\int_0^1 [\frac{1}{2}Y'(x)^2 + Y'(x) + Y(x)\,Y'(x) + Y(x)]\,dx$ among all functions Y of class $\mathcal{C}^1$ on $[0, 1]$ which satisfy the conditions $Y(0) = 0$, $Y(1) = 0$. Prove directly that the integral is actually minimized.

4.2. John Bernoulli's Brachistochrone Problem, and Brachistochrones Through the Earth

We first consider the problem of finding the least time of descent that can be achieved by varying the shape of a wire down which a small bead slides under gravity from one point to a nearby lower point, as illustrated in Figure 1 of Chapter 1. We have seen in Section 1.3 that if we neglect friction and assume that the earth's gravitational force is constant near the surface of the earth and if we represent the wire as a curve γ given as

$$\gamma:\quad y = Y(x), \qquad x_0 \leq x \leq x_1, \tag{4.2.1}$$

then the problem reduces to that of minimizing the time functional $T = T(Y)$ of (1.3.3) among all continuously differentiable functions $Y = Y(x)$ satisfying the constraints

$$Y(x_0) = y_0, \qquad Y(x_1) = y_1, \tag{4.2.2}$$

where the two end points of the wire are located at the specified points $P_0 = (x_0, y_0)$ and $P_1 = (x_1, y_1)$.

In the notation of Section 4.1 the problem is to find a minimum vector in $D[K_0 = y_0, K_1 = y_1]$ for the functional T of (1.3.3), where D is the entire vector space $\mathcal{C}^1[x_0, x_1]$ and where the functionals K_0 and K_1 are defined for any vector $Y = Y(x)$ by $K_0(Y) = Y(x_0)$ and $K_1(Y) = Y(x_1)$. Hence we need only solve the Euler-Lagrange equation (4.1.11) subject to the constraints (4.2.2), where the function F appearing in the Euler-Lagrange equation is given in this case as [see (4.1.13)] $F = \sqrt{(1 + z^2)/2g(y_0 - y)}$. Since $F = F(y, z)$ is independent of x, we may use the simpler equation (4.1.15) instead of

(4.1.11). In this case we find with (2.4.16) that

$$F_{Y'}(Y(x), Y'(x)) = \frac{Y'(x)}{\sqrt{2g[y_0 - Y(x)][1 + Y'(x)^2]}},$$

so that (4.1.15) implies for any extremum function $Y(x)$ that

$$\sqrt{\frac{1 + Y'(x)^2}{y_0 - Y(x)}} - \frac{Y'(x)^2}{\sqrt{[y_0 - Y(x)][1 + Y'(x)^2]}} = \sqrt{2g}C,$$

or, after some simplification, that

$$[y_0 - Y(x)][1 + Y'(x)^2] = A,$$

with $A^{-1} = 2gC^2$. If we solve this equation for $Y'(x)$, we find that

$$Y'(x) = -\sqrt{\frac{A - [y_0 - Y(x)]}{y_0 - Y(x)}}, \tag{4.2.3}$$

where we have taken the negative square root since we expect the slope of the graph of $Y(x)$ to be negative, as indicated in Figure 6 of Chapter 1 (recall from Section 1.3 that $x_0 < x_1$ while $y_0 > y_1$).

Hence in this case the Euler-Lagrange equation furnishes the differential equation (4.2.3), which we hope to solve for the extremum function $Y(x)$. This differential equation can be most easily solved by first introducing a new function $\theta = \theta(x)$ through the relation†

$$y_0 - Y(x) = A\left[\sin\frac{\theta(x)}{2}\right]^2. \tag{4.2.4}$$

If we insert (4.2.4) into (4.2.3), we find for $\theta(x)$ after a short calculation that

$$A \sin\frac{\theta(x)}{2}\cos\frac{\theta(x)}{2}\theta'(x) = \frac{\cos[\theta(x)/2]}{\sin[\theta(x)/2]},$$

where $\theta'(x) = d\theta(x)/dx$. Hence we find that

$$A\left[\sin\frac{\theta}{2}\right]^2\frac{d\theta}{dx} = 1,$$

which can be integrated‡ using the trigonometric identity

$$2\left[\sin\frac{\theta}{2}\right]^2 = 1 - \cos\theta \tag{4.2.5}$$

†Some readers may object here to the use of the tricky substitution given by (4.2.4). A less miraculous approach for the integration of the differential equation (4.2.3) is outlined in Exercise 1.

‡See George B. Thomas, Jr., *Calculus and Analytic Geometry* (Reading, Mass.: Addison-Wesley Publishing Company, Inc., 1968), p. 693.

to give

$$x = x_0 + \frac{A}{2}(\theta - \sin \theta). \tag{4.2.6}$$

Here x_0 is a constant of integration which we have taken to be the x-coordinate of the left end point of the wire. Equation (4.2.6) can in principle be solved to give $\theta = \theta(x)$ as a function of x along the extremum curve, and then (4.2.4) will yield the extremum function $Y(x)$. Actually it is more convenient in this case to represent the extremum curve γ parametrically in terms of the parameter θ rather than in terms of x. Indeed, if we use (4.2.1), (4.2.4), (4.2.5), and (4.2.6), we find for any such extremum curve γ that (explain)

$$\gamma: \begin{cases} x = x_0 + \frac{A}{2}(\theta - \sin \theta) \\ y = y_0 - \frac{A}{2}(1 - \cos \theta) \qquad \text{for } \theta_0 \leq \theta \leq \theta_1, \end{cases} \tag{4.2.7}$$

where θ_0 and θ_1 are the values of θ which correspond to the end points $P_0 = (x_0, y_0)$ and $P_1 = (x_1, y_1)$ of the wire.

Equations (4.2.7) are the parametric equations of a **cycloid,**† which is generated by the motion of a fixed point located on the rim of a wheel of diameter A as the wheel rolls on the underside of the line $y = y_0$, as shown in Figure 6. The parameter θ increases from θ_0 to θ_1 as the point $P = (x, y)$ travels along the cycloid from $P_0 = (x_0, y_0)$ to $P_1 = (x_1, y_1)$. We take the initial value of the parameter to be $\theta_0 = 0$, and then at $\theta = \theta_1$ (4.2.7) gives (explain)

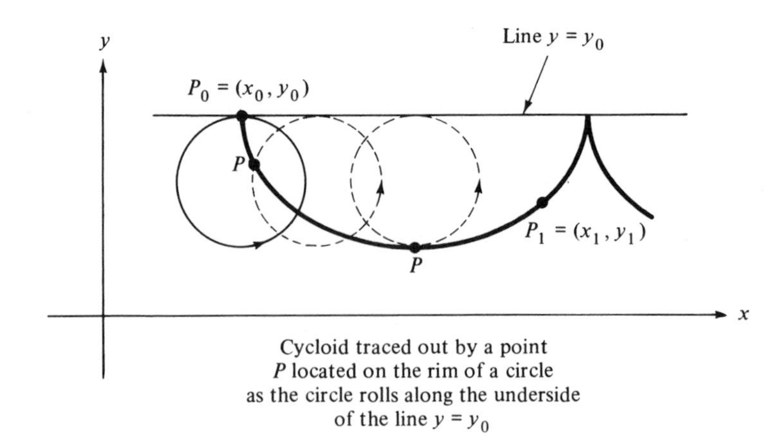

Cycloid traced out by a point
P located on the rim of a circle
as the circle rolls along the underside
of the line $y = y_0$

Figure 6

†For some interesting historical remarks on the cycloid, see E. A. Whitman, "Some Historical Notes on the Cycloid," *American Mathematical Monthly*, **50** (1943), 309–315.

$$\begin{aligned} A(\theta_1 - \sin\theta_1) &= 2(x_1 - x_0) \\ A(1 - \cos\theta_1) &= -2(y_1 - y_0). \end{aligned} \tag{4.2.8}$$

It can be shown† that this pair of equations determines unique values for the two constants A and θ_1 (with $0 < \theta_1 < 2\pi$) in terms of the given values x_0, x_1, y_0, and y_1. Hence the constants A and θ_1 can always be adjusted so that the resulting cycloid of (4.2.7) (with $\theta_0 = 0$) connects the two given points P_0 and P_1. Moreover, it can be shown‡ that the resulting cycloid does, in fact, give the *minimum time of descent* as compared with all other connecting curves. If we use (4.2.1) and (4.2.7) to change the variable of integration in (1.3.3) from x to θ, we find the value of this minimum descent time to be

$$\begin{aligned} T_{\text{minimum}} &= \int_{x_0}^{x_1} \sqrt{\frac{1 + Y'(x)^2}{2g[y_0 - Y(x)]}}\, dx = \int_0^{\theta_1} \sqrt{\frac{(dx/d\theta)^2 + (dy/d\theta)^2}{2g(y_0 - y)}}\, d\theta \\ &= \frac{A}{2} \int_0^{\theta_1} \sqrt{\frac{2(1 - \cos\theta)}{gA(1 - \cos\theta)}}\, d\theta = \sqrt{\frac{A}{2g}}\,\theta_1, \end{aligned} \tag{4.2.9}$$

where the constants A and θ_1 are determined by (4.2.8). We also used the chain rule of differentiation $[dy/d\theta = (dy/dx)(dx/d\theta)]$ in (4.2.9). This completes our analysis of John Bernoulli's brachistochrone problem.

It is interesting to note that Isaac Newton learned of the statement of John Bernoulli's brachistochrone problem on an evening in 1697, and he then solved the problem between dinner and bedtime that very night. At that same time many of the greatest mathematicians of Europe had been struggling for 6 months to obtain the solution.§ In 1703 Newton was elected president of the Royal Society of London, the leading British learned society, and in 1712 he sponsored the election of John Bernoulli into the Royal Society as one of a small number of foreign members.

The cycloid has another remarkable property which was first discovered by Christian Huygens (1629–1695). Huygens showed that if a bead starts from rest *at any point* and slides under gravity down a cycloid, the resulting time required for the bead to reach the lowest point is *independent of the starting point.* Hence three beads which start at the same time from points

†Bliss, *op. cit.*, p. 55.

‡*Ibid.*, p. 68.

§Louis Trenchard More, *Isaac Newton* (New York: Dover Publications, Inc., 1934, 1962), pp. 474–475. It was Lagrange who said of Newton "If you wish to see the human mind truly great, enter Newton's study when he is decomposing white light or unveiling the system of the world." The author learned of this quotation of Lagrange from E. T. Bell, *Men of Mathematics* (New York: Simon and Schuster, 1937, 1965), p. 168.

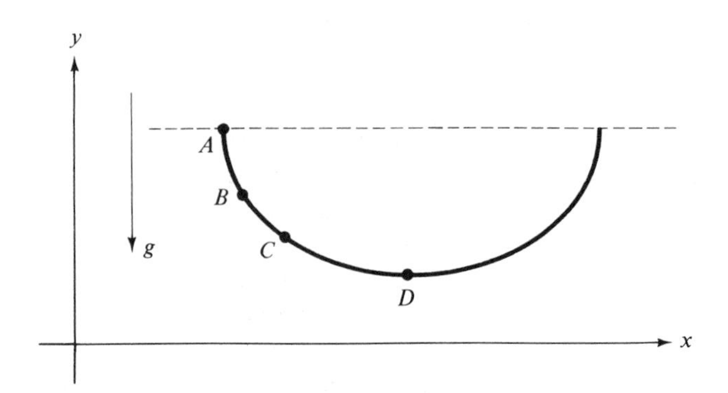

Figure 7

A, B, and C (see Figure 7) will reach D at the same time.† Therefore the cycloid is also a **tautochrone** ("the same time") as well as a brachistochrone. Huygens discovered this result in his studies on pendulum clocks and used this property of the cycloid to show how to construct *tautochronic pendulum clocks* which have periods independent of the amplitude of swing.‡

When John Bernoulli later discovered that the cycloid was also a brachistochrone, he made the following remarks: "With justice we may admire Huygens because he first discovered that a heavy particle falls on a cycloid in the same time always, no matter what the starting point may be. But you will be petrified with astonishment when I say that exactly this same cycloid, the tautochrone of Huygens, is the brachistochrone we are seeking."§

Finally, it is interesting to note that the cycloid continues to reappear in new situations. For example, the cycloid occurs prominently in one of the relativistic models of the universe which were first discovered in 1922 by A. Friedmann¶ using Albert Einstein's general theory of relativity. The particular model in question has been called the **cycloidal universe** by the British cosmologist William Bonnor.||

In this model of the universe the distance R between any two galaxies A and B varies with time t in such a way that the graph of the function $R = R(t)$

†See Thomas, *op. cit.*, pp. 385–386, and Morris Kline, *Calculus*, Part Two (New York: John Wiley & Sons, Inc., 1967), pp. 35–40.

‡See Lancelot Hogben, *Mathematics in the Making* (Garden City, N.Y.: Doubleday & Company, Inc., 1960), p. 175, for a picture of a cycloidal pendulum clock constructed by Huygens in 1673.

§The author learned of this quotation from Bell, *op. cit.*, p. 134.

¶A. Friedmann, "Über die Krümmung des Raumes," *Zeitschrift für Physik*, **10** (1922), 377–386, and "Über die Möglichkeit einer Welt mit konstanter negativer Krümmung des Raumes," *Zeitschrift für Physik*, **21** (1924), 326–332.

||William Bonnor, *The Mystery of the Expanding Universe* (New York: The Macmillan Company, 1964).

is an upright cycloid. The two galaxies initially recede from each other with $R(t)$ increasing during an initial period of expansion of the universe, and then later the galaxies are drawn back together under the force of gravity with $R(t)$ decreasing during a period of contraction of the universe. We refer the reader to Bonnor's book for further details.

Brachistochrones Through the Earth. We turn now to a consideration of brachistochrone curves through the interior of the earth. We let A and B be two fixed given points on the surface of the earth, as indicated in Figure 8, and we let γ be a plane curve connecting A and B passing through the earth's interior. We suppose that a tunnel can be dug through the earth from A to B along the path γ, and we then consider the *time of motion T* required for a bead to slide without friction through the tunnel from A to B, given by (1.3.1) as

$$T = \int_{\gamma} \frac{ds}{v}, \tag{4.2.10}$$

where s measures arc length along γ, ds/dt is the rate of change of arc length with respect to time t during the motion, and the instantaneous speed of motion v is given as $v = ds/dt$. *We seek the particular tunnel γ that yields the least value for T*, where, however, we can no longer use (1.3.2) to give the

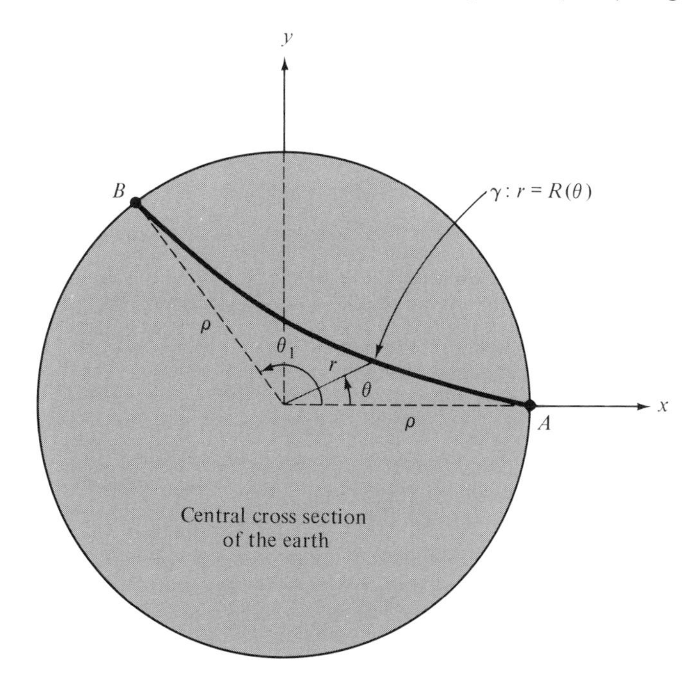

Figure 8

speed of motion of the bead along γ since the potential energy of a bead located *inside* the earth cannot be given by the formula used in Section 1.3 in deriving (1.3.2). In fact, the potential energy of a bead of mass m located inside the earth is found with Newton's law of gravitation to be†

$$\text{potential energy} = mg\frac{r^2}{2\rho},$$

where r is the distance of the bead from the center of the earth, ρ is the radius of the earth, and g is the acceleration due to the earth's gravity at the surface of the earth. Conservation of energy requires that the sum of the potential and kinetic energies of the bead remains constant during the motion (we neglect friction), so that

$$\text{kinetic energy} + \text{potential energy} = \text{constant},$$

where the kinetic energy of motion is given by

$$\text{kinetic energy} = \tfrac{1}{2}mv^2.$$

If the bead starts from rest with $v = 0$ and $r = \rho$ at $t = 0$, it follows that during the motion the relation

$$\frac{1}{2}mv^2 + \frac{1}{2}mg\frac{r^2}{\rho} = \frac{1}{2}mg\rho$$

must hold. This relation can be solved for the speed of motion to give

$$v = \sqrt{\frac{g(\rho^2 - r^2)}{\rho}}, \tag{4.2.11}$$

where ρ is the fixed radius of the earth and r is the instantaneous distance of the bead from the center of the earth $(0 \leq r \leq \rho)$.

We place a Cartesian (x, y)-coordinate plane with the origin at the center of the earth and with the positive x-axis passing through the point A, as shown in Figure 8, and we then let r and θ be the usual plane polar coordinates of the point (x, y), defined by

$$\begin{aligned} x &= r\cos\theta, & r &= \sqrt{x^2 + y^2} \\ y &= r\sin\theta, & \theta &= \tan^{-1}\frac{y}{x}. \end{aligned} \tag{4.2.12}$$

We then represent the curve γ in terms of polar coordinates by giving the

†See Morris Kline, *Calculus*, Part One (New York: John Wiley & Sons, Inc., 1967), pp. 507–512, and J. M. A. Danby, *Fundamentals of Celestial Mechanics* (New York: The Macmillan Company, 1962), pp. 92–95.

radial distance r as a function of θ along γ as

$$\gamma\colon\quad r = R(\theta) \qquad \text{for } 0 \leq \theta \leq \theta_1 \tag{4.2.13}$$

for some suitable function $R(\theta)$. The angle θ_1 is the fixed central angle (in radians) determined by the given points A and B, as indicated in Figure 8; hence θ_1 is given by

$$\rho\theta_1 = S_{AB}, \tag{4.2.14}$$

where S_{AB} is the known arc length between the given points A and B measured along the surface of the earth and ρ is again the radius of the earth. From (4.2.12) and (4.2.13) we may give the x and y Cartesian coordinates of any point of γ as

$$\gamma\colon\quad \begin{cases} x = R(\theta)\cos\theta \\ y = R(\theta)\sin\theta, \end{cases} \tag{4.2.15}$$

and the differential element of arc length along γ then becomes†

$$\begin{aligned} ds &= \sqrt{\left(\frac{dx}{d\theta}\right)^2 + \left(\frac{dy}{d\theta}\right)^2}\, d\theta \\ &= \sqrt{R(\theta)^2 + R'(\theta)^2}\, d\theta, \end{aligned}$$

where $R'(\theta) = dR(\theta)/d\theta$. It follows from this result and (4.2.10), (4.2.11), and (4.2.13) that the time of motion T of the bead from A to B can be written as (explain)

$$T = \sqrt{\frac{\rho}{g}} \int_0^{\theta_1} \sqrt{\frac{R(\theta)^2 + R'(\theta)^2}{\rho^2 - R(\theta)^2}}\, d\theta$$

or

$$T(R) = \int_0^{\theta_1} F(R(\theta), R'(\theta))\, d\theta, \tag{4.2.16}$$

where

$$F(r, z) = \sqrt{\frac{\rho}{g}} \sqrt{\frac{r^2 + z^2}{\rho^2 - r^2}} \tag{4.2.17}$$

for any suitable numbers r and z.

We seek to minimize the functional (4.2.16) among all continuously differentiable functions $R(\theta)$ on the fixed interval $0 \leq \theta \leq \theta_1$ which satisfy the constraints

$$R(0) = \rho, \qquad R(\theta_1) = \rho. \tag{4.2.18}$$

In the notation of Section 4.1 the problem is to find a minimum vector $R(\theta)$

†See John G. Hocking, *Calculus with an Introduction to Linear Algebra* (New York: Holt, Rinehart and Winston, Inc., 1970), pp. 677–678.

in $D[K_0 = \rho, K_1 = \rho]$ for the time functional T of (4.2.16), where K_0 and K_1 are defined for any vector $R = R(\theta)$ in D by

$$K_0(R) = R(0), \qquad K_1(R) = R(\theta_1),$$

and where D is the entire vector space $\mathcal{C}^1[0, \theta_1]$ consisting of all continuously differentiable functions $R = R(\theta)$ on the interval $0 \leq \theta \leq \theta_1$. It follows from the results of Section 4.1 that the extremum function $R(\theta)$ must satisfy the Euler-Lagrange equation [see (4.1.11)]

$$F_R(R(\theta), R'(\theta)) - \frac{d}{d\theta} F_{R'}(R(\theta), R'(\theta)) = 0$$

for $0 \leq \theta \leq \theta_1$. Since $F = F(r, z)$ is independent of the first coordinate θ, we may replace this last Euler-Lagrange equation with the simpler equation [see (4.1.15)]

$$F(R(\theta), R'(\theta)) - R'(\theta)F_{R'}(R(\theta), R'(\theta)) = C \tag{4.2.19}$$

for some constant of integration C, where

$$F_{R'}(R(\theta), R'(\theta)) = \left.\frac{\partial F(r, z)}{\partial z}\right|_{\substack{r=R(\theta) \\ z=R'(\theta)}}. \tag{4.2.20}$$

If we use (4.2.17) and (4.2.20) in (4.2.19), we find for the extremum function $R(\theta)$ the differential equation

$$\sqrt{\frac{R(\theta)^2 + R'(\theta)^2}{\rho^2 - R(\theta)^2}} - \frac{R'(\theta)^2}{\sqrt{[\rho^2 - R(\theta)^2][R(\theta)^2 + R'(\theta)^2]}} = \sqrt{\frac{g}{\rho}}\,C,$$

which can be simplified after a short calculation to give

$$R'(\theta)^2 = \frac{\rho^2}{r_1^2}\frac{R(\theta)^2 - r_1^2}{\rho^2 - R(\theta)^2}R(\theta)^2, \tag{4.2.21}$$

where we have introduced a new constant of integration r_1 related to the earlier constant C through the relation

$$r_1 = \frac{\rho}{\sqrt{1 + (\rho C^2/g)}}. \tag{4.2.22}$$

We may solve (4.2.21) for $R'(\theta) = dR/d\theta$ and find

$$\frac{dR}{d\theta} = \begin{cases} -\dfrac{\rho}{r_1} R\sqrt{\dfrac{R^2 - r_1^2}{\rho^2 - R^2}} & \text{for } 0 \leq \theta \leq \dfrac{1}{2}\theta_1 \\[2ex] +\dfrac{\rho}{r_1} R\sqrt{\dfrac{R^2 - r_1^2}{\rho^2 - R^2}} & \text{for } \dfrac{1}{2}\theta_1 \leq \theta \leq \theta_1, \end{cases} \tag{4.2.23}$$

where we have taken the negative square root for $0 \leq \theta \leq \frac{1}{2}\theta_1$ since we expect the extremum function $R = R(\theta)$ to *decrease* [with $R'(\theta) \leq 0$] as θ increases for $0 \leq \theta \leq \frac{1}{2}\theta_1$, as indicated in Figure 8.

The differential equation (4.2.23) for $R = R(\theta)$ which is given by the Euler-Lagrange equation can be most easily solved by introducing a new function $\varphi = \varphi(\theta)$ through the relation [compare with (4.2.3) and (4.2.4)]†

$$R(\theta)^2 = \frac{\rho^2 + r_1^2}{2} + \frac{\rho^2 - r_1^2}{2}\cos\frac{2\rho\varphi(\theta)}{\rho - r_1}. \tag{4.2.24}$$

If we use (4.2.24) in (4.2.23), we find after a short calculation (which the reader should carry through) the following differential equation for $\varphi = \varphi(\theta)$,

$$\frac{d\varphi}{d\theta} = \frac{(\rho^2 + r_1^2) + (\rho^2 - r_1^2)\cos[2\rho\varphi/(\rho - r_1)]}{r_1(\rho + r_1)\{1 - \cos[2\rho\varphi/(\rho - r_1)]\}},$$

or, equivalently (explain),

$$\left\{-1 + \frac{2\rho^2}{(\rho^2 + r_1^2) + (\rho^2 - r_1^2)\cos[2\rho\varphi/(\rho - r_1)]}\right\}\frac{d\varphi}{d\theta} = \frac{\rho - r_1}{r_1} \tag{4.2.25}$$

for $0 \leq \theta \leq \theta_1$. Equation (4.2.25) can be integrated to give $\varphi = \varphi(\theta)$ along the extremum curve γ, and then (4.2.24) will give the extremum function $R(\theta)$. Along with the differential equation (4.2.25) we have the following initial condition obtained from (4.2.24) and the first constraint of (4.2.18) ($R = \rho$ at $\theta = 0$):

$$\varphi = 0 \qquad \text{at } \theta = 0. \tag{4.2.26}$$

Morevoer, (4.2.25) shows that φ *increases* as θ increases since r_1 satisfies $0 < r_1 \leq \rho$ [see (4.2.22)]. Hence in addition to (4.2.26) we have the following additional condition obtained from (4.2.24) and the remaining constraint of (4.2.18) ($R = \rho$ at $\theta = \theta_1$):

$$\varphi = \frac{\rho - r_1}{\rho}\pi \qquad \text{at } \theta = \theta_1. \tag{4.2.27}$$

If we now integrate both sides of (4.2.25) with respect to θ and use the condition (4.2.26), we find that

$$2\rho^2\int_0^{\varphi}\frac{d\varphi}{(\rho^2 + r_1^2) + (\rho^2 - r_1^2)\cos[2\rho\varphi/(\rho - r_1)]} = \varphi + \frac{\rho - r_1}{r_1}\theta. \tag{4.2.28}$$

†Those readers who objected earlier to the tricky substitution (4.2.4) will probably throw up their arms in despair here at the use of this miraculous substitution (4.2.24). Unfortunately, the author knows of no simpler way to solve the differential equation (4.2.23).

The integral on the left-hand side of (4.2.28) can be evaluated with the aid of the formula†

$$\int \frac{d\varphi}{a + b\cos\varphi} = \frac{2}{\sqrt{a^2 - b^2}} \tan^{-1} \frac{\sqrt{a^2 - b^2}\tan\frac{1}{2}\varphi}{a + b},$$

which is valid for $a^2 > b^2$. In this way we find from (4.2.28) that

$$\frac{r_1}{\rho}\tan\frac{\rho\varphi}{\rho - r_1} = \tan\left(\theta + \frac{r_1\varphi}{\rho - r_1}\right), \tag{4.2.29}$$

relating φ and θ. The remaining constant r_1 is determined by imposing the remaining condition (4.2.27) in (4.2.29); we find that

$$\theta_1 + \frac{r_1}{\rho}\pi = \pi$$

or

$$r_1 = \rho\left(1 - \frac{\theta_1}{\pi}\right),$$

where the known central angle θ_1 is understood to satisfy $0 < \theta_1 < \pi$. Note that this choice of r_1 implies with (4.2.27) that $\varphi = \theta_1$ when $\theta = \theta_1$. Hence as θ increases from 0 to θ_1, φ also increases from 0 to θ_1. We can use (4.2.14) to rewrite the expression for r_1 in terms of the known arc length S_{AB} between the given points A and B as

$$r_1 = \rho - \frac{S_{AB}}{\pi}. \tag{4.2.30}$$

Equations (4.2.29) and (4.2.30) can in principle be solved to give $\varphi = \varphi(\theta)$ as a function of θ along the extremum curve, and then (4.2.24) will yield the extremum function $R(\theta)$. It is more convenient, however, to represent the extremum curve γ by giving the Cartesian coordinates x, y of any point on γ parametrically in terms of the parameter φ. For example, the trigonometric identity $\tan(a + b) = (\tan a + \tan b)/(1 - \tan a \tan b)$ can be used with $a = \theta$ and $b = r_1\varphi/(\rho - r_1)$ in (4.2.29) to calculate $\tan\theta$ explicitly in terms of φ, and it is then a simple matter to obtain $\cos\theta$ and $\sin\theta$ as functions of φ along the extremum curve γ. These results can be used in (4.2.15) along with $R(\theta)$ from (4.2.24) to find the following parametric representation for γ,

$$\gamma: \begin{cases} x = \dfrac{\rho + r_1}{2}\cos\varphi + \dfrac{\rho - r_1}{2}\cos\dfrac{\rho + r_1}{\rho - r_1}\varphi \\ y = \dfrac{\rho + r_1}{2}\sin\varphi - \dfrac{\rho - r_1}{2}\sin\dfrac{\rho + r_1}{\rho - r_1}\varphi, \qquad 0 \le \varphi \le \theta_1, \end{cases} \tag{4.2.31}$$

†*Standard Mathematical Tables*, 10th ed. (Cleveland: The Chemical Rubber Publishing Company, 1956), p. 286.

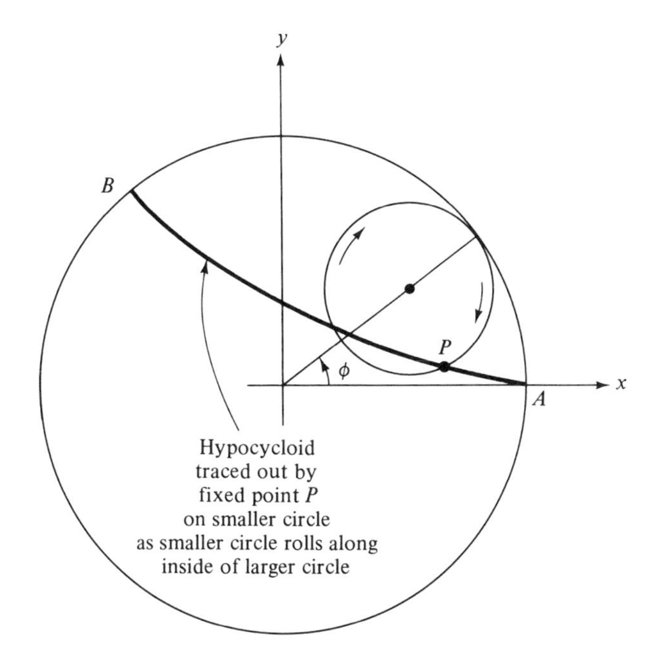

Figure 9

where r_1 is given by (4.2.30) and θ_1 is given by (4.2.14). We leave the derivation of (4.2.31) from (4.2.15), (4.2.24), and (4.2.29) as an exercise for the foolhardy! (The algebra involved is horrendous.) However, it is easy to check that the quantity $x^2 + y^2$ as calculated from (4.2.31) agrees with the quantity $R(\theta)^2 = r^2 = x^2 + y^2$ as given by (4.2.24), (4.2.12), and (4.2.13).

Equations (4.2.31) are the parametric equations of a **hypocycloid**, which is generated by the motion of a fixed point on a circle of radius $\rho - r_1 = S_{AB}/\pi$ as the circle rolls on the inside of the larger circle of radius ρ, as shown in Figure 9. The resulting hypocycloid connecting A and B is the desired brachistochrone curve through the interior of the earth. The resulting *minimum transit time* of a bead in the tunnel can be calculated easily from (4.2.16), (4.2.21), (4.2.24), and (4.2.25). We find that

$$T_{\text{minimum}} = \theta_1 \sqrt{\frac{\rho(\rho + r_1)}{g(\rho - r_1)}},$$

or, with (4.2.14) and (4.2.30),

$$T_{\text{minimum}} = \sqrt{\frac{2\pi\rho S_{AB} - S_{AB}^2}{\rho g}}.$$

This formula can be used to verify the assertion in Section 1.1 concerning the minimum transit time through the earth from San Diego to San Francisco if we use for the earth the values $\rho = 4000$ miles and $g = 32$ feet per square second $= \frac{32}{5280}$ miles per square second.†

Exercise

1. Show how to solve the differential equation (4.2.3) using separation of variables. *Hints:* To get rid of the nasty square root sign, make the substitution $(y_0 - Y)/[A - (y_0 - Y)] = u^2$, with $dY = -[2Au/(1 + u^2)^2]du$. The resulting integral is a standard one which can be evaluated by letting $u = \tan \varphi$. The author is indebted to Professor Jerry L. Kazdan for this more plausible approach to the integration of (4.2.3).

4.3. Geodesic Curves

We consider a given fixed surface S in 3-space, as shown in Figure 10, and we let P_0 and P_1 be any two given fixed points on S. We then consider the problem of finding a curve γ which has the **shortest length** among all curves which lie on the surface S and which connect P_0 and P_1. Stated somewhat differently, we seek the *shortest distance between P_0 and P_1 on S* that can be achieved by varying the path connecting the two points on S. Any such curve giving the minimum distance between two points of S is called a **geodesic curve**‡ on the surface S and is determined by the intrinsic properties of the given surface. Such geodesic curves can be constructed by stretching

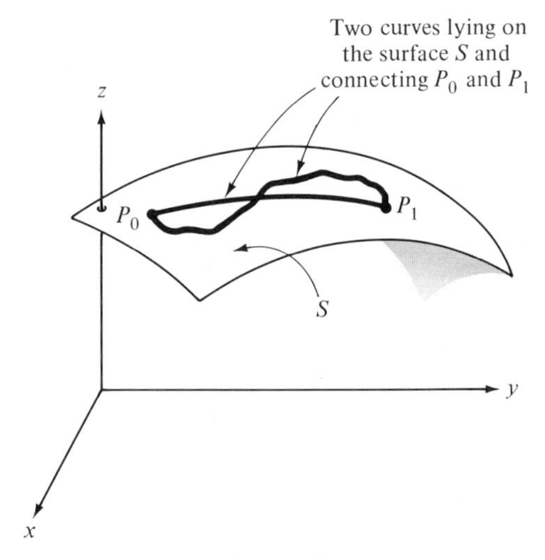

Figure 10

†See also Kline, *op. cit.*, Part One, pp. 263–264.

‡The word **geodesy** has Greek roots and means literally "the dividing of the earth." *Geodetic surveying* consists of determining points and areas on the spherical surface of the earth over large distances for which the earth's curvature must be taken into account.

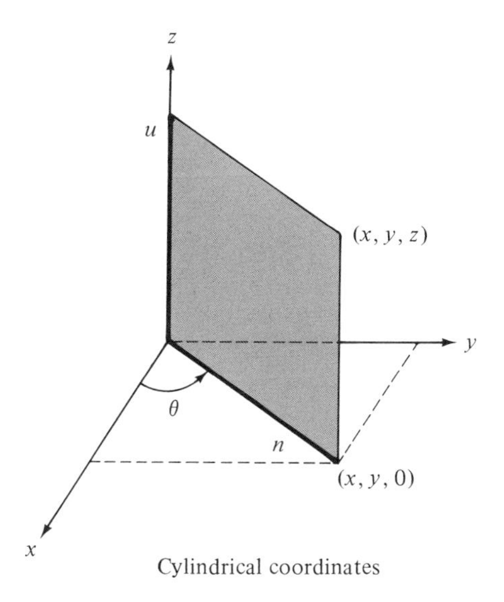

Cylindrical coordinates

Figure 11

a rubber band between two nearby points over a model of the given surface. The simplest case occurs if the surface S is a plane, in which case the geodesic curve connecting any two points on S is just the usual straight-line segment between the points. Two other somewhat simple cases occur if S is either a *right circular cylinder* or a *sphere*. We shall consider here only these two special cases. The interested reader may see Robert Weinstock, *Calculus of Variations* (New York: McGraw-Hill Book Company, 1952), for an indication of the situation obtaining in more general cases.

Geodesics on a Right Circular Cylinder. Suppose first that S is a right circular cylinder. We shall find it useful to represent S parametrically,† and this can be most easily done in this case by introducing *cylindrical coordinates*. We let x, y, z be the usual Cartesian coordinates in 3-space, and then cylindrical coordinates r, θ, u may be defined by (see Figure 11)

$$x = r\cos\theta, \qquad r = \sqrt{x^2 + y^2}$$

$$y = r\sin\theta, \qquad \theta = \tan^{-1}\frac{y}{x}$$

$$z = u, \qquad u = z.$$

†The use of parametric representations in the study of surfaces is due to Carl Friedrich Gauss (1777–1855), who made outstanding contributions in number theory, algebra, geometry, mathematical foundations, analysis, statistics, complex number theory, analytic function theory, mathematical astronomy, capillarity, optics, electromagnetism, and geodesy. Gauss and the physicist Wilhelm Weber (1804–1891) together invented an electromagnetic telegraph in 1833. E. T. Bell (*op. cit.*) ranks Archimedes, Newton, and Gauss as the three greatest mathematicians. See also G. Waldo Dunnington, *Carl Friedrich Gauss: Titan of Science* (New York: Hafner Publishing Company, Inc., 1955).

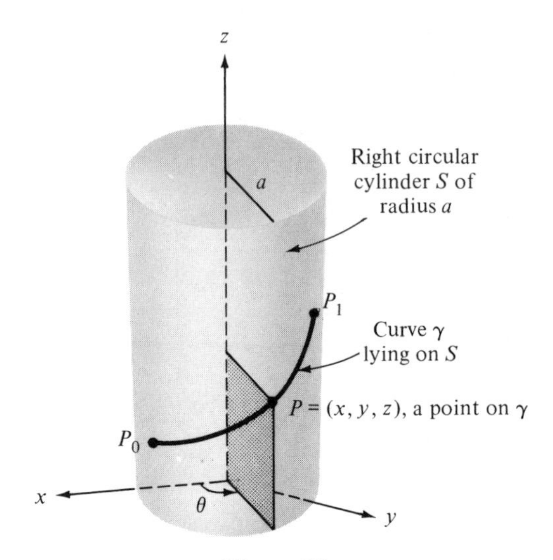

Figure 12

We let the central axis of the cylinder S coincide with the z-axis as shown in Figure 12, and then the surface S can be given parametrically in terms of the parameters u and θ as

$$S: \begin{cases} x = a\cos\theta \\ y = a\sin\theta \\ z = u \qquad \text{for } 0 \leq \theta \leq 2\pi,\ -\infty < u < \infty, \end{cases}$$

where the quantity a is a fixed positive constant which gives the value of the radius of the cylinder and x, y, z are the Cartesian coordinates of a representative point $P = (x, y, z)$ on the cylinder.

We now consider an arbitrary curve γ *lying on the surface of the cylinder* S. We represent γ parametrically in terms of the cylindrical angle θ as (see Figure 12)

$$\gamma: \begin{cases} x = a\cos\theta \\ y = a\sin\theta \\ z = U(\theta) \qquad \text{for } \theta_0 \leq \theta \leq \theta_1 \end{cases} \tag{4.3.1}$$

for some suitable function $U(\theta)$ relating the parameters u and θ along γ. The constants θ_0 and θ_1 are the θ-coordinates of the specified end points $P_0 = (x_0, y_0, z_0)$ and $P_1 = (x_1, y_1, z_1)$ of γ, which satisfy

$$\begin{aligned} x_0 &= a\cos\theta_0, \qquad y_0 = a\sin\theta_0 \\ x_1 &= a\cos\theta_1, \qquad y_1 = a\sin\theta_1. \end{aligned} \tag{4.3.2}$$

We also assume that θ_0 and θ_1 satisfy the condition $\theta_0 \leq \theta_1$, which can always be achieved simply be interchanging the roles of P_0 and P_1 if necessary. Finally, since we are interested in finding the *shortest* curve γ connecting P_0 and P_1, there will be no loss in assuming the additional condition $\theta_1 - \theta_0 \leq \pi$, which can be achieved by choosing the angular difference $\theta_1 - \theta_0$ to be the smallest possible cylindrical angle between the points P_0 and P_1. Hence we shall assume the following conditions:

$$0 \leq \theta_1 - \theta_0 \leq \pi. \tag{4.3.3}$$

There is one exceptional case in which the parametric representation (4.3.1) is *not* possible, namely the case in which the points P_0 and P_1 lie directly above and below each other on the vertical cylindrical surface. In this case P_0 and P_1 will have the same x- and y-coordinates and $\theta_0 = \theta_1$ will hold [see (4.3.2) and (4.3.3)]. Of course, in this exceptional case the shortest curve connecting P_0 and P_1 on S is obvious. [One would use u rather than θ as the parameter along γ in this case by specifying $\theta = \Theta(u)$ for some suitable function $\Theta(u)$.]

The length L of any parametric curve γ such as in (4.3.1) is given by the integral†

$$L = \int_{\theta_0}^{\theta_1} \sqrt{\left(\frac{dx}{d\theta}\right)^2 + \left(\frac{dy}{d\theta}\right)^2 + \left(\frac{dz}{d\theta}\right)^2}\, d\theta, \tag{4.3.4}$$

where the derivatives $dx/d\theta$, $dy/d\theta$, and $dz/d\theta$ are calculated from (4.3.1). In this way we find for the length of any such curve γ lying on the surface of the circular cylinder that

$$L = \int_{\theta_0}^{\theta_1} \sqrt{a^2 + U'(\theta)^2}\, d\theta \tag{4.3.5}$$

or

$$L(U) = \int_{\theta_0}^{\theta_1} F(U'(\theta))\, d\theta, \tag{4.3.6}$$

where the function $F = F(w)$ is defined by

$$F(w) = \sqrt{a^2 + w^2} \tag{4.3.7}$$

for any number w, and where we have written $L(U)$ in (4.3.6) to indicate that the length of γ is a functional which depends on the particular function $U = U(\theta)$ used in the parametric representation (4.3.1).

Finally, then, the problem of finding the geodesic curve connecting $P_0 = (x_0, y_0, z_0)$ and $P_1 = (x_1, y_1, z_1)$ on the cylinder can be reduced to the problem of finding the particular function $U = U(\theta)$ which minimizes the

†See Hocking, *op. cit.*, pp. 677–678.

length functional $L(U)$ of (4.3.6) subject to the constraints (explain)

$$U(\theta_0) = z_0, \qquad U(\theta_1) = z_1. \tag{4.3.8}$$

It follows then from the results of Section 4.1 that the minimizing function $U(\theta)$ must satisfy the following Euler-Lagrange equation:

$$F_U(U'(\theta)) - \frac{d}{d\theta} F_{U'}(U'(\theta)) = 0$$

for $\theta_0 < \theta < \theta_1$. Since $F = F(U')$ is independent of U, we may replace this equation with the simpler equation

$$F_{U'}(U'(\theta)) = C$$

for some constant of integration C, where

$$F_{U'}(U'(\theta)) = \left.\frac{\partial F(w)}{\partial w}\right|_{w=U'(\theta)}.$$

If we use the last two equations along with (4.3.7), we find for the extremum function $U = U(\theta)$ the differential equation

$$\frac{U'(\theta)}{\sqrt{a^2 + U'(\theta)^2}} = C,$$

which can be simplified to give

$$U'(\theta) = A$$

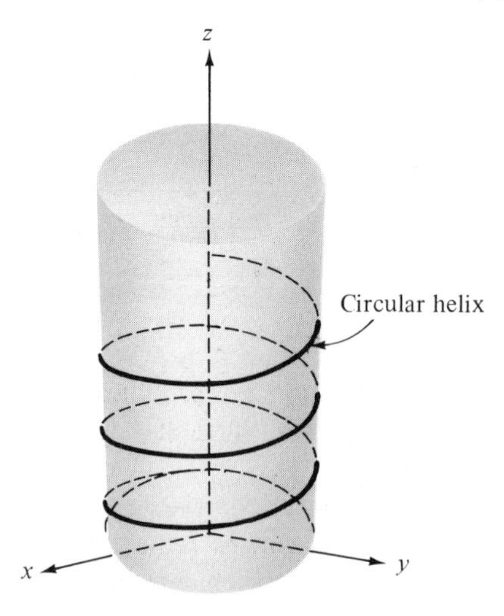

Figure 13

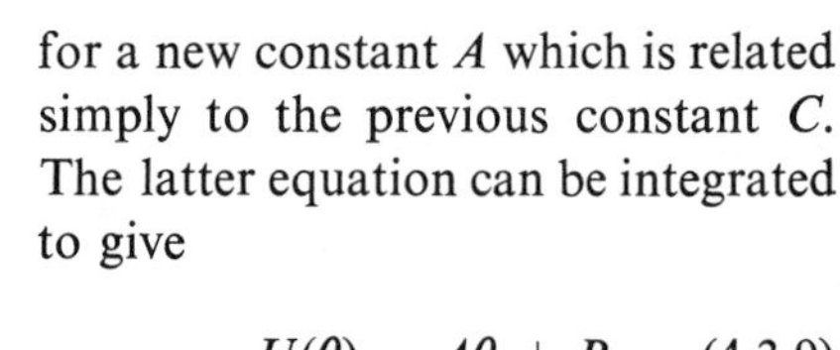

for a new constant A which is related simply to the previous constant C. The latter equation can be integrated to give

$$U(\theta) = A\theta + B, \tag{4.3.9}$$

where B is a second constant of integration.

Equations (4.3.1) and (4.3.9) are the parametric equations of an arc of a **circular helix** which winds around the cylinder S, as shown in Figure 13. The circular helix intersects lines parallel to the z-axis at a constant angle which is determined by the choice of the constant A. The constants θ_0, θ_1, A, and B appearing in (4.3.1) and (4.3.9) can always be chosen so that the resulting helix connects the given points $P_0 = (x_0, y_0, z_0)$ and $P_1 = (x_1, y_1, z_1)$ for $\theta_0 \leq \theta \leq \theta_1$. Indeed, (4.3.2) and (4.3.3) determine θ_0 and θ_1, and then A and B are found from (4.3.8) and (4.3.9) to be

$$A = \frac{z_1 - z_0}{\theta_1 - \theta_0}, \qquad B = \frac{\theta_1 z_0 - \theta_0 z_1}{\theta_1 - \theta_0}. \tag{4.3.10}$$

It can be shown that the resulting arc of the helix connecting P_0 and P_1 is, in fact, the *geodesic curve* connecting P_0 and P_1 on the cylinder S. The resulting *minimum length* among all curves connecting P_0 and P_1 on S is found from (4.3.5), (4.3.9), and (4.3.10) to be

$$L_{\text{minimum}} = \sqrt{a^2(\theta_1 - \theta_0)^2 + (z_1 - z_0)^2}, \tag{4.3.11}$$

where the angular difference $\theta_1 - \theta_0$ is determined by the following conditions [which follow from (4.3.2) and (4.3.3); explain]:

$$\cos(\theta_1 - \theta_0) = \frac{x_0 x_1 + y_0 y_1}{a^2}, \qquad 0 \leq \theta_1 - \theta_0 \leq \pi. \tag{4.3.12}$$

It is interesting to note from (4.3.11) that the **Pythagorean theorem**† holds for the *right geodesic triangle* QP_0P_1 on S, as shown in Figure 14, where the hypotenuse P_0P_1 as well as the other two sides QP_0 and QP_1 are all given by the appropriate geodesic curves on S. Indeed, (4.3.11) implies that

$$L^2 = (a\,\Delta\theta)^2 + (\Delta z)^2, \tag{4.3.13}$$

where we have set $L = L_{\text{minimum}}$, $\Delta z = z_1 - z_0$, and $\Delta\theta = \theta_1 - \theta_0$, and where Δz is the length of the straight-line geodesic connecting Q and P_1, $a\Delta\theta$ is the length of the circular geodesic arc connecting Q and P_0 (explain), and L is the length of the geodesic helix connecting P_0 and P_1. The two geodesics QP_0 and QP_1 meet in a right angle at Q. The result (4.3.13) is then a statement of the Pythagorean theorem for right

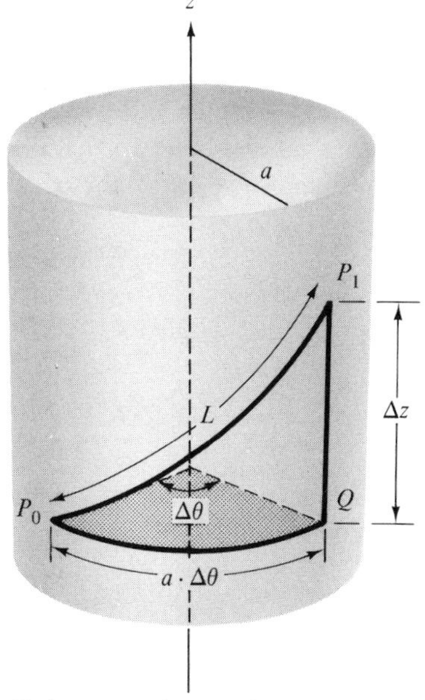

Pythagorean theorem for a right geodesic triangle: $L^2 = (a \cdot \Delta\theta)^2 + (\Delta z)^2$

Figure 14

†*The square of the length of the hypotenuse equals the sum of the squares of the lengths of the other two sides of a right triangle.*

geodesic triangles on the cylinder. This result can be obtained alternatively from the usual Pythagorean theorem by a geometrical construction which involves the cutting and rolling out of the cylinder. Indeed, if the cylinder S is sliced along one of the lines of intersection of the cylinder and a plane containing the central axis of the cylinder, and if the resulting sliced cylinder is then rolled out flat onto a plane surface, the original right geodesic triangle QP_0P_1 on the cylinder will roll out onto an ordinary right triangle $Q'P'_0P'_1$ on the plane surface. The three geodesic curves which form the sides of QP_0P_1 on the cylinder will all roll out onto the corresponding straight-line geodesic arcs which form the sides of $Q'P'_0P'_1$ on the flat plane surface. (One can show that any circular helix on the cylinder will roll out onto a straight line.) It is reasonably clear (and true) that distances along the geodesic curves between corresponding points will be preserved during the "rolling out" process. In this way one can show that (4.3.13) will follow from the usual Pythagorean theorem (for right triangles lying in plane surfaces). The reader may wish to consider whether or not the Pythagorean theorem holds in general on a right circular cylinder for *every* right geodesic triangle formed by any three suitably intersecting geodesic curves two of which meet in a right angle.† The case considered above is somewhat special, since only one of the three geodesic sides of QP_0P_1 is formed by an arc of a helix which winds around and up the cylinder.

Geodesics on a Sphere. We turn now to a consideration of geodesic curves on a sphere. In this case it is convenient to introduce *spherical polar coordinates* r, θ, ϕ by the usual relations (see Figure 15)

$$\begin{aligned} x &= r\sin\phi\cos\theta, & r &= \sqrt{x^2+y^2+z^2} \\ y &= r\sin\phi\sin\theta, & \theta &= \tan^{-1}\frac{y}{x} \\ z &= r\cos\phi, & \phi &= \cos^{-1}\frac{z}{r}, \end{aligned} \tag{4.3.14}$$

and then the surface of a sphere S of radius a centered at the origin can be given parametrically in terms of the angles ϕ and θ as

$$S: \begin{cases} x = a\sin\phi\cos\theta \\ y = a\sin\phi\sin\theta \\ z = a\cos\phi \qquad \text{for } 0 \le \phi \le \pi,\ 0 \le \theta \le 2\pi. \end{cases}$$

Finally, we let γ be a curve lying on the surface of the sphere S as shown in Figure 16, and we represent γ parametrically in terms of the single parameter

†See Barrett O'Neill, *Elementary Differential Geometry* (New York: Academic Press, Inc., 1966), where many interesting questions concerning surfaces and curves are pursued.

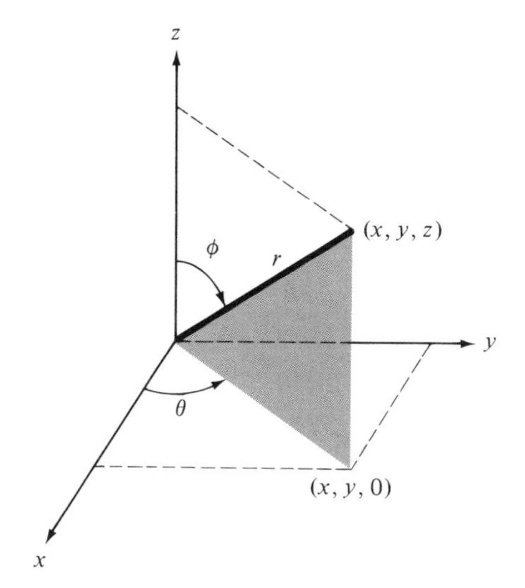

Spherical coordinates

Figure 15

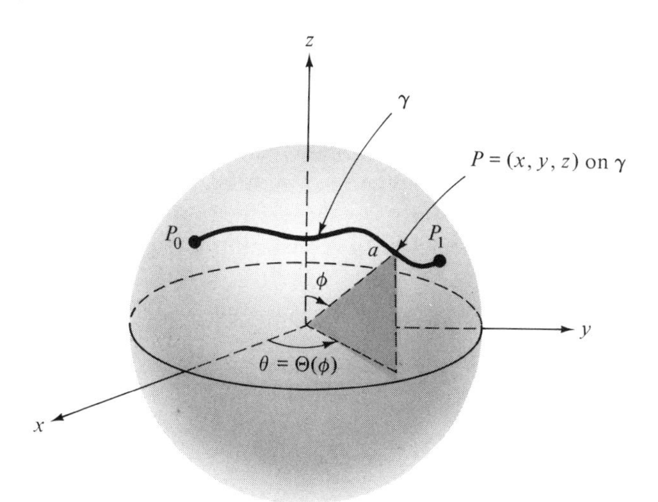

Figure 16

ϕ as

$$\gamma: \quad \begin{cases} x = a \sin \phi \cos \Theta(\phi) \\ y = a \sin \phi \sin \Theta(\phi) \\ z = a \cos \phi \qquad \text{for } \phi_0 \le \phi \le \phi_1 \end{cases} \tag{4.3.15}$$

for some suitable function $\Theta(\phi)$ relating the angles θ and ϕ on γ. The constants ϕ_0 and ϕ_1 are the ϕ-coordinates of the end points $P_0 = (x_0, y_0, z_0)$ and $P_1 = (x_1, y_1, z_1)$ of γ, which satisfy [see (4.3.14)]

$$z_0 = \sqrt{x_0^2 + y_0^2 + z_0^2} \cos \phi_0, \qquad z_1 = \sqrt{x_1^2 + y_1^2 + z_1^2} \cos \phi_1,$$

with $0 \leq \phi_0 \leq \pi$ and $0 \leq \phi_1 \leq \pi$. The length L of any parametric curve γ such as in (4.3.15) is given as [compare with (4.3.4)]

$$L = \int_{\phi_0}^{\phi_1} \sqrt{\left(\frac{dx}{d\phi}\right)^2 + \left(\frac{dy}{d\phi}\right)^2 + \left(\frac{dz}{d\phi}\right)^2}\, d\phi,$$

and then this formula along with (4.3.15) leads to

$$L = a \int_{\phi_0}^{\phi_1} \sqrt{1 + \Theta'(\phi)^2 \sin^2 \phi}\, d\phi, \tag{4.3.16}$$

where $\Theta'(\phi) = d\Theta/d\phi$. We may rewrite this result as

$$L(\Theta) = \int_{\phi_0}^{\phi_1} F(\phi, \Theta'(\phi))\, d\phi, \tag{4.3.17}$$

where the function $F = F(\phi, w)$ is defined by

$$F(\phi, w) = a\sqrt{1 + w^2 \sin^2 \phi} \tag{4.3.18}$$

for any numbers ϕ and w.

The problem of finding the geodesic curve connecting $P_0 = (x_0, y_0, z_0)$ and $P_1 = (x_1, y_1, z_1)$ on the sphere can now be reduced to the problem of finding the particular function $\Theta = \Theta(\phi)$ which minimizes the length functional $L(\Theta)$ of (4.3.17) subject to the constraints [see (4.3.15)]

$$\tan \Theta(\phi_0) = \frac{y_0}{x_0}, \qquad \tan \Theta(\phi_1) = \frac{y_1}{x_1}.$$

It follows directly from the results of Section 4.1 that the minimizing function $\Theta(\phi)$ must satisfy the Euler-Lagrange equation

$$F_\Theta(\phi, \Theta'(\phi)) - \frac{d}{d\phi} F_{\Theta'}(\phi, \Theta'(\phi)) = 0,$$

and since F is independent of Θ, we may replace this equation with the simpler equation

$$F_{\Theta'}(\phi, \Theta'(\phi)) = C$$

for some constant of integration C, where

$$F_{\Theta'}(\phi, \Theta'(\phi)) = \left.\frac{\partial F(\phi, w)}{\partial w}\right|_{w=\Theta'(\phi)}.$$

If we use the last two equations with (4.3.18), we find for the extremum function $\Theta(\phi)$ the differential equation

$$\Theta'(\phi) = \frac{A}{\sin\phi\sqrt{\sin^2\phi - A^2}}$$

for some constant A which is related to the previous constant C. It is convenient† to replace A with yet a different constant α through the relation $A = \sin\alpha$, in which case the previous differential equation becomes

$$\frac{d\Theta}{d\phi} = \frac{\sin\alpha}{\sin\phi\sqrt{\sin^2\phi - \sin^2\alpha}} \tag{4.3.19}$$

for some constant α. This equation can be most easily integrated by introducing a new independent variable u through the substitution‡

$$\tan\phi = \frac{1}{u}, \tag{4.3.20}$$

and then equation (4.3.19) becomes

$$\frac{d\Theta}{du} = \frac{-\tan\alpha}{\sqrt{1 - u^2\tan^2\alpha}}.$$

Finally, the last equation can be integrated to give

$$\Theta + \beta = \cos^{-1}(u\tan\alpha),$$

where β is a constant of integration. Hence, using (4.3.20), we find that

$$\Theta(\phi) = -\beta + \cos^{-1}\left(\frac{\tan\alpha}{\tan\phi}\right), \tag{4.3.21}$$

where the constants α and β can be chosen so that the resulting extremum curve (4.3.15) connects the given two points P_0 and P_1 on S.

It is convenient in this case to transform the extremum relation (4.3.21) directly back into Cartesian coordinates. To this end we write (4.3.21) as

$$\cos[\Theta(\phi) + \beta] = \frac{\tan\alpha}{\tan\phi},$$

and we then multiply this equation on both sides by $a\sin\phi$. In this way,

†See Charles Fox, *An Introduction to the Calculus of Variations* (London: Oxford University Press, 1954), pp. 22–23.

‡*Ibid.*

with (4.3.15) we easily find that

$$x \cos \beta - y \sin \beta = z \tan \alpha, \tag{4.3.22}$$

which must hold for any point $P = (x, y, z)$ on the geodesic curve γ connecting P_0 and P_1. Equation (4.3.22) is the equation of a *plane passing through the origin* in 3-space, and it follows then that *any geodesic curve on a sphere must lie along the intersection of the sphere and a plane through its center.* Hence the geodesic curves are arcs of **great circles** on the sphere.

The *minimum distance* between P_0 and P_1 on the sphere can now be easily calculated. Indeed, the geodesic curve connecting P_0 and P_1 will be a suitable arc of the circle of radius a centered at the center of the sphere and lying in the plane passing through P_0, P_1 and the center of the sphere. The length of such a plane circular arc is given by

$$L = a\zeta, \tag{4.3.23}$$

where a is the radius of the circle and ζ is the central angle in radians determined by the end points $P_0 = (x_0, y_0, z_0)$ and $P_1 = (x_1, y_1, z_1)$ of the circular arc, as shown in Figure 17. In the present case the central angle ζ will be determined by

$$\cos \zeta = \frac{x_0 x_1 + y_0 y_1 + z_0 z_1}{a^2}, \qquad 0 \leq \zeta \leq \pi, \tag{4.3.24}$$

which follows directly from (4.3.12) in the special case in which P_0 and P_1 both lie in the (x, y)-plane $z = 0$ (i.e., $z_0 = 0$ and $z_1 = 0$); the general case can be reduced to this special case by rotating the sphere so as to place the points P_0 and P_1 in the plane $z = 0$. Hence the length of the geodesic arc connecting P_0 and P_1 on the sphere is given by (4.3.23) and (4.3.24).

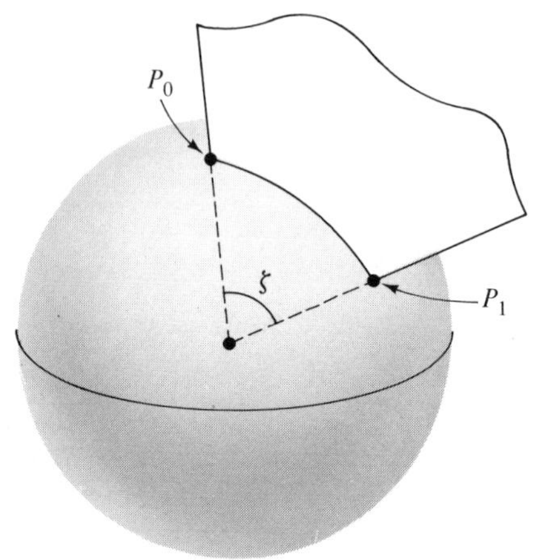

Figure 17

Exercise

1. Let S be the right circular *cone* with constant slope m given parametrically in terms of plane polar coordinates r and θ as

$$S: \begin{cases} x = r \cos \theta \\ y = r \sin \theta \\ z = mr \quad \text{for } 0 \leq \theta \leq 2\pi, \ r \geq 0, \end{cases}$$

and let γ be a curve on S given parametrically in terms of θ as

$$\gamma: \begin{cases} x = R(\theta)\cos\theta \\ y = R(\theta)\sin\theta \\ z = mR(\theta) \qquad \text{for } \theta_0 \leq \theta \leq \theta_1 \end{cases}$$

for some suitable function $R(\theta)$ which relates r and θ along γ. If γ is a geodesic curve on S, show that the function $R = R(\theta)$ must satisfy the differential equation $(1 + m^2)r_0^2 R'(\theta)^2 = R(\theta)^2[R(\theta)^2 - r_0^2]$ for some suitable constant r_0 $[0 \leq r_0 \leq R(\theta)]$, where $R'(\theta) = dR(\theta)/d\theta$. Show how to integrate this equation to get

$$R(\theta) = \frac{r_0}{\sin[(\theta + \beta)/\sqrt{1 + m^2}]}$$

for any geodesic curve which winds around and up the cone [with $R(\theta)$ *nonconstant*], where r_0 and β are constants of integration which can be chosen so as to obtain the particular geodesic curve connecting any two suitable points P_0 and P_1 on S.

4.4. Problems with Variable End Points

The functional $J = J(Y)$ of (4.1.1) can be considered to be a function which is defined for certain *curves* γ in the (x, y)-plane, where γ is to be given in the form

$$\gamma: \ y = Y(x), \qquad x_0 \leq x \leq x_1, \tag{4.4.1}$$

for some suitable function $Y(x)$. In Section 4.1 we studied the case where all such curves γ considered were required to connect two *given fixed* points $P_0 = (x_0, y_0)$ and $P_1 = (x_1, y_1)$.

It sometimes happens that one wishes to maximize or minimize a functional of the type (4.1.1) for all curves γ with one or both end points of γ allowed to vary on some given locus of points. For example, we might want to choose the steering angle of a boat so as to *minimize the transit time* of the boat crossing a river from a given initial point P_0 *among all possible paths* γ *terminating on the other side of the river*, as shown in Figure 18. In this case the terminal point P_1 is *not* given but rather must be determined along with the curve $y = Y(x)$ connecting it with P_0 so as to give the least transit time among *all* paths starting at P_0.† Or, as another example, we might consider a modified form of John Bernoulli's brachistochrone problem in which the initial point $P_0 = (x_0, y_0)$ remains fixed, while the terminal point P_1 is only required to have a specified x-coordinate given as $x = x_1$, as illustrated in Figure 19. The problem would be to find the curve (4.4.1) giving a minimum

†The solution to this steering problem may be intuitively obvious to some readers. See Exercise 3 at the end of this section.

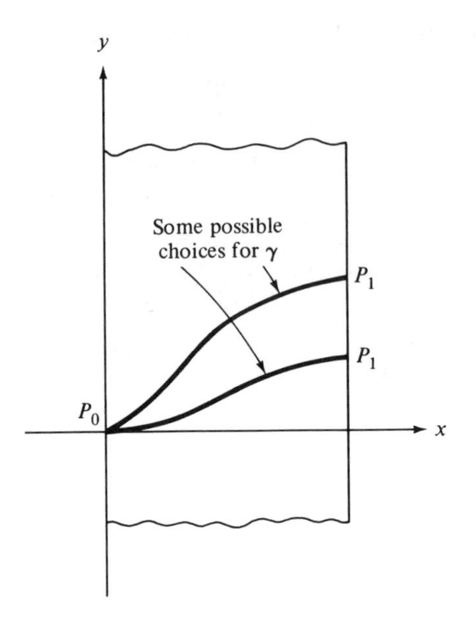

Figure 18

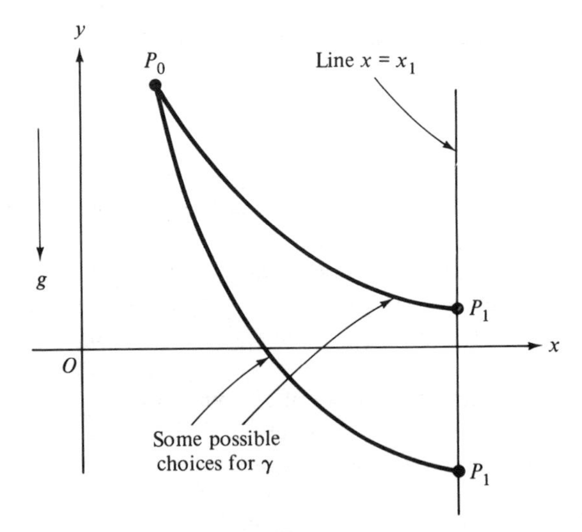

Figure 19

value to the functional $T = T(Y)$ of (1.3.3) subject to the single constraint [compare with (4.2.2)]

$$Y(x_0) = y_0, \tag{4.4.2}$$

where it is required also that γ terminate on the line C given as

$$C: \quad x = x_1, \qquad \text{all } y. \tag{4.4.3}$$

This modified brachistochrone problem was posed by James Bernoulli in 1697, one year after his younger brother John had challenged the mathematicians of the day to solve the fixed end-point brachistochrone problem. We call this modified problem **James Bernoulli's brachistochrone problem.**

More generally the problem may be to minimize the time functional T of (1.3.3) among all curves γ starting at $P_0 = (x_0, y_0)$ and ending on some specified line C given as

$$C: \quad y = ax + b, \qquad \text{all } x, \tag{4.4.4}$$

where a and b are given numbers (see Figure 20). In the latter case the x-coordinate of the terminal point $P_1 = (x_1, y_1)$ is not even specified. It is only required that P_1 satisfy the relation $y_1 = ax_1 + b$.

Even more generally the problem may be to minimize or maximize the functional J of (4.1.1) among all curves γ starting at $P_0 = (x_0, y_0)$ and terminating on some given arbitrary curve C, as shown in Figure 21, where the curve C might be described in the form

$$C: \quad \Phi(x, y) = 0, \tag{4.4.5}$$

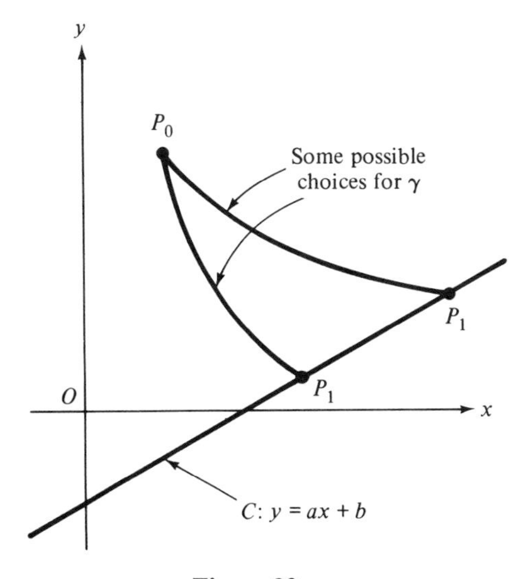

Figure 20

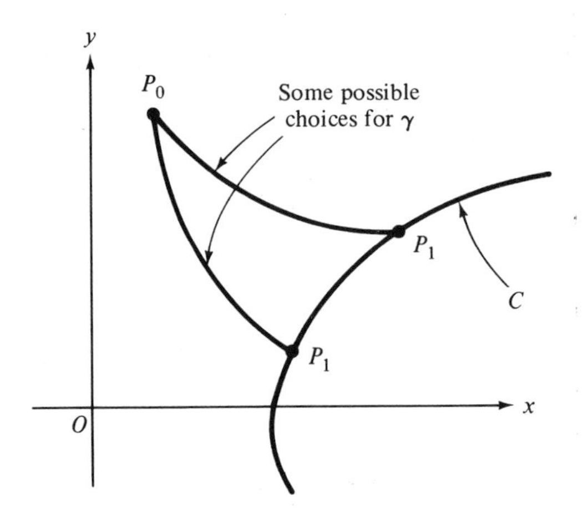

Figure 21

where Φ is a given function of x and y. For example, (4.4.5) includes (4.4.3) if we define Φ by

$$\Phi(x, y) = x - x_1,$$

while (4.4.5) includes (4.4.4) if we take

$$\Phi(x, y) = y - ax - b.$$

Hence we consider the general problem of minimizing or maximizing the functional

$$J = \int_{x_0}^{x_1} F(x, Y(x), Y'(x))\, dx \tag{4.4.6}$$

among all curves γ given by (4.4.1) which satisfy the initial constraint (4.4.2) and the terminal constraint [see (4.4.1) and (4.4.5)]

$$\Phi(x_1, Y(x_1)) = 0. \tag{4.4.7}$$

Since in this case the terminal x-coordinate x_1 appearing in (4.4.6) is in general free to vary provided only that (4.4.7) holds, we must consider the functional (4.4.6) to be a function of x_1 as well as a function of $Y = Y(x)$. (Note, however, that x_0 is fixed throughout.) Hence we take as our basic vector space $\mathfrak{X}$ the set of all pairs $(x_1, Y) = (x_1, Y(x))$, *where x_1 may be any arbitrary number in the vector space $\mathfrak{R}$ of all numbers and where $Y = Y(x)$ may be any arbitrary continuously differentiable function on $\mathfrak{R}$ which vanishes for all large $|x|$*. [It is convenient to require $Y(x)$ to vanish for large $|x|$ so that the right-hand side of (4.4.8) below will be finite.] If (x_1, Y) and (x_1^*, Y^*) are any two vectors in $\mathfrak{X}$, then we define their sum by

$$(x_1, Y) + (x_1^*, Y^*) = (x_1 + x_1^*, Y + Y^*),$$

which again gives a vector in $\mathfrak{X}$. Similarly, we define the product $a(x_1, Y)$ by

$$a(x_1, Y) = (ax_1, aY),$$

which gives a vector in $\mathfrak{X}$ for any number a in $\mathcal{R}$ and any vector (x_1, Y) in $\mathfrak{X}$. One easily checks that all the rules listed in Section 1.2 for vector spaces are satisfied by $\mathfrak{X}$ with these definitions of addition and multiplication by numbers. Finally, we equip $\mathfrak{X}$ with the norm $\|\cdot\|$ defined by

$$\|(x_1, Y)\| = |x_1| + \max_{\text{all } x \text{ in } \mathcal{R}} |Y(x)| + \max_{\text{all } x \text{ in } \mathcal{R}} |Y'(x)| \tag{4.4.8}$$

for any vector $(x_1, Y) = (x_1, Y(x))$ in $\mathfrak{X}$; one easily checks that this function $\|\cdot\|$ satisfies all the properties of a norm on $\mathfrak{X}$ as given in Section 1.4. We emphasize that the normed vector space $\mathfrak{X}$ consists of *all* pairs (x_1, Y) for all numbers x_1 and for all continuously differentiable functions $Y = Y(x)$. There need *not* be any special relation between the number x_1 and the continuously differentiable function $Y = Y(x)$ occurring in a pair (x_1, Y). In particular, every such pair is in $\mathfrak{X}$ whether or not the pair satisfies the relation (4.4.7).

The extremum problem considered for the functional (4.4.6) can now be stated as follows. We seek a vector $(x_1, Y) = (x_1, Y(x))$ in some given open set D in $\mathfrak{X}$ that will maximize or minimize in D the functional J defined as

$$J(x_1, Y) = \int_{x_0}^{x_1} F(x, Y(x), Y'(x))\,dx, \tag{4.4.9}$$

where the admissible vectors $(x_1, Y) = (x_1, Y(x))$ are also required to satisfy the constraints of (4.4.2) and (4.4.7). The domain D of the functional (4.4.9) is taken to be some specified open subset of the normed vector space $\mathfrak{X}$. For example, D might consist of all vectors (x_1, Y) in $\mathfrak{X}$ with x_1 restricted to lie in some given open interval in $\mathcal{R}$ and with $Y = Y(x)$ restricted to be in some suitable open set in the normed vector space $\mathcal{C}^1(\mathcal{R})$.

If we define functionals K_0 and K_1 on D by

$$K_0(x_1, Y) = Y(x_0) \tag{4.4.10}$$

and

$$K_1(x_1, Y) = \Phi(x_1, Y(x_1)), \tag{4.4.11}$$

then the present problem is to find extremum vectors in $D[K_0 = y_0, K_1 = 0]$ for the functional J of (4.4.9).† We emphasize that the number x_0 appearing in (4.4.9) and (4.4.10) is *given* and kept fixed throughout, although we could consider a more general problem in which x_0 as well as x_1 were allowed to vary. [For such a more general problem we would take the vector space $\mathfrak{X}$ to be the set of all *triples* (x_0, x_1, Y), where x_0 and x_1 could be any numbers

†This approach to variable end-point problems was developed by the author in 1967.

and where $Y = Y(x)$ could be any suitable continuously differentiable function.]

To proceed in the present case, we need to calculate the variations of the functionals J, K_0, and K_1, where we allow x_1 as well as $Y = Y(x)$ to vary (but not x_0). From (2.3.4) we have

$$\delta J(x_1, Y; \Delta x_1, \Delta Y) = \frac{d}{d\epsilon} J(x_1 + \epsilon \Delta x_1, Y + \epsilon \Delta Y)\Big|_{\epsilon=0} \tag{4.4.12}$$

and a similar equation for K_0 and K_1, where $(\Delta x_1, \Delta Y)$ may be *any* vector in the vector space $\mathfrak{X}$ and $(x_1, Y) + \epsilon(\Delta x_1, \Delta Y) = (x_1 + \epsilon \Delta x_1, Y + \epsilon \Delta Y)$ for any two vectors (x_1, Y) and $(\Delta x_1, \Delta Y)$ and for any number ϵ. We use (4.4.9) to find that

$$J(x_1 + \epsilon \Delta x_1, Y + \epsilon \Delta Y) = \int_{x_0}^{x_1 + \epsilon \Delta x_1} F(x, Y(x) + \epsilon \Delta Y(x), Y'(x) + \epsilon \Delta Y'(x))dx. \tag{4.4.13}$$

The required derivative of this expression with respect to the parameter ϵ [see (4.4.12)] can be calculated using [compare with (2.4.5)]

$$\frac{d}{d\epsilon} \int_{x_0}^{\xi(\epsilon)} f(x;\epsilon)\,dx = f(\xi(\epsilon);\epsilon)\xi'(\epsilon) + \int_{x_0}^{\xi(\epsilon)} \frac{\partial f(x;\epsilon)}{\partial \epsilon}\,dx, \tag{4.4.14}$$

which holds for any smooth function $f = f(x;\epsilon)$ and any continuously differentiable function $\xi = \xi(\epsilon)$ depending on ϵ, where $\xi'(\epsilon) = d\xi(\epsilon)/d\epsilon$. [We include a proof of equation (4.4.14) in Section A6 of the Appendix.] If we take $\xi(\epsilon) = x_1 + \epsilon \Delta x_1$ and $f(x;\epsilon) = F(x, Y(x) + \epsilon \Delta Y(x), \; Y'(x) + \epsilon \Delta Y'(x))$ in (4.4.14), we find with (4.4.12) and (4.4.13) that (explain)

$$\delta J(x_1, Y; \Delta x_1, \Delta Y) = F(x_1, Y(x_1), Y'(x_1))\,\Delta x_1 + \int_{x_0}^{x_1} \frac{\partial}{\partial \epsilon} F(x, Y(x) + \epsilon \Delta Y(x), Y'(x) + \epsilon \Delta Y'(x))\Big|_{\epsilon=0} dx,$$

which with (2.4.15) implies that

$$\delta J(x_1, Y; \Delta x_1, \Delta Y) = F(x_1, Y(x_1), Y'(x_1))\,\Delta x_1 + \int_{x_0}^{x_1} [F_Y(x, Y(x), Y'(x))\,\Delta Y(x) + F_{Y'}(x, Y(x), Y'(x))\,\Delta Y'(x)]\,dx \tag{4.4.15}$$

for any vector $(\Delta x_1, \Delta Y)$ in the vector space $\mathfrak{X}$.

Now that we have the variation of J in hand, we turn to the functionals K_0 and K_1. Similarly to (4.4.12), for K_0 we have

$$\delta K_0(x_1, Y; \Delta x_1, \Delta Y) = \frac{d}{d\epsilon} K_0(x_1 + \epsilon \Delta x_1, Y + \epsilon \Delta Y)\Big|_{\epsilon=0},$$

and since (4.4.10) implies that

$$K_0(x_1 + \epsilon\,\Delta x_1,\ Y + \epsilon\,\Delta Y) = Y(x_0) + \epsilon\,\Delta Y(x_0),$$

we find that

$$\delta K_0(x_1,\ Y; \Delta x_1, \Delta Y) = \Delta Y(x_0) \tag{4.4.16}$$

for any vector $(\Delta x_1, \Delta Y)$ in $\mathfrak{X}$. [It is important that the reader understand fully the derivation of equation (4.4.16).] To calculate the variation of K_1, we again have

$$\delta K_1(x_1, Y; \Delta x_1, \Delta Y) = \frac{d}{d\epsilon} K_1(x_1 + \epsilon\,\Delta x_1, Y + \epsilon\,\Delta Y)\bigg|_{\epsilon=0}, \tag{4.4.17}$$

where in this case we find from (4.4.11) that

$$\begin{aligned}&K_1(x_1 + \epsilon\,\Delta x_1, Y + \epsilon\,\Delta Y)\\ &\quad = \Phi(x_1 + \epsilon\,\Delta x_1, Y(x_1 + \epsilon\,\Delta x_1) + \epsilon\,\Delta Y(x_1 + \epsilon\,\Delta x_1)).\end{aligned}$$

[It is important to understand the last equation.] If $x = x(\epsilon)$ and $y = y(\epsilon)$ are any two continuously differentiable functions of the parameter ϵ, then for any continuously differentiable function $\Phi = \Phi(x, y)$ the chain rule of differential calculus gives

$$\frac{d}{d\epsilon}\Phi(x(\epsilon), y(\epsilon)) = \frac{\partial}{\partial x}\Phi(x(\epsilon), y(\epsilon))\frac{dx}{d\epsilon} + \frac{\partial}{\partial y}\Phi(x(\epsilon), y(\epsilon))\frac{dy}{d\epsilon}.$$

If we take

$$\begin{aligned} x(\epsilon) &= x_1 + \epsilon\,\Delta x_1\\ y(\epsilon) &= Y(x_1 + \epsilon\,\Delta x_1) + \epsilon\,\Delta Y(x_1 + \epsilon\,\Delta x_1)\end{aligned} \tag{4.4.18}$$

with (explain)

$$\frac{dx}{d\epsilon} = \Delta x_1$$

$$\frac{dy}{d\epsilon} = Y'(x_1 + \epsilon\,\Delta x_1)\,\Delta x_1 + \Delta Y(x_1 + \epsilon\,\Delta x_1) + \epsilon\,\Delta Y'(x_1 + \epsilon\,\Delta x_1)\Delta x_1,$$

then with the last few equations we find that

$$\begin{aligned}\frac{d}{d\epsilon}K_1(x_1 + \epsilon\,\Delta x_1, Y + \epsilon\,\Delta Y) &= \frac{\partial}{\partial x}\Phi(x(\epsilon), y(\epsilon))\,\Delta x_1\\ &\quad + \frac{\partial}{\partial y}\Phi(x(\epsilon), y(\epsilon))[Y'(x_1 + \epsilon\,\Delta x_1)\,\Delta x_1 + \Delta Y(x_1 + \epsilon\,\Delta x_1)\\ &\quad + \epsilon\,\Delta Y'(x_1 + \epsilon\,\Delta x_1)\,\Delta x_1],\end{aligned}$$

where $x(\epsilon)$ and $y(\epsilon)$ are given by (4.4.18). If we now put $\epsilon = 0$ in the last equation, we find with (4.4.17) and (4.4.18) that

$$\delta K_1(x_1, Y; \Delta x_1, \Delta Y) = \frac{\partial}{\partial x}\Phi(x_1, Y(x_1))\,\Delta x_1 + \frac{\partial}{\partial y}\Phi(x_1, Y(x_1))[Y'(x_1)\,\Delta x_1 + \Delta Y(x_1)] \tag{4.4.19}$$

for any vector $(\Delta x_1, \Delta Y)$ in the vector space $\mathfrak{X}$.

Suppose now that the vector $(x_1, Y) = (x_1, Y(x))$ is a local extremum vector in $D[K_0 = y_0, K_1 = 0]$ for J. By the Euler-Lagrange multiplier theorem of Section 3.6 we find the necessary condition

$$\delta J(x_1, Y; \Delta x_1, \Delta Y) = \lambda_0\,\delta K_0(x_1, Y; \Delta x_1, \Delta Y) + \lambda_1\,\delta K_1(x_1 Y; \Delta x_1, \Delta Y) \tag{4.4.20}$$

for suitable constants λ_0 and λ_1, *for all numbers* Δx_1, *and for all continuously differentiable functions* $\Delta Y = \Delta Y(x)$ on the interval $x_0 \le x \le x_1$. [Every such function $\Delta Y(x)$ can be *extended* so as to be defined and continuously differentiable on $\mathfrak{R}$ and so as to vanish for all large $|x|$; hence $(\Delta x_1, \Delta Y)$ may be considered to be in $\mathfrak{X}$.] If we use (4.4.15), (4.1.8), (4.4.16), and (4.4.19), we may rewrite (4.4.20) as [compare with (4.1.9)]

$$\begin{aligned}\int_{x_0}^{x_1} &[F_Y(x, Y(x), Y'(x)) - \frac{d}{dx}F_{Y'}(x, Y(x), Y'(x))]\,\Delta Y(x)\,dx \\ &= [\lambda_0 + F_{Y'}(x_0, y_0, Y'(x_0))]\,\Delta Y(x_0) \\ &\quad + \left[\lambda_1 \frac{\partial}{\partial y}\Phi(x_1, Y(x_1)) - F_{Y'}(x_1, Y(x_1), Y'(x_1))\right]\Delta Y(x_1) \\ &\quad + \Big\{-F(x_1, Y(x_1), Y'(x_1)) + \lambda_1\Big[\frac{\partial}{\partial x}\Phi(x_1, Y(x_1)) \\ &\quad + Y'(x_1)\frac{\partial}{\partial y}\Phi(x_1, Y(x_1))\Big]\Big\}\,\Delta x_1,\end{aligned} \tag{4.4.21}$$

which must hold for all numbers Δx_1 *and all continuously differentiable functions* $\Delta Y = \Delta Y(x)$ if $(x_1, Y) = (x_1, Y(x))$ is a local extremum vector in $D[K_0 = y_0, K_1 = 0]$ for J. In (4.4.21) we have used the constraint (4.4.2) which must be satisfied by any such extremum vector (x_1, Y).

If we first choose $\Delta x_1 = 0$ in (4.4.21) and consider only functions $\Delta Y(x)$ which vanish at the end points x_0 and x_1 (it is important to understand that this is permitted), we find the condition

$$\int_{x_0}^{x_1} [F_Y(x, Y(x), Y'(x)) - \frac{d}{dx}F_{Y'}(x, Y(x), Y'(x))]\,\Delta Y(x)\,dx = 0, \tag{4.4.22}$$

which must hold for all continuously differentiable functions $\Delta Y(x)$ on $[x_0, x_1]$ satisfying $\Delta Y(x_0) = 0$, $\Delta Y(x_1) = 0$. Similarly, as in Section 4.1, it follows necessarily from the last result and the lemma of Section A4 of the Appendix that *the Euler-Lagrange equation* (4.1.11) *must be satisfied by the function* $Y = Y(x)$ for $x_0 < x < x_1$ if $(x_1, Y(x))$ is a local extremum vector in $D[K_0 = y_0, K_1 = 0]$ for J.

But then if we use the Euler-Lagrange equation (4.1.11) in (4.4.21), it follows that

$$\begin{aligned} 0 = {} & [\lambda_0 + F_{Y'}(x_0, y_0, Y'(x_0))]\,\Delta Y(x_0) \\ & + \left[\lambda_1 \frac{\partial}{\partial y}\Phi(x_1, Y(x_1)) - F_{Y'}(x_1, Y(x_1), Y'(x_1))\right]\Delta Y(x_1) \\ & + \Bigg\{-F(x_1, Y(x_1), Y'(x_1)) \\ & + \lambda_1\left[\frac{\partial}{\partial x}\Phi(x_1, Y(x_1)) + Y'(x_1)\frac{\partial}{\partial y}\Phi(x_1, Y(x_1))\right]\Bigg\}\,\Delta x_1 \end{aligned} \tag{4.4.23}$$

must also hold for all numbers Δx_1 and for all continuously differentiable functions $\Delta Y(x)$ on $[x_0, x_1]$. [It is important that the reader understand the validity of (4.4.23).] If we set $\Delta x_1 = 0$ and first take

$$\Delta Y(x) = \frac{x_1 - x}{x_1 - x_0}$$

in (4.4.23) and then

$$\Delta Y(x) = \frac{x - x_0}{x_1 - x_0},$$

we find in turn the following necessary conditions:

$$\lambda_0 + F_{Y'}(x_0, y_0, Y'(x_0)) = 0$$

and

$$\lambda_1 \frac{\partial}{\partial y}\Phi(x_1, Y(x_1)) - F_{Y'}(x_1, Y(x_1), Y'(x_1)) = 0. \tag{4.4.24}$$

Using the last two results in (4.4.23) and taking $\Delta x_1 = 1$, we also find the following additional condition:

$$\begin{aligned} F(x_1, Y(x_1), Y'(x_1)) = \lambda_1 \Bigg[& \frac{\partial}{\partial x}\Phi(x_1, Y(x_1)) \\ & + Y'(x_1)\frac{\partial}{\partial y}\Phi(x_1, Y(x_1))\Bigg]. \end{aligned} \tag{4.4.25}$$

We can eliminate the Euler-Lagrange multiplier λ_1 between (4.4.24) and (4.4.25) by multiplying the latter equation on both sides by $\partial\Phi/\partial y$ and then using the former equation to find that

$$\frac{\partial}{\partial y}\Phi(x_1, Y(x_1))F(x_1, Y(x_1), Y'(x_1))$$
$$= F_{Y'}(x_1, Y(x_1), Y'(x_1))\left[\frac{\partial}{\partial x}\Phi(x_1, Y(x_1)) + Y'(x_1)\frac{\partial}{\partial y}\Phi(x_1, Y(x_1))\right] \tag{4.4.26}$$

which is then a **natural boundary condition** which must hold at the variable end-point $x = x_1$ for any local extremum vector $(x_1, Y(x))$. This natural boundary condition has a simple geometric interpretation in the special case in which the integrand function $F = F(x, y, z)$ of (4.4.9) has the form

$$F(x, y, z) = f(x, y)\sqrt{1 + z^2} \tag{4.4.27}$$

for some given function $f = f(x, y)$. (For example, the integrand function of the brachistochrone functional (1.3.3) is of this form, with $f(x, y) = [2g(y_0 - y)]^{-1/2}$.) If (4.4.27) holds, then according to (2.4.16), we find that

$$F_{Y'}(x, Y(x), Y'(x)) = \frac{f(x, Y(x))Y'(x)}{\sqrt{1 + Y'(x)^2}},$$

so that (4.4.26) simplifies with (4.4.27) to give

$$\frac{\partial}{\partial y}\Phi(x_1, Y(x_1)) = Y'(x_1)\frac{\partial}{\partial x}\Phi(x_1, Y(x_1)), \tag{4.4.28}$$

which must hold at the end point $x = x_1$ [except possibly if $f(x_1, Y(x_1)) = 0$; explain]. Condition (4.4.28) simply requires that *the extremum curve* γ *of* (4.4.1) *must intersect the given curve C of* (4.4.5) *orthogonally* (at right angles), as indicated in Figure 22. Indeed, it is known from differential calculus that the slope $y' = y'_C$ of the curve C of (4.4.5) is given as

$$y'_C = -\frac{\partial\Phi/\partial x}{\partial\Phi/\partial y}$$

[as can be checked by differentiating (4.4.5) implicitly with respect to x], so that (4.4.28) requires the slope $y' = Y'(x)$ of the extremum curve γ to satisfy

$$Y'(x_1) = -\frac{1}{y'_C(x_1)} \tag{4.4.29}$$

from which the asserted result follows.†

We should mention that if the terminal curve C happens to be a vertical line with $\Phi(x, y) = x - x_1$ in (4.4.5) for some *given fixed* x_1, then we must

†See Thomas, *op. cit.*, p. 8.

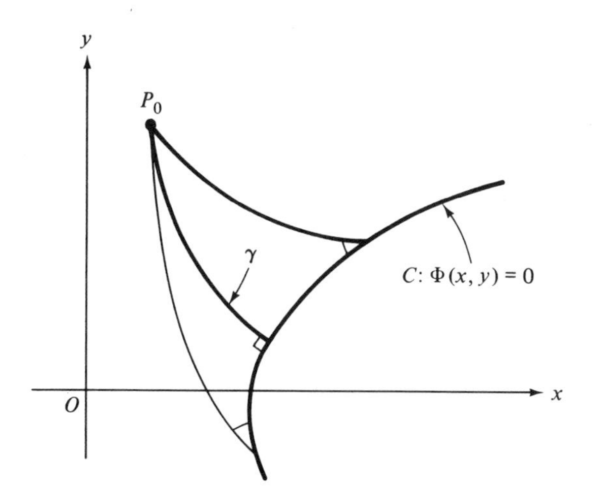

Figure 22

consider the functional J of (4.4.9) to be a function only of $Y = Y(x)$ on the *fixed* interval $x_0 \leq x \leq x_1$. In this case the resulting natural boundary condition at the terminal point is also found to be given by (4.4.26). The calculation is similar to the calculation in Section 4.1 for the fixed end-point problem except that the multiplier λ_1 of (4.1.12) must be set equal to zero (why?). This gives the natural boundary condition

$$F_{Y'}(x_1, Y(x_1), Y'(x_1)) = 0 \tag{4.4.30}$$

at $x = x_1$, which agrees with (4.4.26) in this case since $\Phi(x, y) = x - x_1$ is independent of y.

Summarizing, we have shown that in searching for those curves γ of the form (4.4.1) which minimize or maximize the functional J among all such curves starting at $P_0 = (x_0, y_0)$ and ending on the curve C given by (4.4.5), *we need only consider those curves γ which satisfy the Euler-Lagrange equation* (4.1.11) *for $x_0 < x < x_1$, along with the constraints*

$$Y(x_0) = y_0$$

and

$$\Phi(x_1, Y(x_1)) = 0,$$

and the natural boundary condition (4.4.26) [*or* (4.4.28) *in the special case* (4.4.27)].

Example. To illustrate the situation, we consider the problem of minimizing the functional

$$J(x_1, Y) = \int_5^{x_1} \frac{\sqrt{1 + Y'(x)^2}}{Y(x)}\, dx \tag{4.4.31}$$

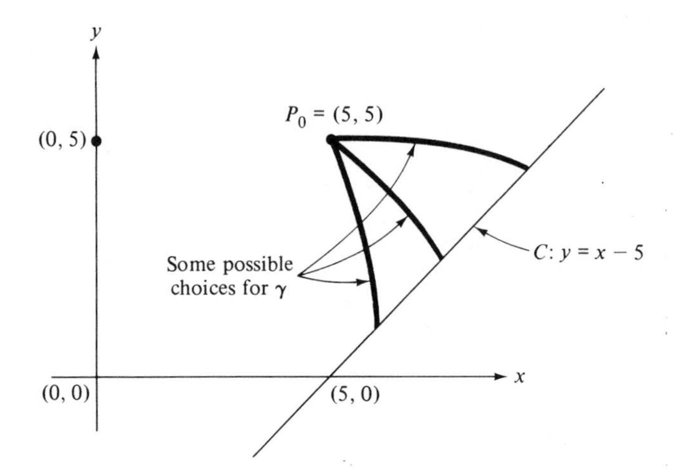

Figure 23

among all curves γ given as

$$\gamma: \quad y = Y(x), \qquad 5 \leq x \leq x_1$$

which join the point $P_0 = (5, 5)$ to the line C defined as

$$C: \quad y = x - 5. \tag{4.4.32}$$

The situation is shown in Figure 23. The functional J of (4.4.31) is of the form (4.4.9) with $x_0 = 5$ and

$$F(x, y, z) = \frac{\sqrt{1 + z^2}}{y}. \tag{4.4.33}$$

If the vector (x_1, Y) furnishes a local minimum in $D[K_0 = y_0, K_1 = 0]$ for J, it follows as above from (4.4.22) that $Y = Y(x)$ must satisfy the Euler-Lagrange equation (4.1.11) for $5 < x < x_1$. Since the function F of (4.4.33) does not depend on x, we can use equation (4.1.15) instead of (4.1.11). We find from (4.4.33) and (2.4.16) that

$$F_{Y'}(Y(x), Y'(x)) = \frac{Y'(x)}{Y(x)\sqrt{1 + Y'(x)^2}},$$

so that in this case (4.1.15) gives the differential equation (explain)

$$Y(x)^2[1 + Y'(x)^2] = A^2$$

for some constant A. Solving for $Y'(x)$, we find that

$$Y'(x) = \pm\frac{\sqrt{A^2 - Y(x)^2}}{Y(x)},$$

which can be integrated by separation of variables as

$$\int \frac{Y(x)Y'(x)\,dx}{\sqrt{A^2 - Y(x)^2}} = \pm \int dx.$$

If we make the substitution $Y(x) = y$ in the integral on the left-hand side of the last result, we find that

$$\int \frac{y\,dy}{\sqrt{A^2 - y^2}} = \pm \int dx,$$

which may be evaluated to give

$$y^2 + (x - B)^2 = A^2$$

for some constant B. Since $y = Y(x)$, we find

$$Y(x) = \sqrt{A^2 - (x - B)^2}, \qquad (4.4.34)$$

which must be satisfied by any possible extremum function $Y = Y(x)$. The two constants of integration A and B and the unknown abscissa x_1 of the terminal point must be chosen so that the resulting curve γ starts at $P_0 = (5, 5)$, terminates on C, and satisfies the required natural boundary condition (4.4.26). In particular, we have taken the *positive* square root in (4.4.34) so as to make the condition $Y(5) = +5$ possible at the point P_0.

If the curve γ is to start at $P_0 = (x_0, y_0) = (5, 5)$, we must require that [see (4.4.34)]

$$A^2 = 25 + (5 - B)^2$$

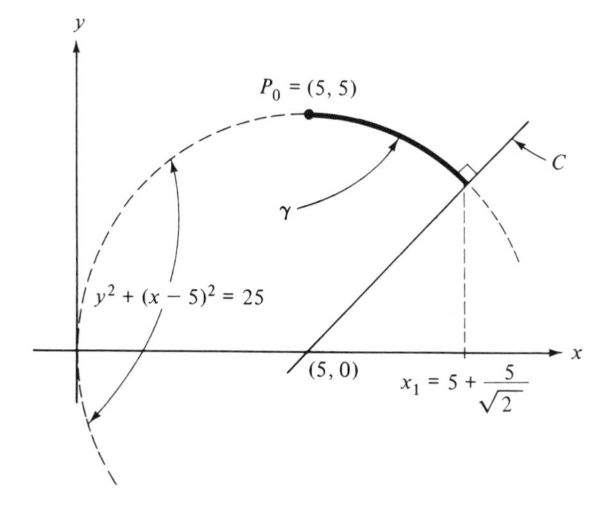

Figure 24

hold, so that (4.4.34) becomes

$$Y(x) = \sqrt{25 + (5 - B)^2 - (x - B)^2}. \tag{4.4.35}$$

On the other hand, we must require that

$$Y(x_1) = x_1 - 5 \tag{4.4.36}$$

in order that γ will terminate on C at $x = x_1$ [see (4.4.32)]. Differentiating (4.4.35), we find that

$$Y'(x) = -\frac{x - B}{Y(x)},$$

which with (4.4.36) gives at $x = x_1$ the condition

$$Y'(x_1) = -\frac{x_1 - B}{x_1 - 5}. \tag{4.4.37}$$

Now the integrand function F of (4.4.33) is of the special form (4.4.27), and so the natural boundary condition (4.4.26) requires that γ must intersect C at right angles [see (4.4.29)]. Since C has constant slope $y'_C = +1$, it follows that any extremum curve γ must intersect C with slope -1,

$$Y'(x_1) = -1.$$

Comparing this result with (4.4.37), we find that we must necessarily make the choice $B = 5$, so that (4.4.35) becomes

$$Y(x) = \sqrt{25 - (x - 5)^2}. \tag{4.4.38}$$

Using (4.4.36) and (4.4.38), for x_1 we find that

$$x_1 = 5 + \frac{5}{\sqrt{2}}. \tag{4.4.39}$$

Hence the extremum curve γ is an arc of the *circle* of radius 5 centered at the point $(x, y) = (5, 0)$, as shown in Figure 24. It can be shown (but we shall not show it here) that this curve γ actually furnishes a minimum value to the functional (4.4.31) among all admissible curves starting at P_0 and terminating on C. The resulting minimum value of J is found from (4.4.31), (4.4.38), and (4.4.39) to be

$$J_{\text{minimum}} = \tfrac{1}{2}\log(3 + 2\sqrt{2}).$$

James Bernoulli's Brachistochrone Problem. In closing this section we shall consider again the brachistochrone problem of James Bernoulli described in the second paragraph at the beginning of this section (see Figure 19). We seek the shape of a curved wire given by (4.4.1) connecting a given point $P_0 = (x_0, y_0)$ with the line $x = x_1$ and along which a bead will slide in minimum time under the force of gravity. Although the abscissa x_1 of the terminal point $P_1 = (x_1, y_1)$ is given, the ordinate y_1 is left *unspecified* so that the system of equations (4.2.8) which was used earlier to specify the solution of John Bernoulli's problem *cannot be used* in this case to determine the extremum curve γ.

In fact, James Bernoulli's brachistochrone problem is just of the type considered here with the terminal curve C described as

$$C\colon \quad \Phi(x, y) = 0$$

with $\Phi(x, y) = x - x_1$. As in John Bernoulli's problem (see Section 4.2) we find again in this case that *any minimizing curve* γ *of the form* (4.4.1) *must satisfy the Euler-Lagrange equation* (4.1.11) *for* $x_0 \leq x \leq x_1$. Hence [see (4.2.7)]

$$\gamma\colon \quad \begin{cases} x = x_0 + \dfrac{A}{2}(\theta - \sin\theta) \\ y = y_0 - \dfrac{A}{2}(1 - \cos\theta) \qquad \text{for } 0 \leq \theta \leq \theta_1 \end{cases} \tag{4.4.40}$$

must hold, where A and θ_1 are unspecified constants and the parameter θ is as in Section 4.2. In particular, the extremum curve $y = Y(x)$ must again be an arc of a cycloid.

To determine the constants A and θ_1, we impose the constraint

$$x = x_1 \qquad \text{at } \theta = \theta_1 \tag{4.4.41}$$

along with the natural boundary condition (4.4.30). Since this natural boundary condition requires the extremum curve γ to intersect C *orthogonally* [see (4.4.29)], we must require that (why?)

$$Y'(x_1) = 0,$$

where along γ the chain rule of differential calculus gives

$$Y'(x) = \frac{dy/d\theta}{dx/d\theta}. \tag{4.4.42}$$

Hence the natural boundary condition implies that

$$\frac{dy}{d\theta} = 0 \qquad \text{at } \theta = \theta_1,$$

which with the second equation of (4.4.40) leads to

$$\theta_1 = \pi. \tag{4.4.43}$$

The first equation of (4.4.40) with (4.4.41) and (4.4.43) now gives (explain)

$$A = \frac{2(x_1 - x_0)}{\pi}. \tag{4.4.44}$$

Hence among *all* curves connecting $P_0 = (x_0, y_0)$ with the line $x = x_1$, the minimum time will be achieved along that particular cycloid generated by the motion of a fixed point on the circumference of a circle of radius $(x_1 - x_0)/\pi$ which rolls on the underside of the line $y = y_0$, as shown in Figure 25. Any other (single-arched) cycloid dropping down from P_0 will intersect C in an angle different from 90° and will yield a greater time of descent for the sliding bead (as will all other noncycloidal curves). The value of the minimum time achieved is found from (1.3.3), (4.4.40), (4.4.42), (4.4.43), and (4.4.44) to be

$$T_{\text{minimum}} = \int_0^{\pi} \sqrt{\frac{(dx/d\theta)^2 + (dy/d\theta)^2}{2g(y_0 - y)}}\, d\theta = \sqrt{\frac{|x_1 - x_0|\pi}{g}}. \tag{4.4.45}$$

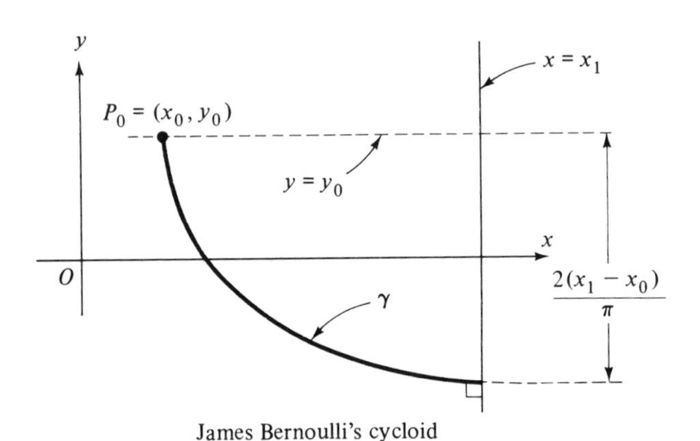

James Bernoulli's cycloid

Figure 25

Exercises

1. Give a careful derivation of the natural boundary condition (4.4.30) for the problem of minimizing the functional $J(Y) = \int_{x_0}^{x_1} F(x, Y(x), Y'(x))dx$ among all continuously differentiable functions $Y = Y(x)$ on the *fixed* interval $x_0 \leq x \leq x_1$ subject only to the single constraint (4.4.2).

2. A boatman wishes to steer his boat so as to minimize the transit time required to cross a river from a fixed initial point, as shown in Figure 18. The situation is described briefly in the second paragraph at the beginning of this section, and the transit time is given by equation (1.3.17). Find the natural boundary condition which must be satisfied at $x = l$ by the extremum curve $\gamma: y = Y(x)$, $0 \leq x \leq l$. Use this natural boundary condition and equation (4.1.20) to show that the extremum curve must satisfy the differential equation $Y'(x) = e(x)$ if (as is customarily the case) the river current speed vanishes at the river banks with [see (1.3.15)]

$$e(0) = e(l) = 0. \tag{4.4.46}$$

3. Use the results of Exercise 2 and the condition (4.4.46) along with equation (4.1.23) to show that the boat will minimize the transit time from P_0 to the opposite bank by steering always *directly toward* the opposite bank, as shown in Figure 26. (Certain animals correctly solve this extremum problem instinctively whenever they swim across a stream.) Show that the resulting minimum transit time is independent of the river current strength and is actually the same as would be obtained by traveling directly across a stagnant stream which has no flow.

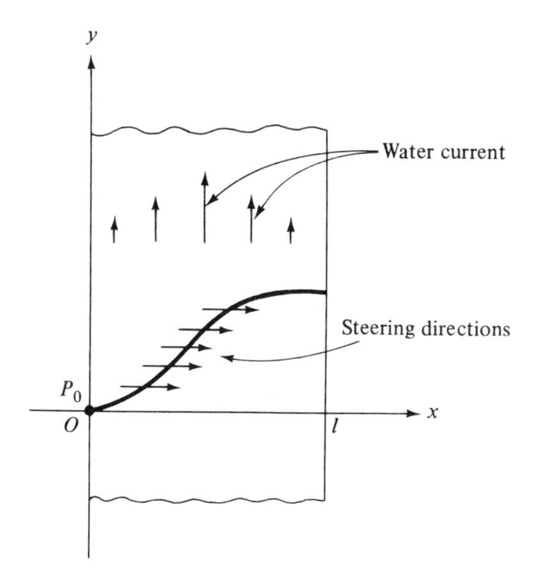

Figure 26

4. A dog is dropped into the ocean at a point labeled as $P_0 = (x_0, y_0) = (4, 4)$, where x and y are Cartesian coordinates measured in the plane surface of the quiet ocean. The region $(x - 9)^2 + y^2 \leq 9$ constitutes an island (see Figure 27). Taking into account the dog's ability and the strength of the ocean currents, it is known that the speed v of the swimming dog at any point (x, y) is given by $v = |y|$. Along what path γ should the dog swim in order to arrive at the island so as to minimize his time T in the water, where $T = \int_\gamma dt = \int_\gamma ds/v$? Show in

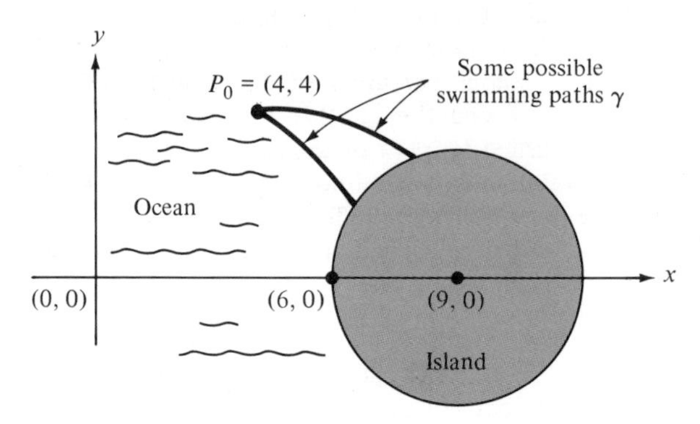

Figure 27

particular that the dog should arrive at the island at the point $P_1 = (x_1, y_1) = (\frac{36}{5}, \frac{12}{5})$ after a transit time in the water given as $T_{\text{minimum}} = \log 3$.

5. Let $P_0 = (x_0, y_0) = (0, y_0)$ be a given *fixed* point on the y-axis, and let $P_1 = (x_1, y_1) = (x_1, 0)$ represent any *variable* point on the x-axis, with $x_1 > 0$ and $y_0 > 0$. Let γ be any curve connecting P_0 and P_1 given as $\gamma\colon y = Y(x)$, $0 \leq x \leq x_1$, as shown in Figure 28. Among all such curves enclosing with the coordinate axes a given constant area A, show that the line $(x/x_1) + (y/y_0) = 1$, $x_1 y_0/2 = A$ generates the *least area* when rotated about the x-axis.† Why does the resulting extremum curve γ *not* intersect the x-axis orthogonally as might seem to be required by the natural boundary condition (4.4.29)? *Hints:* The area enclosed between γ and the coordinate axes is given by the integral $\int_0^{x_1} Y(x)\, dx$, while

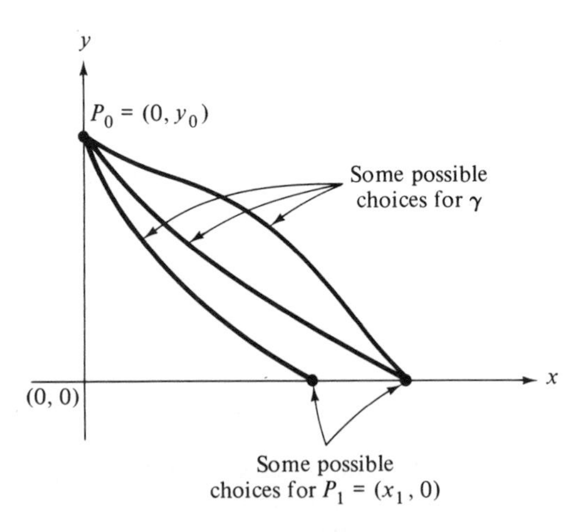

Figure 28

†This problem is related to a special case of one of the many variational problems solved by Euler in 1744. See D. J. Struik, *A Source Book In Mathematics, 1200–1800* (Cambridge, Mass.: Harvard University Press, 1969), p. 406.

the surface area generated when γ is rotated about the x-axis is given as $2\pi \int_0^{x_1} Y(x)\sqrt{1 + Y'(x)^2}\,dx$. You should be able to use the Euler-Lagrange multiplier theorem with three constraints to show that the extremum curve $y = Y(x)$ must satisfy the differential equation $1 + Y'(x)^2 = [Y(x)/(aY(x) + b)]^2$ for some suitable constants a and b. Using this equation and one of the constraints, you should then be able to conclude that $b = 0$, after which the solution is apparent.

6. Find the minimum value of the integral $\int_0^1 [\frac{1}{2}Y'(x)^2 + Y'(x) + Y(x)Y'(x) + Y(x)]\,dx$ among all functions Y of class $\mathcal{C}^1$ on $[0, 1]$ which satisfy the single condition $Y(0) = 0$. Compare the result with that of Exercise 8 of Section 4.1.

7. Find the minimum value of the integral $\int_1^2 x^2 Y'(x)^2\,dx$ subject to the condition $Y(1) = 1$. The solution may be obvious. (Compare with Exercise 6 of Section 4.1.)

8. Find the minimum value of the integral $\int_0^\pi [Y'(x)^2 + 2Y(x)\sin x]\,dx$ subject to the single condition $Y(0) = 0$. Compare the result with that of Exercise 7 of Section 4.1.

9. Among all curves γ that have length l and begin and end on the parabola $y = x^2$, it is desired to find such a curve that bounds the *greatest possible area* between itself and the given parabola.† If γ^* is any such extremum curve, show that γ^* must be an appropriate arc of the circle of radius r centered at the point $(0, -b)$, where r is determined by the condition

$$2r\left(\sin\frac{l}{2r}\right)^2 = \cos\frac{l}{2r} \qquad \text{with } 0 < \frac{l}{2r} \leq \pi \tag{4.4.47}$$

and where b is then given as

$$b = \frac{\sqrt{1 + 16r^2} - 1}{8}. \tag{4.4.48}$$

Hints: You should be able to use the Euler-Lagrange multiplier theorem with three constraints to show that any extremum curve γ given as $\gamma\colon y = Y(x)$, $x_0 \leq x \leq x_1$, must satisfy the differential equation

$$\frac{d}{dx}\left(\frac{Y'(x)}{\sqrt{1 + Y'(x)^2}}\right) = \text{constant} \qquad \text{for } x_0 < x < x_1 \tag{4.4.49}$$

and the natural boundary conditions

$$Y'(x_0) = -\frac{1}{2x_0}, \qquad Y'(x_1) = -\frac{1}{2x_1} \tag{4.4.50}$$

along with the specified constraints

$$Y(x_0) = x_0^2, \qquad Y(x_1) = x_1^2 \tag{4.4.51}$$

†Professor Polya has observed that Queen Dido would have faced a similar problem if she had bargained with natives living on a peninsula. See G. Polya, *Induction and Analogy in Mathematics* (Princeton, N. J.: Princeton University Press, 1954), p. 180.

and

$$\int_{x_0}^{x_1} \sqrt{1 + Y'(x)^2}\, dx = l. \tag{4.4.52}$$

Equation (4.4.49) can be integrated to give $(x + a)^2 + (y + b)^2 = r^2$ for $y = Y(x)$ and for suitable constants a, b, and r. The latter equation implies with the natural conditions (4.4.50) that both x_0 and x_1 must satisfy the quadratic equation $u^2 + 2au - b = 0$ for $u = x_0$ and for $u = x_1$. If you require x_0 and x_1 to be distinct, you should be able to conclude that $a^2 + b \neq 0$ and that $x_0 = -a - \sqrt{a^2 + b}$, $x_1 = -a + \sqrt{a^2 + b}$. Finally, you should be able to use (4.4.51) to conclude that $a = 0$, giving $x_1 = \sqrt{b} = -x_0$, with which the stated equation (4.4.48) can be shown to hold. Condition (4.4.47) can be shown to follow from the length constraint (4.4.52). The interested reader should be able to show that there is always a unique value of r satisfying (4.4.47) for any given positive number l.

4.5. *How To Design a Thrilling Chute-the-Chute*

An amusement park wishes to build a chute-the-chute which will be more thrilling than that of its competitor down the street. The designer of the new chute knows that a cycloid† gives the quickest descent from one point down to a nearby lower point, but he suspects that a cycloidal chute-the-chute would be *too thrilling* since it would start *vertically downward*! (Would you try it? See Figure 29.) Hence the designer decides to consider a *composite* chute whose path will consist of a piece of a *circular arc* with a *horizontal initial slope* for the sake of safety, followed by a piece of some other arc suitably chosen so as to make a thrilling slide, as shown in Figure 30. Therefore the designer is led to seek the quickest descent that can be achieved along any such composite curve γ connecting a fixed initial point to a given terminal vertical line as shown in Figure 31, where γ is to consist of an initial piece of a given circle joined to a piece of any other arc.

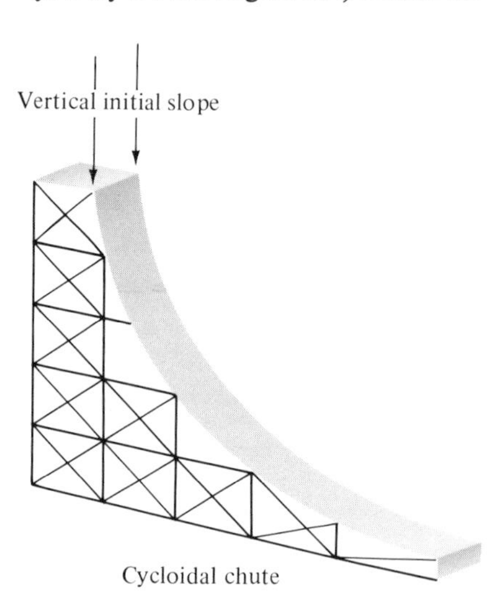

Figure 29

We introduce Cartesian coordinates with the origin of coordinates located at the center of the given

†The author is indebted to Professor Edna E. Kramer for making this connection between the cycloid and an amusement park. See Edna E. Kramer, *The Main Stream of Mathematics* (New York: Oxford University Press, Inc., 1951), p. 180.

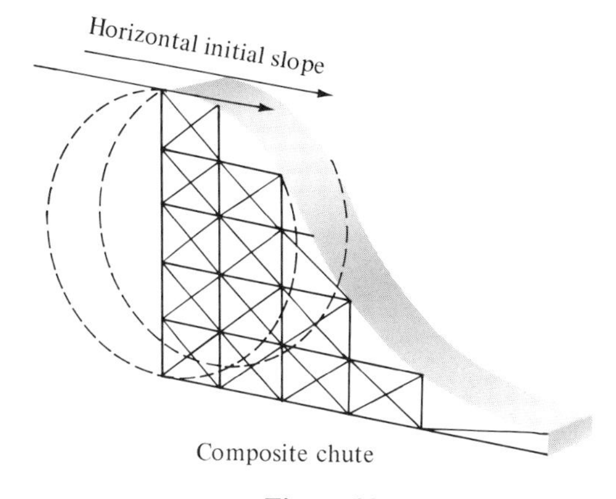

Figure 30

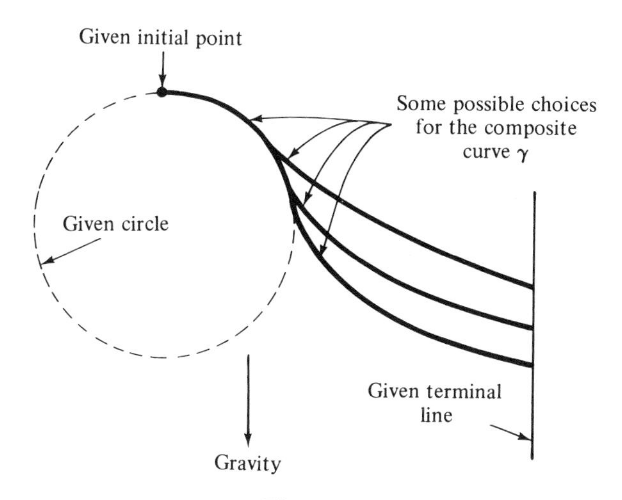

Figure 31

circle and with the given terminal line given by

$$x = x_1, \tag{4.5.1}$$

as shown in Figure 32. The given initial point is labeled $P_0 = (0, y_0)$, and then the equation of the given circle can be given as

$$x^2 + y^2 = y_0^2. \tag{4.5.2}$$

We assume that the earth's gravitational force acts down along the negative y-direction.

The path of the chute can be represented by a composite curve γ given

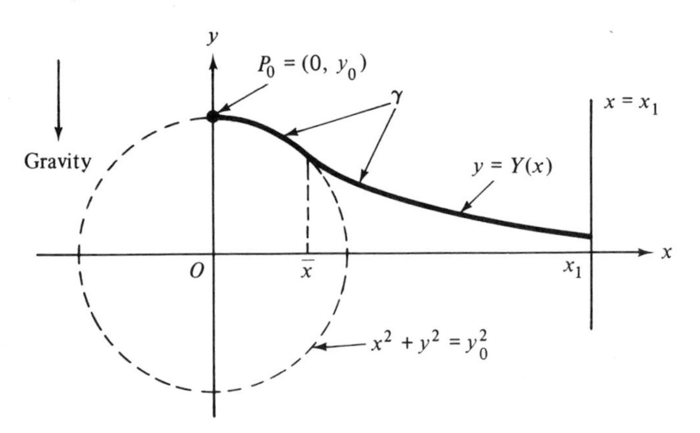

Figure 32

as [see (4.5.2) and Figure 32]

$$\gamma:\ \ y = \begin{cases} \sqrt{y_0^2 - x^2} & \text{for } 0 \leq x \leq \bar{x} \\ Y(x) & \text{for } \bar{x} \leq x \leq x_1 \end{cases} \tag{4.5.3}$$

for some suitable number $\bar{x}$ which is to be the x-coordinate of the point at which the two arcs comprising γ join together ($0 < \bar{x} < x_1$) and for some suitable function $Y(x)$ which is to satisfy the constraint

$$Y(\bar{x}) = \sqrt{y_0^2 - \bar{x}^2}. \tag{4.5.4}$$

This constraint simply guarantees that the two arcs actually join together at $x = \bar{x}$.

The time T required for a person to slide along γ is given by (1.3.1) and can be written as the sum of the corresponding times along each subarc as

$$T = \int_0^{\bar{x}} \frac{\sqrt{1 + (dy/dx)^2}}{v(x)}\, dx + \int_{\bar{x}}^{x_1} \frac{\sqrt{1 + (dy/dx)^2}}{v(x)}\, dx, \tag{4.5.5}$$

where $v = v(x)$ is the speed of motion of the person, which is given as the rate of change of arc length with respect to time ($v = ds/dt$), and where we have used the formula $ds = \sqrt{1 + (dy/dx)^2}\, dx$ for the differential element of arc length along γ.

To calculate the velocity $v(x)$ required in (4.5.5), we recall that the potential energy of a person of mass m located at a height y (near the surface of the earth) is given as

$$\text{potential energy} = mgy,$$

where g is the constant acceleration at the earth's surface due to the gravitational field of the earth. Hence from (4.5.3) we find for a person located at a point (x, y) on γ that

$$\text{potential energy} = \begin{cases} mg\sqrt{y_0^2 - x^2} & \text{for } 0 \le x \le \bar{x} \\ mg\,Y(x) & \text{for } \bar{x} \le x \le x_1. \end{cases}$$

On the other hand, the person's kinetic energy of motion is given as

$$\text{kinetic energy} = \tfrac{1}{2}mv^2.$$

Conservation of energy† requires that the sum of the kinetic and potential energies remains constant during the motion of a person sliding down the chute (we neglect friction). It then follows that a person who starts from *rest* at P_0 and slides along γ will travel with a speed $v = v(x)$ given as [explain; see, for example, the calculation leading to equation (4.2.11)]

$$v(x) = \begin{cases} [2g(y_0 - \sqrt{y_0^2 - x^2})]^{1/2} & \text{if } 0 \le x \le \bar{x} \\ [2g(y_0 - Y(x))]^{1/2} & \text{if } \bar{x} \le x \le x_1. \end{cases}$$

If we insert this result into (4.5.5) and also use (4.5.3), we find for the time T that

$$T(\bar{x}, Y) = \int_0^{\bar{x}} \frac{y_0\,dx}{\sqrt{2g(y_0^2 - x^2)(y_0 - \sqrt{y_0^2 - x^2})}} + \int_{\bar{x}}^{x_1} \sqrt{\frac{1 + Y'(x)^2}{2g(y_0 - Y(x))}}\,dx,$$

where we have written $T(\bar{x}, Y)$ to indicate that the time T depends explicitly on $\bar{x}$ as well as on $Y = Y(x)$ for $\bar{x} \le x \le x_1$. For brevity we rewrite the last equation as

$$T(\bar{x}, Y) = \int_0^{\bar{x}} f(x)\,dx + \int_{\bar{x}}^{x_1} F(Y(x), Y'(x))\,dx, \qquad (4.5.6)$$

where

$$f(x) = y_0[2g(y_0^2 - x^2)(y_0 - \sqrt{y_0^2 - x^2})]^{-1/2} \qquad (4.5.7)$$

and

$$F(y, z) = \sqrt{\frac{1 + z^2}{2g(y_0 - y)}}. \qquad (4.5.8)$$

We take the underlying vector space $\mathfrak{X}$ corresponding to the functional $T = T(\bar{x}, Y)$ to be the set of *all* pairs $(\bar{x}, Y) = (\bar{x}, Y(x))$, where $\bar{x}$ may be any arbitrary number in $\mathfrak{R}$ and where $Y = Y(x)$ may be any arbitrary continuously differentiable function on the fixed interval $0 \le x \le x_1$. For the moment there need not be any special relation between the number $\bar{x}$ and the func-

†See Section 4.8.

tion $Y = Y(x)$ occurring in the pair $(\bar{x}, Y)$. In particular, the relation (4.5.4) need *not* hold even if $\bar{x}$ happens to be in the interval $[0, x_1]$ (which it may not). If $(\bar{x}_1, Y_1)$ and $(\bar{x}_2, Y_2)$ are any two vectors in $\mathfrak{X}$, we define their sum by

$$(\bar{x}_1, Y_1) + (\bar{x}_2, Y_2) = (\bar{x}_1 + \bar{x}_2, Y_1 + Y_2),$$

which gives a vector in $\mathfrak{X}$. Similarly, we define the product $a(\bar{x}, Y)$ by

$$a(\bar{x}, Y) = (a\bar{x}, aY),$$

which gives a vector in $\mathfrak{X}$ for any number a in $\mathbb{R}$ and any vector $(\bar{x}, Y)$. One easily checks that all the rules listed in Section 1.2 for vector spaces are satisfied by $\mathfrak{X}$. We equip $\mathfrak{X}$ with the norm $\|\cdot\|$ defined by

$$\|(\bar{x}, Y)\| = |\bar{x}| + \max_{0 \leq x \leq x_1} |Y(x)| + \max_{0 \leq x \leq x_1} |Y'(x)|$$

for any vector $(\bar{x}, Y)$ in $\mathfrak{X}$. Finally, we take the domain D of the functional $T = T(\bar{x}, Y)$ to be the subset of $\mathfrak{X}$ consisting of all vectors $(\bar{x}, Y)$ in $\mathfrak{X}$ for which $0 < \bar{x} < x_1$. One easily checks that D is an open set in $\mathfrak{X}$.

We now seek a vector $(\bar{x}, Y)$ in D that will minimize in D the functional T of (4.5.6) *subject to the variable constraint* (4.5.4). This constraint (4.5.4) now gives a relation between $\bar{x}$ and $Y = Y(x)$ which is to be imposed on all admissible pairs $(\bar{x}, Y)$. If $(\bar{x}, Y)$ is a minimizing vector for T in D subject to (4.5.4), then the corresponding curve γ giving the path of quickest descent for the chute is given by (4.5.3).

If we define a functional K on D by [see (4.5.4)]

$$K(\bar{x}, Y) = Y(\bar{x}) - \sqrt{y_0^2 - \bar{x}^2}, \tag{4.5.9}$$

then *the problem is to find a minimum vector in* $D[K = 0]$ *for the functional* T *of* (4.5.6). We can use the Euler-Lagrange multiplier theorem to find such a minimum vector; for this purpose we need the variations of the functionals T and K.

If we compare (4.4.11) and (4.5.9), it follows that the variation of K can be given by (4.4.19) with (explain)

$$\Phi(x, y) = y - \sqrt{y_0^2 - x^2}.$$

Hence we find that

$$\delta K(\bar{x}, Y; \Delta\bar{x}, \Delta Y) = \left[\frac{\bar{x}}{\sqrt{y_0^2 - \bar{x}^2}} + Y'(\bar{x})\right]\Delta\bar{x} + \Delta Y(\bar{x}) \tag{4.5.10}$$

for any vector $(\bar{x}, Y)$ in D and any vector $(\Delta\bar{x}, \Delta Y)$ in the vector space $\mathfrak{X}$. [The reader should carry out the calculation leading to (4.5.10).] It only remains, then, to find the variation of T. If we use (4.5.6) and the same type of calculation as was used earlier in obtaining (4.4.15), we find in the present case that

$$\begin{aligned}\delta T(\bar{x}, Y; \Delta\bar{x}, \Delta Y) = [f(\bar{x}) - F(Y(\bar{x}), Y'(\bar{x}))]\,\Delta\bar{x} \\ + \int_{\bar{x}}^{x_1} [F_Y(Y(x), Y'(x))\,\Delta Y(x) \\ + F_{Y'}(Y(x), Y'(x))\,\Delta Y'(x)]\,dx,\end{aligned} \tag{4.5.11}$$

which the reader should be able to verify. Using (4.1.8), this result can be rewritten as

$$\begin{aligned}\delta T(\bar{x}, Y; \Delta\bar{x}, \Delta Y) = \int_{\bar{x}}^{x_1} \left[F_Y(Y(x), Y'(x)) - \frac{d}{dx} F_{Y'}(Y(x), Y'(x))\right]\Delta Y(x)\,dx \\ + F_{Y'}(Y(x_1), Y'(x_1))\,\Delta Y(x_1) - F_{Y'}(Y(\bar{x}), Y'(\bar{x}))\,\Delta Y(\bar{x}) \\ + [f(\bar{x}) - F(Y(\bar{x}), Y'(\bar{x}))]\,\Delta\bar{x}.\end{aligned} \tag{4.5.12}$$

It can be verified from (4.5.10) and (4.5.12) that *all the hypotheses of the Euler-Lagrange multiplier theorem* (as stated in Section 3.3) *are satisfied by* K *and* T. Moreover, if we take $\Delta\bar{x} = 0$ and $\Delta Y = \Delta Y(x) = 1$ (for all x) in (4.5.10), we find that

$$\delta K(\bar{x}, Y; \Delta\bar{x}, \Delta Y) = 1,$$

so that (3.3.1) does *not* hold in this case. Hence the second possibility (3.3.2) of the multiplier theorem must hold, so that if $(\bar{x}, Y)$ is any extremum vector in $D[K = 0]$ for T, there must be a multiplier λ such that

$$\delta T(\bar{x}, Y; \Delta\bar{x}, \Delta Y) = \lambda\,\delta K(\bar{x}, Y; \Delta\bar{x}, \Delta Y) \tag{4.5.13}$$

holds for *all* numbers $\Delta\bar{x}$ and for *all* continuously differentiable functions $\Delta Y = \Delta Y(x)$ on $[0, x_1]$.

Using (4.5.10) and (4.5.12) along with the constraint (4.5.4), we may rewrite the necessary condition (4.5.13) as

$$\begin{aligned}\int_{\bar{x}}^{x_1} \left[F_Y(Y(x), Y'(x)) - \frac{d}{dx} F_{Y'}(Y(x), Y'(x))\right]\Delta Y(x)\,dx \\ = -F_{Y'}(Y(x_1), Y'(x_1))\,\Delta Y(x_1) + [F_{Y'}(\sqrt{y_0^2 - \bar{x}^2}, Y'(\bar{x})) + \lambda]\,\Delta Y(\bar{x}) \\ + \left\{F(\sqrt{y_0^2 - \bar{x}^2}, Y'(\bar{x})) - f(\bar{x}) + \lambda\left[\frac{\bar{x}}{\sqrt{y_0^2 - \bar{x}^2}} + Y'(\bar{x})\right]\right\}\Delta\bar{x},\end{aligned} \tag{4.5.14}$$

which must hold for *all numbers* $\Delta\bar{x}$ *and all continuously differentiable functions* $\Delta Y = \Delta Y(x)$. A standard argument based on the lemma of Section A4 of the Appendix and the arbitrariness of $\Delta\bar{x}$ and ΔY in (4.5.14) [see the discussion following equation (4.4.21)] can now be used to obtain from (4.5.14) the necessary conditions

$$F_Y(Y(x), Y'(x)) - \frac{d}{dx} F_{Y'}(Y(x), Y'(x)) = 0 \qquad \text{for } \bar{x} \leq x \leq x_1 \tag{4.5.15}$$

$$F_{Y'}(Y(x_1), Y'(x_1)) = 0 \tag{4.5.16}$$

$$F_{Y'}(\sqrt{y_0^2 - \bar{x}^2}, Y'(\bar{x})) + \lambda = 0, \tag{4.5.17}$$

and

$$F(\sqrt{y_0^2 - \bar{x}^2}, Y'(\bar{x})) - f(\bar{x}) + \lambda\left[\frac{\bar{x}}{\sqrt{y_0^2 - \bar{x}^2}} + Y'(\bar{x})\right] = 0, \tag{4.5.18}$$

which must be satisfied by any local extremum vector $(\bar{x}, Y)$ in $D[K = 0]$ for the time functional T of (4.5.6). We leave the derivation of these conditions from (4.5.14) as an exercise for the reader.

The multiplier λ can be eliminated from equations (4.5.17) and (4.5.18) to give the single *natural condition*

$$\begin{aligned} F(\sqrt{y_0^2 - \bar{x}^2}, Y'(\bar{x})) &- Y'(\bar{x})F_{Y'}(\sqrt{y_0^2 - \bar{x}^2}, Y'(\bar{x})) \\ &= f(\bar{x}) + \frac{\bar{x}}{\sqrt{y_0^2 - \bar{x}^2}} F_{Y'}(\sqrt{y_0^2 - \bar{x}^2}, Y'(\bar{x})), \end{aligned} \tag{4.5.19}$$

which must then hold at the *variable point* $x = \bar{x}$. Finally, if we use (4.5.7) and (4.5.8) along with [see (2.4.16)]

$$F_{Y'}(\sqrt{y_0^2 - \bar{x}^2}, Y'(\bar{x})) = \left.\frac{\partial F(y, z)}{\partial z}\right|_{\substack{y=\sqrt{y_0^2 - \bar{x}^2} \\ z = Y'(\bar{x})}},$$

it follows that (4.5.19) implies that

$$\sqrt{y_0^2 - \bar{x}^2} = y_0\sqrt{1 + Y'(\bar{x})^2} + \bar{x}\,Y'(\bar{x}),$$

from which follows

$$Y'(\bar{x}) = -\frac{\bar{x}}{\sqrt{y_0^2 - \bar{x}^2}}. \tag{4.5.20}$$

This natural condition (4.5.20) simply requires that the minimizing curve γ be *tangent* to the circle (4.5.2) at $x = \bar{x}$, as the reader should be able to verify.

A similar calculation shows that the natural boundary condition (4.5.16) implies that

$$Y'(x_1) = 0, \tag{4.5.21}$$

which requires that γ intersect the line $x = x_1$ *orthogonally* (at right angles).

We can now show that the Euler-Lagrange equation (4.5.15) along with the constraint (4.5.4) and the natural conditions (4.5.20) and (4.5.21) can be used to determine the extremum curve γ. In fact, (4.5.15) and (4.5.8) imply that [see (4.2.3)]

$$Y'(x) = -\sqrt{\frac{A - [y_0 - Y(x)]}{y_0 - Y(x)}}$$

for some constant of integration A. This differential equation has cycloidal solutions which can be given parametrically as [see (4.2.7)]

$$\begin{aligned} x &= x_0 + \frac{A}{2}(\theta - \sin\theta) \\ Y &= y_0 - \frac{A}{2}(1 - \cos\theta) \qquad \text{for } \bar{\theta} \leq \theta \leq \theta_1, \end{aligned} \tag{4.5.22}$$

where the first equation here can be solved in principle to give $\theta = \theta(x)$ along γ, and then the second equation will yield $Y = Y(x)$ (for $\bar{x} \leq x \leq x_1$). The constant x_0 appearing in (4.5.22) is an undetermined constant which fixes the initial point of the cycloid when $\theta = 0$, while the constants $\bar{\theta}$ and θ_1 are positive constants which must be chosen so that the appropriate constraints and natural conditions are all satisfied. In particular, θ_1 must be chosen so that the cycloid terminates on the line $x = x_1$ of (4.5.1) when $\theta = \theta_1$; this requires with (4.5.22) the condition

$$x_1 = x_0 + \frac{A}{2}(\theta_1 - \sin\theta_1). \tag{4.5.23}$$

Also, when $\theta = \bar{\theta}$, we require that the cycloid intersect the given circle (4.5.2) with $x = \bar{x}$ and with [see (4.5.4)] $Y = \sqrt{y_0^2 - \bar{x}^2}$; this requires with (4.5.22) the conditions

$$\bar{x} = x_0 + \frac{A}{2}(\bar{\theta} - \sin\bar{\theta}) \tag{4.5.24}$$

and

$$\sqrt{y_0^2 - \bar{x}^2} = y_0 - \frac{A}{2}(1 - \cos\bar{\theta}). \tag{4.5.25}$$

We can get two additional conditions from the natural conditions (4.5.20) and (4.5.21). In fact, with (4.5.22) the chain rule of differential calculus gives

$$Y'(x) = \frac{dY/d\theta}{dx/d\theta} = \frac{-\sin\theta}{1 - \cos\theta}, \tag{4.5.26}$$

so that the natural condition (4.5.20) at $\theta = \bar{\theta}$ gives the requirement

$$\frac{\sin\bar{\theta}}{1 - \cos\bar{\theta}} = \frac{\bar{x}}{\sqrt{y_0^2 - \bar{x}^2}},$$

which can be solved for $\bar{x}$ as

$$\bar{x} = \frac{y_0 \sin \bar{\theta}}{\sqrt{2(1 - \cos \bar{\theta})}}. \tag{4.5.27}$$

Finally, the natural condition (4.5.21) and (4.5.26) at $\theta = \theta_1$ give the requirement $\sin \theta_1 = 0$, which will be satisfied with

$$\theta_1 = \pi.$$

Inserting this result into (4.5.23), we find that

$$x_0 = x_1 - \frac{\pi}{2} A. \tag{4.5.28}$$

Hence we have the four equations (4.5.24), (4.5.25), (4.5.27), and (4.5.28) from which to determine the remaining four constants $\bar{x}$, x_0, A, and $\bar{\theta}$ in terms of the given numbers x_1 and y_0. In fact, we can use (4.5.27) and (4.5.28) to eliminate $\bar{x}$ and x_0 from equations (4.5.24) and (4.5.25) to find the two equations

$$\frac{y_0 \sin \bar{\theta}}{\sqrt{2(1 - \cos \bar{\theta})}} - x_1 + \frac{\pi}{2} A = \frac{A}{2}(\bar{\theta} - \sin \bar{\theta})$$

and

$$y_0\left(1 - \sqrt{\frac{1 - \cos \bar{\theta}}{2}}\right) = \frac{A}{2}(1 - \cos \bar{\theta}), \tag{4.5.29}$$

and then the constant A may be eliminated between the last two equations to give the single equation

$$(\pi - \bar{\theta})\sqrt{\frac{1 - \cos \bar{\theta}}{2}} = \pi - \bar{\theta} + \sin \bar{\theta} - \frac{x_1}{y_0}(1 - \cos \bar{\theta}), \tag{4.5.30}$$

which can be solved for $\bar{\theta}$ ($0 < \bar{\theta} < \pi$) in terms of the given numbers x_1 and y_0. In practice, equation (4.5.30) must be solved numerically using, perhaps, Newton's method or the iterative method of Picard.† Once $\bar{\theta}$ is obtained from (4.5.30), then the corresponding values for A, $\bar{x}$, and x_0 can be found directly from (4.5.29), (4.5.27), and (4.5.28).

Example. As an example, we consider the special case in which the data x_1 and y_0 satisfy

$$x_1 = \left[1 + \frac{\pi}{2}\left(1 - \frac{1}{\sqrt{2}}\right)\right] y_0 \approx 1.46 y_0, \tag{4.5.31}$$

†Such numerical methods are discussed, for example, in Peter Henrici, *Elements of Numerical Analysis* (New York: John Wiley & Sons, Inc., 1964).

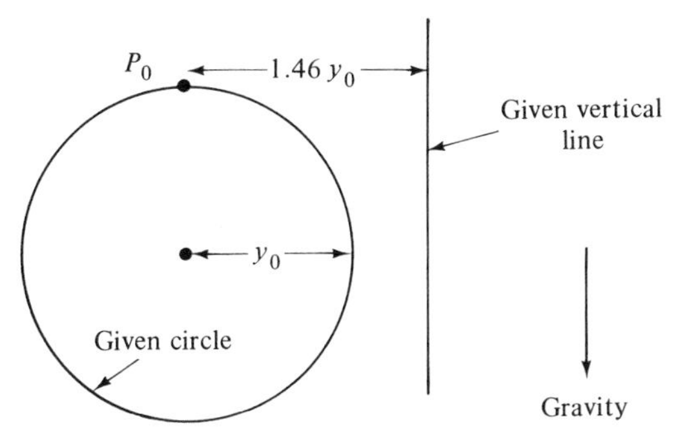

Figure 33

so that the terminal line is located at a distance about 1.46 times the radius of the given circle from the given initial point, as shown in Figure 33. Using (4.5.31) in (4.5.30), for $\bar{\theta}$ we find that

$$\bar{\theta} = \frac{\pi}{2},$$

and with (4.5.29) this result gives

$$A = 2\left(1 - \frac{1}{\sqrt{2}}\right) y_0 \approx .59 y_0.$$

The last two results can now be used in (4.5.27) and (4.5.28) to give

$$\bar{x} = \frac{1}{\sqrt{2}} y_0 \approx .71\, y_0$$

and

$$x_0 = \left[1 - \frac{\pi}{2}\left(1 - \frac{1}{\sqrt{2}}\right)\right] y_0 \approx .54\, y_0,$$

provided, of course, that the relation (4.5.31) holds.

Hence in the special case in which x_1 and y_0 are related by (4.5.31), the most thrilling chute-the-chute of the type considered here will follow a composite path γ consisting of the circular arc

$$y = \sqrt{y_0^2 - x^2} \qquad \text{for } 0 \le x \le \frac{1}{\sqrt{2}} y_0 \approx .71\, y_0$$

followed by the cycloidal arc

$$\begin{aligned} x &= y_0\left[1 - \frac{\pi}{2}\left(1 - \frac{1}{\sqrt{2}}\right) + \left(1 - \frac{1}{\sqrt{2}}\right)(\theta - \sin\theta)\right] \\ &\approx y_0[.54 + .295(\theta - \sin\theta)] \end{aligned}$$

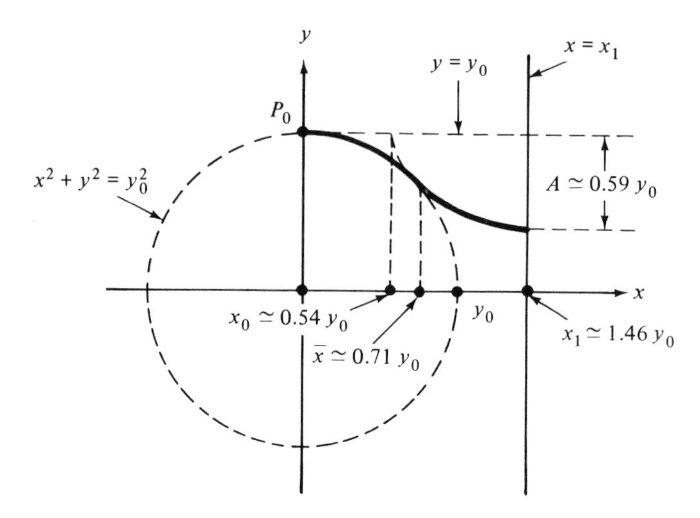

Figure 34

$$y = y_0\left[1 - \left(1 - \frac{1}{\sqrt{2}}\right)(1 - \cos\theta)\right]$$

$$\approx y_0[1 - .295(1 - \cos\theta)] \qquad \text{for } \frac{\pi}{2} \leq \theta \leq \pi,$$

which extends from $x = \bar{x} = (1/\sqrt{2})y_0 \approx .71y_0$ to $x = x_1 \approx 1.46y_0$. The given cycloid is generated by the motion of a fixed point on the rim of a circle of diameter $A \approx .59y_0$ which rolls under the line $y = y_0$ starting at $x = x_0 \approx .54y_0$, as shown in Figure 34.

Exercises

1. Give a careful derivation of (4.5.11) for the variation of the functional T defined by (4.5.6).

2. Show how to obtain the necessary conditions (4.5.15), (4.5.16), (4.5.17), and (4.5.18) from (4.5.14).

3. Show that the maximum *slope* of the extremum composite chute obtained above in the special case (4.5.31) is 45° and occurs at $x = \bar{x}$.

4.6. Functionals Involving Several Unknown Functions

It sometimes happens that one wishes to minimize or maximize a functional J of the form

$$J = \int_{x_0}^{x_1} F(x, Y_1(x), Y_2(x), \ldots, Y_n(x), Y_1'(x), Y_2'(x), \ldots, Y_n'(x))\, dx, \tag{4.6.1}$$

which depends on n unknown functions $Y_1, Y_2, \ldots, Y_n$ of class $\mathcal{C}^1$ on the interval $[x_0, x_1]$. The function $F = F(x, y_1, y_2, \ldots, y_n, z_1, z_2, \ldots, z_n)$ is a given function of the $2n + 1$ variables $x, y_1, y_2, \ldots, y_n, z_1, z_2, \ldots, z_n$, which in (4.6.1) is evaluated along $y_i = Y_i(x)$ and $z_i = Y_i'(x)$ for $i = 1, 2, \ldots, n$. The problem is to find those particular functions $Y_1, Y_2, \ldots, Y_n$ which furnish an appropriate extreme value to J subject to any specified constraints. For example, we might consider a fixed end-point problem in which the values of $Y_1, Y_2, \ldots, Y_n$ are specified at the end points as

$$Y_i(x_0) = a_i \qquad \text{and} \qquad Y_i(x_1) = b_i \qquad \text{for } i = 1, 2, \ldots, n, \tag{4.6.2}$$

where $a_1, a_2, \ldots, a_n, b_1, b_2, \ldots, b_n$ are given constants. In its most general setting the problem of finding geodesic curves is of this type, as are certain other geometric extremum problems. In fact, even the simple problem of finding the curve γ of shortest length which connects two given points $P_0 = (a_1, a_2, a_3)$ and $P_1 = (b_1, b_2, b_3)$ in three-dimensional Euclidean space is of this type. For example, we might represent a variable point $P = (y_1, y_2, y_3)$ on such a curve γ parametrically as

$$\gamma: \begin{cases} y_1 = Y_1(x) \\ y_2 = Y_2(x) \\ y_3 = Y_3(x) \qquad \text{for } x_0 \le x \le x_1 \end{cases}$$

for suitable functions $Y_1(x)$, $Y_2(x)$, and $Y_3(x)$, where x may be any suitable parameter along γ. Then the length of γ may be given as [see (4.3.4)]

$$L = \int_{x_0}^{x_1} \sqrt{Y_1'(x)^2 + Y_2'(x)^2 + Y_3'(x)^2}\, dx,$$

and the problem is to minimize this length functional subject to fixed end-point conditions. Also, we shall see in Section 4.8 that the motions of dynamical systems as governed by Newton's laws of motion can be characterized through such a minimization problem involving several unknown functions.

For notational purposes it is convenient in such problems to combine the functions $Y_1, Y_2, \ldots, Y_n$ into a single vector function $\mathbf{Y}$ as

$$\mathbf{Y} = \mathbf{Y}(x) = (Y_1(x), Y_2(x), \ldots, Y_n(x)),$$

and then the previous functional (4.6.1) can be written more briefly as

$$J(\mathbf{Y}) = \int_{x_0}^{x_1} F(x, \mathbf{Y}(x), \mathbf{Y}'(x))\, dx. \tag{4.6.3}$$

The end-point conditions (4.6.2) can be rewritten similarly as

$$\mathbf{Y}(x_0) = \mathbf{a}, \qquad \mathbf{Y}(x_1) = \mathbf{b},$$

where $\mathbf{a} = (a_1, a_2, \ldots, a_n)$ and $\mathbf{b} = (b_1, b_2, \ldots, b_n)$. We take the domain D of the functional $J = J(\mathbf{Y})$ to be the entire vector space $\mathfrak{X}$ which consists of all vector functions $\mathbf{Y} = \mathbf{Y}(x) = (Y_1(x),\ Y_2(x), \ldots,\ Y_n(x))$ whose components $Y_i(x)$ are of class $\mathcal{C}^1$ on $[x_0, x_1]$ for all $i = 1, 2, \ldots, n$. If $\mathbf{Y} = (Y_1, Y_2, \ldots, Y_n)$ and $\mathbf{Z} = (Z_1, Z_2, \ldots, Z_n)$ are any two vectors in $\mathfrak{X}$, we define their sum by

$$\mathbf{Y} + \mathbf{Z} = (Y_1 + Z_1,\ Y_2 + Z_2, \ldots,\ Y_n + Z_n),$$

which again gives a vector in $\mathfrak{X}$. Similarly, we define the product $a\mathbf{Y}$ by

$$a\mathbf{Y} = (aY_1, aY_2, \ldots, aY_n),$$

which gives a vector in $\mathfrak{X}$ for any number a in $\mathfrak{R}$ and any vector $\mathbf{Y}$ in $\mathfrak{X}$. Finally, we equip $\mathfrak{X}$ with the norm $\|\cdot\|$ defined by

$$\|\mathbf{Y}\| = \sum_{i=1}^{n} \{\max_{x_0 \leq x \leq x_1} |Y_i(x)| + \max_{x_0 \leq x \leq x_1} |Y_i'(x)|\}$$

for any vector $\mathbf{Y} = (Y_1, Y_2, \ldots, Y_n)$ in $\mathfrak{X}$, although there are other possible norms which are also acceptable.

If we define functionals K_i and L_i on $\mathfrak{X}$ by

$$K_i(\mathbf{Y}) = Y_i(x_0), \qquad L_i(\mathbf{Y}) = Y_i(x_1) \qquad \text{for } i = 1, 2, \ldots, n \tag{4.6.4}$$

for any vector $\mathbf{Y} = (Y_1, Y_2, \ldots, Y_n)$ in $\mathfrak{X}$, then the fixed end-point extremum problem mentioned above is to find local extremum vectors in $D[K_i = a_i, L_i = b_i$ for $i = 1, 2, \ldots, n]$ for the functional J of (4.6.3) where $D = \mathfrak{X}$. We shall see below that all the hypotheses of the Euler-Lagrange multiplier theorem are satisfied, and we shall be able to conclude that any local extremum vector $\mathbf{Y}$ in $D[K_i = a_i, L_i = b_i$ for $i = 1, 2, \ldots, n]$ for J must satisfy the necessary condition [see equation (3.6.3)]

$$\delta J(\mathbf{Y}; \Delta\mathbf{Y}) = \sum_{i=1}^{n} \{\lambda_i\, \delta K_i(\mathbf{Y}; \Delta\mathbf{Y}) + \mu_i\, \delta L_i(\mathbf{Y}; \Delta\mathbf{Y})\} \tag{4.6.5}$$

for suitable Euler-Lagrange multipliers $\lambda_1, \lambda_2, \ldots, \lambda_n$ and $\mu_1, \mu_2, \ldots, \mu_n$ and for all vectors $\Delta\mathbf{Y} = (\Delta Y_1, \Delta Y_2, \ldots, \Delta Y_n)$ in $\mathfrak{X}$.

To proceed, we must now calculate the variations of all the functionals involved. We easily find from (4.6.4) and (2.3.4) that (explain)

$$\begin{aligned} \delta K_i(\mathbf{Y}; \Delta\mathbf{Y}) &= \Delta Y_i(x_0) \\ \delta L_i(\mathbf{Y}; \Delta\mathbf{Y}) &= \Delta Y_i(x_1) \qquad \text{for } i = 1, 2, \ldots, n \end{aligned} \tag{4.6.6}$$

for any vector $\Delta\mathbf{Y} = (\Delta Y_1, \Delta Y_2, \ldots, \Delta Y_n)$ in $\mathfrak{X}$, so that it only remains to find the variation of J. From (4.6.3) we find that

$$J(\mathbf{Y} + \epsilon\,\Delta\mathbf{Y}) = \int_{x_0}^{x_1} F(x, \mathbf{Y}(x) + \epsilon\,\Delta\mathbf{Y}(x), \mathbf{Y}'(x) + \epsilon\,\Delta\mathbf{Y}'(x))\,dx,$$

and then (2.3.4) leads to

$$\delta J(\mathbf{Y}; \mathbf{\Delta Y}) = \int_{x_0}^{x_1} \frac{\partial}{\partial \epsilon} F(x, \mathbf{Y}(x) + \epsilon\, \mathbf{\Delta Y}(x), \mathbf{Y}'(x) + \epsilon\, \mathbf{\Delta Y}'(x))\bigg|_{\epsilon=0} dx.$$

But now the same type of calculation which led earlier to equation (2.4.15) can be used in the present case to obtain (explain)

$$\frac{\partial}{\partial \epsilon} F(x, \mathbf{Y}(x) + \epsilon\, \mathbf{\Delta Y}(x), \mathbf{Y}'(x) + \epsilon\, \mathbf{\Delta Y}'(x))\bigg|_{\epsilon=0}$$
$$= \sum_{i=1}^{n} \{F_{Y_i}(x, \mathbf{Y}(x), \mathbf{Y}'(x))\, \Delta Y_i(x) + F_{Y_i'}(x, \mathbf{Y}(x), \mathbf{Y}'(x))\, \Delta Y_i'(x)\},$$

where [compare with (2.4.16)]

$$\begin{aligned} F_{Y_i}(x, \mathbf{Y}(x), \mathbf{Y}'(x)) &= \frac{\partial}{\partial y_i} F(x, \mathbf{y}, \mathbf{z})\bigg|_{\substack{\mathbf{y}=\mathbf{Y}(x)\\ \mathbf{z}=\mathbf{Y}'(x)}} \\ F_{Y_i'}(x, \mathbf{Y}(x), \mathbf{Y}'(x)) &= \frac{\partial}{\partial z_i} F(x, \mathbf{y}, \mathbf{z})\bigg|_{\substack{\mathbf{y}=\mathbf{Y}(x)\\ \mathbf{z}=\mathbf{Y}'(x)}}. \end{aligned} \tag{4.6.7}$$

[We have used the notation $\mathbf{y} = (y_1, y_2, \ldots, y_n)$ and $\mathbf{z} = (z_1, z_2, \ldots, z_n)$ in (4.6.7).] Hence the variation of J can be given by

$$\begin{aligned} \delta J(\mathbf{Y}; \mathbf{\Delta Y}) = \sum_{i=1}^{n} \int_{x_0}^{x_1} &\{F_{Y_i}(x, \mathbf{Y}(x), \mathbf{Y}'(x))\, \Delta Y_i(x) \\ &+ F_{Y_i'}(x, \mathbf{Y}(x), \mathbf{Y}'(x))\, \Delta Y_i'(x)\}\, dx, \end{aligned}$$

or using (4.1.8), we find that

$$\begin{aligned} \delta J(\mathbf{Y}; \mathbf{\Delta Y}) = &\sum_{i=1}^{n} \int_{x_0}^{x_1} \left\{F_{Y_i}(x, \mathbf{Y}(x), \mathbf{Y}'(x)) - \frac{d}{dx} F_{F_i'}(x, \mathbf{Y}(x), \mathbf{Y}'(x))\right\} \Delta Y_i(x)\, dx \\ &+ \sum_{i=1}^{n} \{F_{Y_i'}(x_1, \mathbf{Y}(x_1), \mathbf{Y}'(x_1))\, \Delta Y_i(x_1) \\ &- F_{Y_i'}(x_0, \mathbf{Y}(x_0), \mathbf{Y}'(x_0))\, \Delta Y_i(x_0)\} \end{aligned} \tag{4.6.8}$$

for any vector $\mathbf{\Delta Y} = (\Delta Y_1, \Delta Y_2, \ldots, \Delta Y_n)$ in $\mathfrak{X}$.

One can now verify that the Euler-Lagrange multiplier theorem does indeed apply to the present fixed end-point problem, and therefore (4.6.5) must hold if $\mathbf{Y}$ is a local extremum vector in $D[K_i = a_i, L_i = b_i$ for $i = 1, 2, \ldots, n]$ for J. (We leave as an exercise for the reader the elimination of the other possibility of the Euler-Lagrange multiplier theorem; see Exercise 1.) If we use (4.6.6) and (4.6.8) in (4.6.5), we find the necessary condition

$$\begin{aligned} \sum_{i=1}^{n} \int_{x_0}^{x_1} &\left\{F_{Y_i}(x, \mathbf{Y}(x), \mathbf{Y}'(x)) - \frac{d}{dx} F_{Y_i'}(x, \mathbf{Y}(x), \mathbf{Y}'(x))\right\} \Delta Y_i(x)\, dx \\ &= \sum_{i=1}^{n} [F_{Y_i'}(x_0, \mathbf{a}, \mathbf{Y}'(x_0)) + \lambda_i]\, \Delta Y_i(x_0) \\ &\quad + \sum_{i=1}^{n} [-F_{Y_i'}(x_1, \mathbf{b}, \mathbf{Y}'(x_1)) + \mu_i]\, \Delta Y_i(x_1), \end{aligned}$$

which must then hold for all continuously differentiable functions $\Delta Y_1(x)$, $\Delta Y_2(x), \ldots, \Delta Y_n(x)$. In particular, if j is any fixed integer $(1 \leq j \leq n)$, we can choose each $\Delta Y_i(x)$ to be *identically zero* for all integers $i \neq j$ so as to find for this j that

$$\int_{x_0}^{x_1} \left[F_{Y_j}(x, \mathbf{Y}(x), \mathbf{Y}'(x)) - \frac{d}{dx} F_{Y_j'}(x, \mathbf{Y}(x), \mathbf{Y}'(x)) \right] \Delta Y_j(x)\, dx$$
$$= [F_{Y_j'}(x_0, \mathbf{a}, \mathbf{Y}'(x_0)) + \lambda_j]\, \Delta Y_j(x_0) + [-F_{Y_j'}(x_1, \mathbf{b}, \mathbf{Y}'(x_1)) + \mu_j]\, \Delta Y_j(x_1),$$

which must still hold for every function ΔY_j of class $\mathcal{C}^1$ on $[x_0, x_1]$. It follows now, just as in Section 4.1, that the following Euler-Lagrange equation must hold [compare with (4.1.11)]:

$$F_{Y_j}(x, \mathbf{Y}(x), \mathbf{Y}'(x)) - \frac{d}{dx} F_{Y_j'}(x, \mathbf{Y}(x), \mathbf{Y}'(x)) = 0 \qquad \text{for } x_0 < x < x_1.$$

Since j is arbitrary here $(j = 1, 2, \ldots, n)$, we get n such equations which must be satisfied simultaneously by the n extremum functions $Y_1(x)$, $Y_2(x)$, $\ldots$, $Y_n(x)$:

$$F_{Y_i}(x, \mathbf{Y}(x), \mathbf{Y}'(x)) - \frac{d}{dx} F_{Y_i'}(x, \mathbf{Y}(x), \mathbf{Y}'(x)) = 0$$
$$\text{for } x_0 < x < x_1 \qquad \text{and} \qquad \text{for } i = 1, 2, \ldots, n. \tag{4.6.9}$$

Example. By way of illustration we consider the following example with $n = 2$ and with F defined as

$$F(x, y_1, y_2, z_1, z_2) = -2y_1^2 + 2y_1 y_2 - z_1^2 + z_2^2. \tag{4.6.10}$$

If we use this function F in (4.6.1), we obtain the functional

$$J(\mathbf{Y}) = \int_{x_0}^{x_1} [-2Y_1(x)^2 + 2Y_1(x)Y_2(x) - Y_1'(x)^2 + Y_2'(x)^2]\, dx$$

for any vector $\mathbf{Y} = \mathbf{Y}(x) = (Y_1(x), Y_2(x))$. We seek to minimize or maximize J among all such vector functions $\mathbf{Y}(x)$ on the fixed interval $[x_0, x_1]$ subject to the fixed end-point conditions

$$Y_1(x_0) = a_1, \qquad Y_2(x_0) = a_2, \qquad Y_1(x_1) = b_1, \qquad Y_2(x_1) = b_2. \tag{4.6.11}$$

From (4.6.7) and (4.6.10) we find that (explain)

$$F_{Y_1}(x, \mathbf{Y}(x), \mathbf{Y}'(x)) = -4Y_1(x) + 2Y_2(x)$$
$$F_{Y_2}(x, \mathbf{Y}(x), \mathbf{Y}'(x)) = 2Y_1(x)$$
$$F_{Y_1'}(x, \mathbf{Y}(x), \mathbf{Y}'(x)) = -2Y_1'(x)$$
$$F_{Y_2'}(x, \mathbf{Y}(x), \mathbf{Y}'(x)) = 2Y_2'(x),$$

and then the Euler-Lagrange equations (4.6.9) imply in this case that

$$\begin{aligned} -2Y_1(x) + Y_2(x) + Y_1''(x) &= 0 \\ Y_1(x) - Y_2''(x) &= 0 \qquad \text{for } x_0 < x < x_1. \end{aligned} \tag{4.6.12}$$

Hence the components $Y_1(x)$ and $Y_2(x)$ of any local extremum vector $\mathbf{Y} = \mathbf{Y}(x) = (Y_1(x),\ Y_2(x))$ for J must satisfy the coupled system of differential equations given by (4.6.12). We shall see that the most general solution of this system of two second-order differential equations will involve *four* arbitrary constants of integration, which can be specified in the present case through the use of the four conditions of (4.6.11).

We shall solve the system (4.6.12) by replacing it with a *single* fourth-order differential equation which can be easily solved. [Alternatively, one might use Laplace transforms to solve (4.6.12).] Indeed, we can use the second equation of (4.6.12) to eliminate $Y_1(x)$ in the first equation, and in this way we find for $Y_2(x)$ the single equation

$$\frac{d^4}{dx^4} Y_2(x) - 2\frac{d^2}{dx^2} Y_2(x) + Y_2(x) = 0.$$

This equation can be *factored* as

$$\left(\frac{d^2}{dx^2} - 1\right)\left(\frac{d^2}{dx^2} - 1\right) Y_2(x) = 0,$$

or, what amounts to the same thing, we can write

$$u''(x) - u(x) = 0, \tag{4.6.13}$$

where $u(x)$ is defined as

$$u(x) = Y_2''(x) - Y_2(x). \tag{4.6.14}$$

The reader can verify that the last two equations are equivalent to the previous fourth-order differential equation for $Y_2(x)$. Now the most general solution of (4.6.13) can be given as†

$$u(x) = Ae^x + Be^{-x}$$

in terms of two arbitrary constants of integration A and B, and then (4.6.14) becomes

$$Y_2''(x) - Y_2(x) = Ae^x + Be^{-x}, \tag{4.6.15}$$

which we can now solve for $Y_2(x)$.

If the right-hand side of (4.6.15) were zero (i.e., if $A = B = 0$), then the

†See, for example, Thomas, *op. cit.*, Chapter 20, or any elementary text on differential equations.

solution of (4.6.15) would be given in the form

$$Y_2(x) = Ce^x + De^{-x} \tag{4.6.16}$$

for arbitrary constants of integration C and D. In the present case, however, the right-hand side of (4.6.15) is in general *not* zero, and so the "solution" (4.6.16) is *not valid*. However, there is a method which was first used by John Bernoulli and Isaac Newton independently in certain special situations and which was later generalized and developed by Euler and then by Lagrange into a general procedure to handle just such situations as we now face. The procedure is called the method of **variation of parameters**, or **variation of constants**. In the present case the idea is to seek a valid solution of (4.6.15) in the *same form* as (4.6.16) but where now we allow the "constants" C and D to vary as functions of x; i.e., we seek $Y_2(x)$ in the form

$$Y_2(x) = C(x)e^x + D(x)e^{-x} \tag{4.6.17}$$

for suitable *functions* $C(x)$ and $D(x)$. We insert this expression into (4.6.15) and seek suitable functions $C(x)$ and $D(x)$ so as to obtain a correct solution. It was shown by Euler and Lagrange that the calculation will be greatly simplified if we require not only that $Y_2(x)$ have the same form in (4.6.17) as in (4.6.16) but also that the derivative $Y_2'(x)$ have the same form as obtained from (4.6.17) as obtained from (4.6.16). Since $Y_2'(x) = Ce^x - De^{-x}$ as calculated from (4.6.16) when C and D are constants, we require now that $C(x)$ and $D(x)$ be chosen in (4.6.17) so that

$$Y_2'(x) = C(x)e^x - D(x)e^{-x}. \tag{4.6.18}$$

Of course, this will be the case for $Y_2(x)$ given by (4.6.17) if and only if (explain)

$$C'(x)e^x + D'(x)e^{-x} = 0. \tag{4.6.19}$$

Differentiating (4.6.18) again and using (4.6.15) and (4.6.17), we find the additional equation

$$C'(x)e^x - D'(x)e^{-x} = Ae^x + Be^{-x}. \tag{4.6.20}$$

Hence we must find functions $C(x)$ and $D(x)$ which satisfy these last two equations, and then $Y_2(x)$ will be given correctly by (4.6.17).

We can view (4.6.19) and (4.6.20) as two algebraic equations for the two unknown expressions $C'(x)$ and $D'(x)$, and these equations can be solved to give

$$C'(x) = \frac{A + Be^{-2x}}{2}, \qquad D'(x) = -\frac{Ae^{2x} + B}{2}.$$

The latter equations can be integrated now to give

$$C(x) = C_0 + \frac{A}{2}x - \frac{B}{4}e^{-2x}$$
$$D(x) = D_0 - \frac{A}{4}e^{2x} - \frac{B}{2}x \tag{4.6.21}$$

for arbitrary constants of integration C_0 and D_0. Hence we have found the required functions $C(x)$ and $D(x)$, and now the solution $Y_2(x)$ can be obtained from (4.6.17) and (4.6.21) as

$$Y_2(x) = \left(C_0 - \frac{A}{4}\right)e^x + \left(D_0 - \frac{B}{4}\right)e^{-x} + \frac{x}{2}(Ae^x - Be^{-x}), \tag{4.6.22}$$

which is indeed the general solution of (4.6.15) depending on the arbitrary constants C_0 and D_0. Of course, if $A = B = 0$ in (4.6.15), then (4.6.22) reduces to the known solution (4.6.16). Finally, we can use (4.6.22) and the last equation of (4.6.12) to find

$$Y_1(x) = \left(C_0 + \frac{3}{4}A\right)e^x + \left(D_0 + \frac{3}{4}B\right)e^{-x} + \frac{x}{2}(Ae^x - Be^{-x}). \tag{4.6.23}$$

One easily checks that (4.6.22) and (4.6.23) do indeed furnish a solution of the system of Euler-Lagrange equations (4.6.12). The appropriate constants A, B, C_0, and D_0 can be determined by imposing the boundary conditions (4.6.11). The resulting vector function $\mathbf{Y}(x) = (Y_1(x), Y_2(x))$ will be the only vector in $\mathfrak{X}$ which satisfies the necessary condition (4.6.5) of the Euler-Lagrange multiplier theorem. Of course, one must investigate further to determine if $\mathbf{Y}$ is a maximum vector or a minimum vector for J, or perhaps a saddle point. We shall not do this here. (See the references given at the end of Section 8.3.)

Note in this example that the system of Euler-Lagrange equations furnishes two second-order differential equations which involve the two unknown functions $Y_1(x)$ and $Y_2(x)$ [see (4.6.12)]. In the general case it is reasonably clear that the Euler-Lagrange system (4.6.9) will furnish n second-order differential equations for the n unknown functions $Y_1(x)$, $Y_2(x)$, . . . , $Y_n(x)$. The general solution will then involve $2n$ arbitrary constants of integration, which can be determined by imposing the $2n$ boundary conditions given by (4.6.2).

We remark finally that it is possible similarly to consider extremum problems involving several unknown functions with *variable* end-point conditions. In this case one obtains several natural boundary conditions corresponding to each variable point. The procedure is similar to that developed in Section 4.4, and we leave the details as an exercise for the reader. (See, for example, Exercise 2.)

Exercises

1. Show that the first possibility of the Euler-Lagrange multiplier theorem of Section 3.6 fails to hold for the fixed end-point problem for the functional $J = J(\mathbf{Y})$ of (4.6.3), and thereby show that (4.6.5) must hold for any local extremum vector.

2. Find the curves γ in 3-space given parametrically as

$$\gamma: \begin{cases} y_1 = Y_1(x) \\ y_2 = Y_2(x) \\ y_3 = x \qquad \text{for} \quad 0 \leq x \leq \dfrac{\pi}{2} \end{cases}$$

for which the functional $J = \int_0^{\pi/2} [Y_1'(x)^2 + Y_2'(x)^2 + 2Y_1(x)Y_2(x)]\,dx$ can have local extreme values subject to the conditions $Y_1(0) = 0$, $Y_2(0) = 0$. The terminal point $(y_1, y_2, y_3) = (Y_1(\pi/2), Y_2(\pi/2), \pi/2)$ is free to vary on the plane $y_3 = \pi/2$. Find the natural boundary conditions at the terminal point.

3. Suppose that the function $F = F(x, \mathbf{y}, \mathbf{z})$ which appears in (4.6.1) does *not* depend on the independent variable x, as happens, for example, in the special case (4.6.10). In any such case with $F = F(\mathbf{y}, \mathbf{z})$ independent of x, show that any possible local extremum vector $\mathbf{Y}(x) = (Y_1(x), Y_2(x), \ldots, Y_n(x))$ for J must satisfy [compare with equation (4.1.15)] $F(\mathbf{Y}(x), \mathbf{Y}'(x)) - \sum_{i=1}^{n} Y_i'(x)F_{Y_i'}(\mathbf{Y}(x), \mathbf{Y}'(x)) = C$ for some constant of integration C. [The latter equation can be viewed as a single first-order differential equation which must be satisfied by the extremum functions $Y_1(x), Y_2(x), \ldots, Y_n(x)$.]

4.7. Fermat's Principle in Geometrical Optics

In 1675 the astronomer Olaf Roemer (1644–1710) verified on the basis of telescopic observations of a satellite of Jupiter that *light travels* (or propagates) *with finite speed.* We might also mention for the sake of perspective that the great John Kepler (1571–1630), one of the founders of modern astronomy, in 1604 published a treatise on light in which he accepted the then-current view that the speed of light is infinite. In 1905 Albert Einstein (1879–1955) asserted the fundamental principle that *light travels in a vacuum always with the same constant speed c, which is independent of the motion of the source which emits the light*; that is, the light speed "forgets" the speed of the source. Edmund Whittaker has observed that the situation is somewhat analogous to that of the motion of water waves, where "the waves created by throwing a stone into a pond move outwards from the point where the stone entered the water, without being affected by the velocity of the stone."† The constant speed of light in vacuum is known to be approximately 300,000 kilometers per second.‡

We shall consider the propagation of light in a transparent medium such as air, glass, water, or plastic, or a combination or composition of any of these or similar media. The medium under consideration is assumed to fill a certain region of three-dimensional space, and we allow for the possibility that the speed of light in the medium may vary from point to point. For example, it is known that light travels faster in the upper regions of the earth's atmosphere (where the air density is lower) than in the lower regions near the earth's surface (where the air density is greater). We suppose that a light ray travels through the medium from a given fixed point $\mathbf{a} = (a_1, a_2, a_3)$ to another point $\mathbf{b} = (b_1, b_2, b_3)$, and we seek to distinguish from among all possible paths connecting **a** and **b** the *actual path along which the light ray travels.*

The simplest case to consider is that of an *optically homogeneous* medium (such as a vacuum) in which the speed of light is everywhere constant. In this case it is known experimentally that a light ray propagates from **a** to **b** directly along the straight-line segment connecting **a** and **b**, or along the path of *shortest distance.* A slightly more complicated case is that of two different media such as water and air which are separated by a plane interface, as occurs, for example, when we look at a fish in a pond. In this case the fish is *not* where it appears since the light ray traveling from the fish to the eye of the observer is bent or *refracted* at the interface, as shown in Figure 35. The

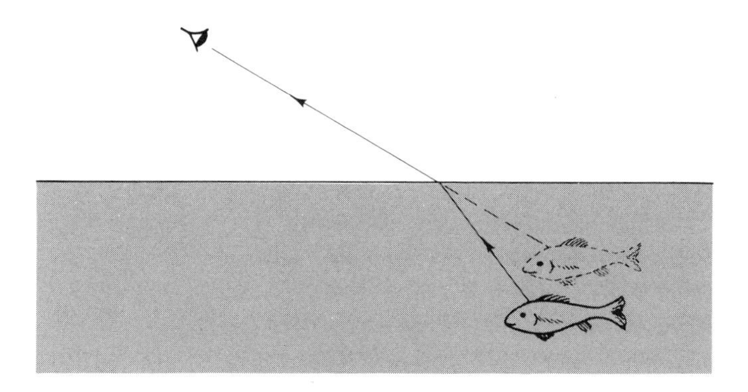

Figure 35

†E. T. Whittaker, *A History of the Theories of Aether and Electricity*, Vol. 2 (London: Thomas Nelson and Sons Ltd., 1953), p. 38. A popular account of the fascinating phenomenon called *light* is given by A. C. S. van Heel and C. H. F. Velzel, *What is Light?*, World University Library (New York: McGraw-Hill Book Company, 1968).

‡See Max Born and Emil Wolf, *Principles of Optics* (Elmsford, N. Y.: Pergamon Press, Inc., 1965), pp. 11–12.

law of refraction for this situation was discovered by Willebrod Snell (1591–1626), who in about 1621† found experimentally that the sines of the angles of incidence and refraction form a ratio with each other which depends only on the two media involved; i e. (see Figure 36),

$$\frac{\sin \theta_1}{\sin \theta_2} = \mu$$

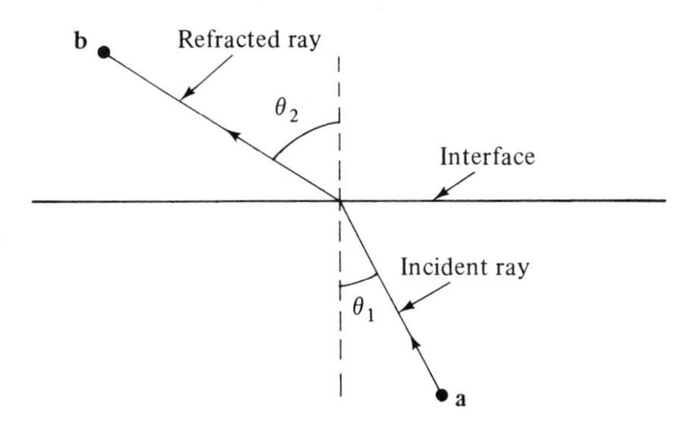

Figure 36

for a fixed constant μ which depends on the given media but not on the angle of incidence of the particular light ray involved. Hence in this case, unlike the former, the light ray refuses to follow the shortest (straight) path.

In 1661, Pierre de Fermat (1601–1665) discovered that the light rays travel along the *quickest paths* in both cases.‡ This is apparent in the former case of a single optically homogeneous medium since the shortest path will give also the quickest path if the speed of motion is everywhere constant. In the latter case of refraction, however, the speed of light changes upon going from one medium to another, and Fermat showed in this case that *the quickest path consists of two line segments connected together but with different slopes which satisfy Snell's law of refraction with* $\mu = v_1/v_2$, where v_1 and v_2 are the speeds of motion of light in the respective media through which the incident and refracted rays travel. Hence Fermat was led to the relation

$$\frac{\sin \theta_1}{v_1} = \frac{\sin \theta_2}{v_2}, \tag{4.7.1}$$

which is often known today as Snell's law of refraction.

†See E. T. Whittaker, *A History of the Theories of Aether and Electricity*, Vol. 1 (London: Thomas Nelson and Sons Ltd., 1910, 1951), p. 10.

‡Fermat followed Descartes in pursuing a theoretical study of the motion of light by conceptually replacing a light ray, whose speed of motion was thought to be infinite, with a moving projectile whose finite speed would change upon going from one medium to another. Hence a certain notion of a finite "speed of light" was introduced even before the establishment of this result by Roemer. See Whittaker, *op. cit.*, Vol. 1, pp. 10–12.

Fermat proposed the general principle that "Nature always acts by the quickest course." We shall formulate this *principle of least time* mathematically for the motion of light rays and discuss its validity and limitations.

As before, we consider any two fixed points $\mathbf{a} = (a_1, a_2, a_3)$ and $\mathbf{b} = (b_1, b_2, b_3)$ located in the medium, and we suppose that a light ray travels through the medium from $\mathbf{a}$ to $\mathbf{b}$ along a path γ given parametrically as

$$\gamma: \quad \mathbf{y} = \mathbf{Y}(\sigma) \qquad \text{for } \sigma_a \leq \sigma \leq \sigma_b \tag{4.7.2}$$

in terms of some suitable vector function $\mathbf{Y}(\sigma) = (Y_1(\sigma), Y_2(\sigma), Y_3(\sigma))$, where σ may be any suitable parameter along γ. The constants σ_a and σ_b are the σ-coordinates of the end points of γ, and the vector function $\mathbf{Y}$ is required to satisfy the conditions

$$\mathbf{Y}(\sigma_a) = \mathbf{a}, \qquad \mathbf{Y}(\sigma_b) = \mathbf{b} \tag{4.7.3}$$

so that γ will connect the given points $\mathbf{a}$ and $\mathbf{b}$. The light signal will travel from $\mathbf{a}$ to $\mathbf{b}$ along γ in time T given as

$$T = \int_0^T dt = \int_\gamma \frac{ds}{v},$$

where s measures the arc length along γ and $ds/dt = v$ is the speed at which the light signal travels. The differential element of arc length along γ can be given in terms of the components of the vector function $\mathbf{Y} = (Y_1, Y_2, Y_3)$ as $ds = [Y_1'(\sigma)^2 + Y_2'(\sigma)^2 + Y_3'(\sigma)^2]^{1/2}\, d\sigma$, where $Y_i'(\sigma) = dY_i/d\sigma$. Hence for the time T we have

$$T(\mathbf{Y}) = \int_{\sigma_a}^{\sigma_b} \frac{\sqrt{Y_1'(\sigma)^2 + Y_2'(\sigma)^2 + Y_3'(\sigma)^2}}{v(\mathbf{Y}(\sigma))}\, d\sigma, \tag{4.7.4}$$

where $v = v(\mathbf{y})$ denotes the speed of light in the medium at the point $\mathbf{y} = (y_1, y_2, y_3)$, and we have taken $\mathbf{y} = \mathbf{Y}(\sigma)$ in (4.7.4). We have written $T = T(\mathbf{Y})$ to indicate that the time is a functional which depends on the choice of the vector function $\mathbf{Y} = \mathbf{Y}(\sigma)$ used in the parametric representation (4.7.2).

Fermat's principle of least time (for the motion of light) now amounts to the assertion that from among all possible functions $\mathbf{Y}$ which satisfy (4.7.3) the *actual* light path γ will correspond to that function $\mathbf{Y}$ which *minimizes* the time functional $T = T(\mathbf{Y})$ of (4.7.4). From Section 4.6 we know that any such minimizing function $\mathbf{Y}$ must satisfy the following Euler-Lagrange equations [explain; see (4.6.9)]

$$\frac{d}{d\sigma}\left[\frac{Y_i'(\sigma)}{v(\mathbf{Y}(\sigma))\sqrt{Y_1'(\sigma)^2 + Y_2'(\sigma)^2 + Y_3'(\sigma)^2}}\right] = -\left[\frac{\partial v(\mathbf{Y}(\sigma))}{\partial y_i}\right]\frac{\sqrt{Y_1'(\sigma)^2 + Y_2'(\sigma)^2 + Y_3'(\sigma)^2}}{v(\mathbf{Y}(\sigma))^2}$$
$$\text{for } \sigma_a < \sigma < \sigma_b \text{ and for } i = 1, 2, 3, \tag{4.7.5}$$

at least provided that the function $v = v(\mathbf{y})$ varies smoothly from point to point in the given medium. Hence, according to Fermat's principle, the differential equations (4.7.5) along with the fixed end-point conditions (4.7.3) should characterize the curve along which a light ray will actually propagate from **a** to **b**. And indeed *this is true*: It was shown at around the beginning of the twentieth century that *Maxwell's electromagnetic theory of light leads to these same equations* (4.7.5) *for the motion of light rays*,† at least for media which satisfy Maxwell's relation‡

$$v = \frac{c}{\sqrt{\epsilon\mu}}.$$

Here c is the speed of light in vacuum, while ϵ is the dielectric constant and μ is the magnetic permeability of the particular medium. The values of both ϵ and μ may vary from point to point in the medium. (They may also depend on *direction*, as occurs, for example, in the case of certain crystals.)

Hence on the basis of Maxwell's electromagnetic theory of light it would appear that Fermat's principle of least time is valid. Of course, a principle of *greatest* time would lead also to the same Euler-Lagrange equations, and there are, in fact, simple cases where a light ray does follow the path of greatest time among all nearby admissible paths. We can obtain such an example by considering the *reflection* of a light ray by a reflecting mirror. The simplest case of reflection occurs when a light ray which is traveling in a homogeneous medium (such as a vacuum) is reflected by a plane mirror. In this case *the angle of reflection is equal to the angle of incidence*, as was known already to Hero of Alexandria.§ It follows that any such light ray traveling from a point **a** to a point **b** via a reflection by a plane mirror will travel along the *shortest possible path* which connects the points via the mirror and hence also along the quickest path. Indeed, if we let **b**′ be the image of **b** in the mirror, as shown in Figure 37, then the length of the actual light path is the same as the length of the straight-line segment connecting **a** and **b**′ (why?). The stated result then follows from the fact that the latter straight-line segment is shorter than any other path which connects **a** and **b**′ (as, for example, the bent-line segment connecting **a** and **b**′ shown in Figure 37). We leave the details of the argument as an exercise for the reader. The conclusion is that Fermat's principle of *least* time is valid in the case of reflection by a plane mirror.

The situation is somewhat different, however, in the case of reflection by a concave *spherical* mirror such as the inside of a reflecting globe. In this case we let y_1, y_2, and y_3 denote the rectangular coordinates of a representative point $\mathbf{y} = (y_1, y_2, y_3)$ in three-dimensional space with the origin of co-

†See Born and Wolf, *op. cit.*, Chapter 3, and Rudolf K. Luneburg, *Mathematical Theory of Optics* (Berkeley: University of California Press, 1964), pp. 25–29.

‡See Born and Wolf, *op. cit.*, pp. 10–14.

§See Whittaker, *op. cit.*, Vol. 1, p. 12.

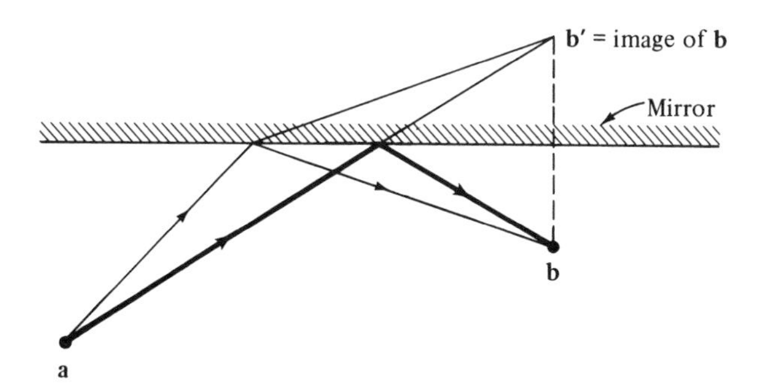

Figure 37

ordinates placed at the center of the spherical globe. Figure 38 shows a cross section of the plane $y_3 = 0$ in which the circle of radius r centered at the origin corresponds to the reflecting mirror. We let l be a positive number less than r, and we consider a light ray traveling from the point $\mathbf{a} = (0, l, 0)$ in a direction toward the point $(r, 0, 0)$. Such a ray will be reflected at $(r, 0, 0)$ back toward the point $\mathbf{b} = (0, -l, 0)$, as shown in Figure 38. We wish to compare the length of the resulting actual light path from **a** to **b** with the lengths of other admissible paths having a single "reflection" off the mirror. (We only consider paths consisting of two line segments meeting at the mirror. Moreover, it can be shown by rotating the coordinate axes suitably about the y_2-axis in three-dimensional space that we need only consider such paths in

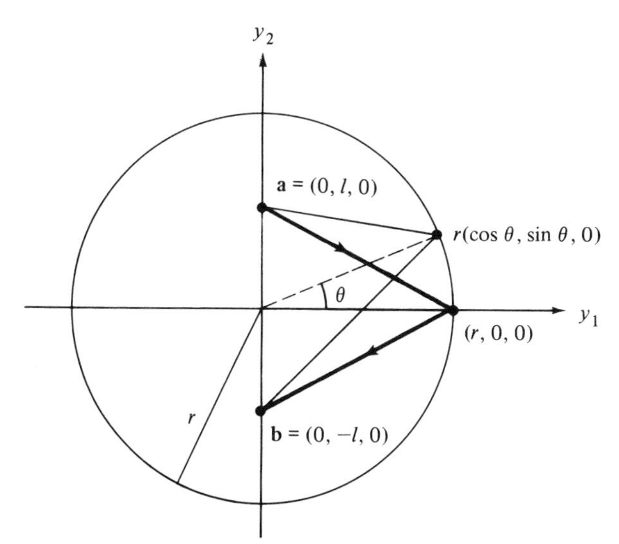

Intersection of a spherical reflecting globe with the plane $y_3 = 0$. ($\mathbf{y} = (y_1, y_2, y_3)$)

Figure 38

the plane $y_3 = 0$.) For convenience we let θ be the angle between the y_1-axis and the radius vector from the origin to the point of reflection for any such admissible path, as shown in Figure 38. Such a path then consists of the line segment from $\mathbf{a} = (0, l, 0)$ to the point $(r \cos \theta, r \sin \theta, 0)$, joined with the line segment from the latter point back to $\mathbf{b} = (0, -l, 0)$. The total length L of this path is given as

$$L = \sqrt{r^2 \cos^2 \theta + (l - r \sin \theta)^2} + \sqrt{r^2 \cos^2 \theta + (l + r \sin \theta)^2}, \tag{4.7.6}$$

and it is easy to verify using the differential calculus that this length is a *maximum* at $\theta = 0$. The conclusion is that the actual light ray follows in this case the path of *greatest* time among all such admissible paths having a single reflection at the mirror. Hence the principle of *least* time is not valid in this case, and indeed Fermat himself was aware of such difficulties involving the reflection of light rays by curved mirrors.†

It is possible to construct similar examples in which actual light rays travel along paths which neither minimize nor maximize the time but which rather furnish *saddle points* for the time functional. (Such an example can be given involving the reflection of a light ray by a right circular cylindrical mirror.) In every case, however, the *variation* of the time functional vanishes for the actual light path $\mathbf{y} = \mathbf{Y}(\sigma)$, that is,

$$\delta T(\mathbf{Y}; \mathbf{\Delta Y}) = 0, \tag{4.7.7}$$

and this is the case for *all* suitable vector functions $\mathbf{\Delta Y}$ for which $\mathbf{Y} + \mathbf{\Delta Y}$ is admissible. It follows that *the actual light path will always satisfy the Euler-Lagrange equations* (4.7.5). Any such vector $\mathbf{Y}$ for which (4.7.7) holds is called a **stationary vector** for the given functional T. Such a stationary vector may be an extremum vector or a saddle point for T. To include all possibilities, therefore, it is customary to refer to Fermat's principle as a principle of *stationary time*. We might mention, however, that the time of motion is indeed always minimized over *successive small segments* of the actual light path. A discussion of the latter result would take us rather far afield, and we shall omit it here.‡

The Motion of Sunlight in the Atmosphere. We shall close this section with an interesting application of Fermat's principle to the motion of sunlight in the earth's atmosphere at sunset. We assume that the speed of light

†*Oeuvres de Fermat*, Vol. 2 (1894), p. 355. The example given above involving the concave spherical mirror is taken from Luneburg, *op. cit.*, p. 87. A similar example which is even simpler is given by Constantin Carathéodory, *Geometrische Optik* (Berlin: Springer-Verlag, 1937), p. 10.

‡See Carathéodory, *op. cit.*, p. 11, and Luneburg, *op. cit.*, p. 87.

at a point in the earth's atmosphere depends only on the *altitude* of the point above the surface of the earth, and, moreover, we shall consider only the simple model of a "flat" earth, as indicated in Figure 39. The qualitative

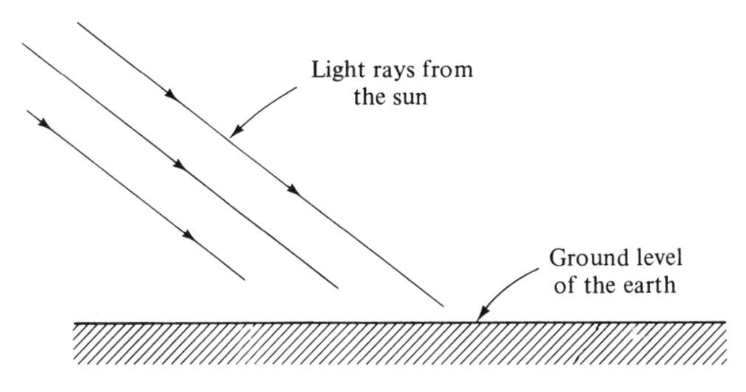

Figure 39

result of interest to us is the same for this model as for the more realistic model of a spherical earth, while the calculation involved is much simpler for a flat earth. We consider light rays traveling in a fixed two-dimensional plane, as indicated in Figure 39. For convenience we introduce a Cartesian (x, y)-coordinate system as shown in Figure 40 with the x-axis coinciding

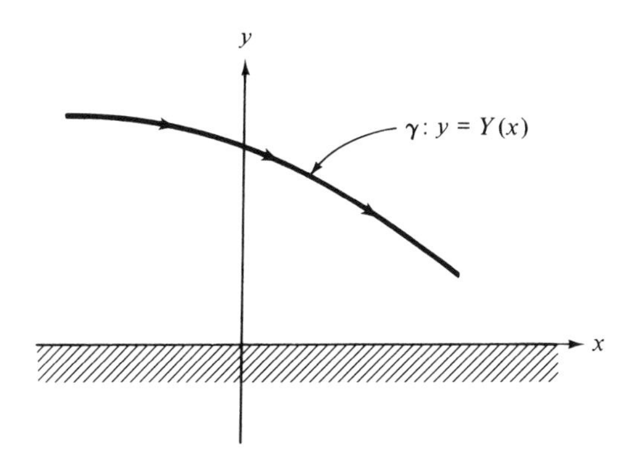

Figure 40

with the earth's surface, and we then represent any such light path as a curve γ given parametrically as

$$\gamma: \quad y = Y(x), \qquad x_0 \leq x \leq x_1, \tag{4.7.8}$$

in terms of some suitable function $Y = Y(x)$. In this situation the light speed v depends only on the y-coordinate, $v = v(y)$, and the time T required for a light ray to travel along γ becomes (explain)

$$T(Y) = \int_{x_0}^{x_1} \frac{\sqrt{1 + Y'(x)^2}}{v(Y(x))}\, dx$$

or

$$T(Y) = \int_{x_0}^{x_1} F(Y(x), Y'(x))\, dx,$$

with $F(y, z) = \sqrt{1 + z^2}/v(y)$. Hence we may use the special form of the Euler-Lagrange equation given by (4.1.15), from which we find that (explain)

$$[1 + Y'(x)^2]v(Y(x))^2 = A,$$

which must hold along any possible extremum (or stationary) curve γ, for some suitable constant of integration A. If we differentiate the last relation on both sides with respect to x, we find after some algebraic manipulation that

$$Y''(x) = -[1 + Y'(x)^2]\frac{v'(Y(x))}{v(Y(x))}, \tag{4.7.9}$$

where v' denotes the derivative of the function $v = v(y)$ with respect to its argument y; i.e.,

$$v'(Y(x)) = \frac{dv(y)}{dy}\bigg|_{y=Y(x)}.$$

Now it is known that the speed of light in the earth's atmosphere *increases* with increasing altitude, so that the function $v = v(y)$ is an increasing function of y. Hence v' is *positive*, and it follows from (4.7.9) on the basis of Fermat's principle that *any actual light path γ given as* (4.7.8) *must satisfy the condition*

$$Y''(x) < 0.$$

Hence all such light rays must curve *concavely toward the earth*, as shown in Figure 40, and therefore *the setting sun appears higher in the sky than it actually is*. The same result can be shown to hold also for a spherical earth, as shown in Figure 41, and, in fact, an observer on the earth continues to "see" the sun for about 5 minutes after it has set.†

Exercises

1. Use Fermat's principle of least time to derive the law of refraction given by (4.7.1).

†See Arnold Sommerfeld, *Optics*, trans. Otto Laporte and Peter A. Moldauer (New York: Academic Press, Inc., 1954), pp. 209 and 340.

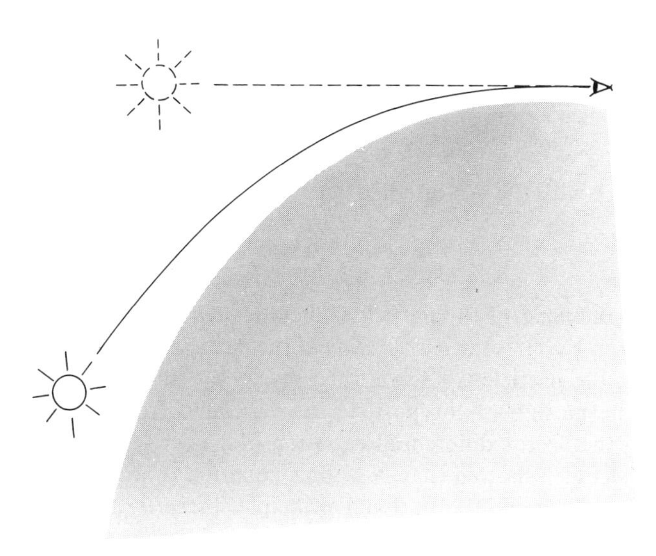

Figure 41

2. Show that an actual light path γ given by (4.7.2) must be a straight line if the speed of light $v = v(\mathbf{y})$ is *everywhere constant* in the given medium. *Hint:* Equations (4.7.5) can be shown in this case to imply that $Y_1'(\sigma) = AY_2'(\sigma) = BY_3'(\sigma)$ for suitable constants A and B, from which the stated result can be shown to follow..

3. Verify carefully that the length $L = L(\theta)$ given by (4.7.6) achieves a maximum value for $\theta = 0$.

4.8. Hamilton's Principle of Stationary Action; an Example on Small Vibrations

The science known today as classical mechanics was founded by such men as Isaac Newton, Christian Huygens, Gottfried Leibniz, and James Bernoulli, and it was further clarified and extended by Leonhard Euler.† These giants interpreted the motions occurring in the natural universe in terms of certain vector quantities such as force and acceleration. For example, *Newton's second law of motion*, which Newton published in 1687 in his great work *Mathematical Principles of Natural Philosophy* [Andrew Motte translation of 1729 revised by Florian Cajori (Berkeley: University of Cali-

†See C. Truesdell, *Essays in the History of Mechanics* (Berlin: Springer-Verlag, 1968) for a sound introduction to the history of mechanics.

fornia Press, 1960)],† states that the total force $\mathbf{F} = (F_1, F_2, F_3)$ which acts on a particle (such as a pendulum bob or a baseball) located at a position $\mathbf{y} = (y_1, y_2, y_3)$ will cause a motion of the particle along a curve γ given as

$$\gamma: \quad \mathbf{y} = \mathbf{Y}(t) \tag{4.8.1}$$

in accordance with the vector equation

$$m\mathbf{Y}'' = \mathbf{F} \tag{4.8.2}$$

where m is the *mass* of the particle, t denotes *time*, and $\mathbf{Y}'' = d^2\mathbf{Y}(t)/dt^2 = (Y_1''(t), Y_2''(t), Y_3''(t))$ is the *acceleration* of the particle. Hence if the total force $\mathbf{F}$ acting on the particle is known, then the vector function $\mathbf{Y} = \mathbf{Y}(t)$ which describes the motion of the particle can be found from Newton's law by integrating the vector differential equation (4.8.2).

In 1744 Euler showed that Newton's equation of motion (4.8.2) can be obtained from a certain variational principle, the *principle of least action.* The roots of this principle along with the roots of Fermat's principle of least time can be traced to a general philosophical principle of economy and harmony in nature. For example, Nicolaus Copernicus (1473–1543) was guided by a strong belief in the simplicity and harmony of nature in his development of the Copernican theory of the universe, which overturned the previously accepted Ptolemaic theory. Similarly, John Kepler believed that nature always follows the simplest and most harmonious course. Euler showed that there was, in fact, a certain quantity which was later called the *action of the motion* and which seemed to be minimized during actual dynamical motions which occur in nature. In 1788, Lagrange showed that a large part of Newtonian dynamics can be derived formally from this principle of least action in the case of *conservative* forces. Later, in 1835, William Hamilton (1805–1865) gave an improved and expanded version of the principle of least action for a wider class of forces, based on a slightly different definition of the *action.* We shall describe Hamilton's version of the principle below, but we shall limit ourselves to the case of conservative forces.

Hence we shall consider only the important special case in which the total force $\mathbf{F} = (F_1, F_2, F_3)$ acting on the particle at the position $\mathbf{y} = (y_1, y_2, y_3)$ can be given as the (negative of the) gradient of a real-valued function

†In this work Newton stated his second law of motion in words only. The law was first used in its now familiar form in terms of fixed rectangular Cartesian coordinates by John Bernoulli in 1742, as pointed out by C. Truesdell on p. 252 of his valuable work "The Rational Mechanics of Flexible or Elastic Bodies, 1638–1788", L. Euleri *Opera Omnia*, (2) **11**$_2$, 428pp. (Zurich: Fussli, 1960). In this work Truesdell includes a careful historical study of the transition from Newton's ideas *as they were understood by Newton's followers* to the eventual realization by Euler in 1750 that "Newton's equations" in Cartesian coordinates *actually provide the equations of motion* for a wide range of complicated mechanical systems.

$V = V(\mathbf{y})$ as

$$\mathbf{F} = -\nabla V, \tag{4.8.3}$$

or $F_i = -\partial V/\partial y_i$ for $i = 1, 2, 3$. We could equally well write $\mathbf{F} = +\nabla U$ with $U = -V$, but it is customary in physics to include the minus sign, as in (4.8.3). For example, the Newtonian gravitational force† exerted on a particle of mass m at $\mathbf{y}$ due to a body of mass M located at the origin can be given as

$$\begin{aligned} \mathbf{F} &= -\frac{GMm\mathbf{y}}{\|\mathbf{y}\|^3} \\ &= -\frac{GMm\mathbf{y}}{(y_1^2 + y_2^2 + y_3^2)^{3/2}}, \end{aligned}$$

which can be written in the form of (4.8.3) with

$$V = -\frac{GMm}{\|\mathbf{y}\|} = -GMm(y_1^2 + y_2^2 + y_3^2)^{-1/2}.$$

Here G is the Newtonian gravitational constant.‡ The function $V = V(\mathbf{y})$ occurring in (4.8.3) was introduced by Lagrange in 1779 and is called the *potential* (or *potential energy*) associated with a particle of mass m located at $\mathbf{y}$ in the presence of the force $\mathbf{F} = -\nabla V$. The *kinetic energy* of such a particle moving along a path γ given as (4.8.1) is denoted as T and is defined by

$$\begin{aligned} T &= \tfrac{1}{2}m\|\mathbf{Y}'\|^2 \\ &= \tfrac{1}{2}m[Y_1'(t)^2 + Y_2'(t)^2 + Y_3'(t)^2]. \end{aligned} \tag{4.8.4}$$

It is a consequence of Newton's second law of motion and the definitions of kinetic and potential energy that *the total energy* $T + V$ *is conserved* (i.e., $T + V$ remains constant) *during the motion of a particle in the presence of a force* $\mathbf{F}$ *which can be given as* $\mathbf{F} = -\nabla V$. Indeed, one can easily show that the derivative of $T + V$ with respect to time t must vanish identically during any such motion, and the stated result then follows. It is for this reason that

†Newton showed that the dynamical motions of such terrestrial objects as pendulums, projectiles, and falling apples as well as the motions of the planetary bodies and satellites in the solar system can all be predicted from the same laws of motion using the single assumption that each particle of matter behaves as if it attracts every other particle with a force which is inversely proportional to the square of the distance between them and directly proportional to the product of their masses. This *universal law of gravitation* was published along with certain laws of motion in 1687 in the single most important work in the history of science: Newton's *Mathematical Principles of Natural Philosophy*.

‡See William Cecil Dampier, *A History of Science*, 4th ed. (London: Cambridge University Press, 1961), p. 178, or J. M. A. Danby, *Fundamentals of Celestial Mechanics* (New York: The Macmillan Company, 1962), pp. 62 and 73, for the actual numerical value of G.

any such force $\mathbf{F}$ which can be given in the form (4.8.3) is said to be *conservative.*

We consider the motion of a particle beginning at a fixed point $\mathbf{a} = (a_1, a_2, a_3)$ at time $t = t_a$ and ending at a point $\mathbf{b} = (b_1, b_2, b_3)$ at time $t = t_b$, and we describe the motion of the particle along its path γ as

$$\gamma: \quad \mathbf{y} = \mathbf{Y}(t) \qquad \text{for } t_a \leq t \leq t_b \tag{4.8.5}$$

in terms of the time t as parameter. The vector function $\mathbf{Y} = \mathbf{Y}(t)$ is required to satisfy the end-point conditions

$$\mathbf{Y}(t_a) = \mathbf{a}, \qquad \mathbf{Y}(t_b) = \mathbf{b}, \tag{4.8.6}$$

and the motion is assumed to be caused by a conservative force $\mathbf{F} = -\nabla V$. The *actual* motion will be determined by the conditions (4.8.6) along with Newton's second law of motion as given by the differential equation (4.8.2). However, we wish to consider other hypothetical motions in addition to the actual motion, so for this purpose *we allow the vector function* $\mathbf{Y} = \mathbf{Y}(t)$ *of* (4.8.5) *to be any continuously differentiable vector function which need not satisfy Newton's equation* (4.8.2). The kinetic energy of motion along any such curve γ is again defined by (4.8.4), while the potential energy at $\mathbf{y} = \mathbf{Y}(t)$ is given as $V = V(\mathbf{Y}(t))$. The **action** of such a hypothetical motion of a particle of mass m along such a path γ is defined (following Hamilton) as

$$A = \int_{t_a}^{t_b} (T - V)\,dt$$
$$= \int_{t_a}^{t_b} [\tfrac{1}{2}m\,||\mathbf{Y}'(t)||^2 - V(\mathbf{Y}(t))]\,dt$$

or

$$A(\mathbf{Y}) = \int_{t_a}^{t_b} L(\mathbf{Y}(t), \mathbf{Y}'(t))\,dt, \tag{4.8.7}$$

where the integrand function $L = L(\mathbf{y}, \mathbf{z})$ is given as

$$L(\mathbf{y}, \mathbf{z}) = T(\mathbf{z}) - V(\mathbf{y}) \tag{4.8.8}$$

and where [see (4.8.4)]

$$T(\mathbf{z}) = \tfrac{1}{2}m\,||\mathbf{z}||^2 = \tfrac{1}{2}m(z_1^2 + z_2^2 + z_3^2)$$

for any vector $\mathbf{z} = (z_1, z_2, z_3)$. The function $L = T - V$ is called the *Lagrangian function* of the motion and is evaluated along $\mathbf{y} = \mathbf{Y}(t)$ and $\mathbf{z} = \mathbf{Y}'(t)$ in (4.8.7). We have written $A = A(\mathbf{Y})$ to indicate that the action is a functional of the motion which depends on the particular choice of the vector function $\mathbf{Y} = \mathbf{Y}(t)$ used in (4.8.5).

Hamilton's principle of least action (for conservative forces) now amounts to the assertion that from among all (actual or hypothetical) motions which begin at **a** at time $t = t_a$ and end at **b** at time $t = t_b$ the particle will actually experience that motion which *minimizes the action*; i.e., the actual motion will correspond to the function $\mathbf{Y} = \mathbf{Y}(t)$ which minimizes the action functional $A = A(\mathbf{Y})$ among all continuously differentiable functions $\mathbf{Y}$ which satisfy (4.8.6). Indeed, we know from Section 4.6 that any such minimizing function $\mathbf{Y}$ for $A = A(\mathbf{Y})$ must satisfy the Euler-Lagrange equations [see (4.6.9)]

$$L_{Y_i}(\mathbf{Y}(t), \mathbf{Y}'(t)) = \frac{d}{dt} L_{Y_i'}(\mathbf{Y}(t), \mathbf{Y}'(t)) \qquad \text{for } i = 1, 2, 3 \tag{4.8.9}$$

for $t_a < t < t_b$. From (4.8.8) we find that

$$\begin{aligned} L_{Y_i'}(\mathbf{Y}(t), \mathbf{Y}'(t)) &= \frac{\partial}{\partial z_i} L(\mathbf{y}, \mathbf{z})\bigg|_{\substack{\mathbf{y}=\mathbf{Y}(t)\\ \mathbf{z}=\mathbf{Y}'(t)}} \\ &= \frac{\partial}{\partial z_i} T(\mathbf{z})\bigg|_{\mathbf{z}=\mathbf{Y}'(t)} = mz_i\bigg|_{\mathbf{z}=\mathbf{Y}'(t)} = mY_i'(t) \end{aligned}$$

and that

$$\begin{aligned} L_{Y_i}(\mathbf{Y}(t), \mathbf{Y}'(t)) &= \frac{\partial}{\partial y_i} L(\mathbf{y}, \mathbf{z})\bigg|_{\substack{\mathbf{y}=\mathbf{Y}(t)\\ \mathbf{z}=\mathbf{Y}'(t)}} \\ &= -\frac{\partial}{\partial y_i} V(\mathbf{y})\bigg|_{\mathbf{y}=\mathbf{Y}(t)} = F_i(\mathbf{Y}(t)), \end{aligned}$$

where $F_i = -\partial V/\partial y_i$ is the ith component of the force vector $\mathbf{F} = -\nabla V$. Hence in the present case the Euler-Lagrange equations (4.8.9) can be rewritten as the vector equation

$$\mathbf{F} = m\mathbf{Y}'',$$

which agrees with Newton's equation of motion (4.8.2).

Hence the principle of least action does indeed lead back to Newton's law of motion. Of course, a principle of *greatest* action will do the same, and, in fact, in 1842 Carl Jacobi (1804–1851) gave an example of a simple dynamical motion for which the action is a *maximum* among all nearby admissible paths.† In any case, however, the actual motion will always satisfy the Euler-Lagrange equations (4.8.9). Hence *the variation of the action functional will always vanish* for the actual motion $\mathbf{y} = \mathbf{Y}(t)$; i.e.,

$$\delta A(\mathbf{Y}; \mathbf{\Delta Y}) = 0$$

†C. G. J. Jacobi, *Vorlesungen über Dynamik* (Berlin: Druck and Verlag von Georg Reimer, 1866), pp. 43–49. A similar example can be found in Magnus R. Hestenes, "Elements of the Calculus of Variations," in *Modern Mathematics for the Engineer* (Edwin F. Beckenback, ed.) (New York: McGraw-Hill Book Company, 1956), p. 77.

for all suitable vectors $\Delta\mathbf{Y}$ for which $\mathbf{Y} + \Delta\mathbf{Y}$ is admissible. Just as in the case of Fermat's principle, therefore, it is customary to refer to Hamilton's principle as a *principle of stationary action.* In fact, the action is always *minimized* over *short* time intervals, but a discussion of this result would take us beyond the intended scope of this book.

A Vibration Problem. Hamilton's principle is actually valid in a much wider context than for the motion of a *single* particle, and we shall close this section with an application of the principle to a problem on the motion of several particles. Specifically, we shall consider the *vibration of beads on an elastic string.* For simplicity, however, we shall consider a case involving only two beads. The beads are attached to a light elastic string of length $4l$ which is stretched at a large tension τ between two fixed points, as indicated in Figure 42. One bead of mass m is located at the position $2l$, and a heavier

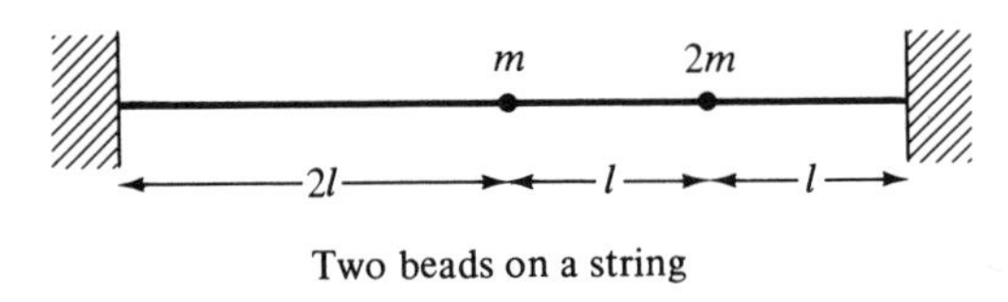

Two beads on a string

Figure 42

bead of mass $2m$ is located at the position $3l$. We consider only transverse vibrations in a fixed vertical plane, and we let $Y_1 = Y_1(t)$ and $Y_2 = Y_2(t)$ be the perpendicular displacements of the two beads from the equilibrium position of the string, as indicated in Figure 43.

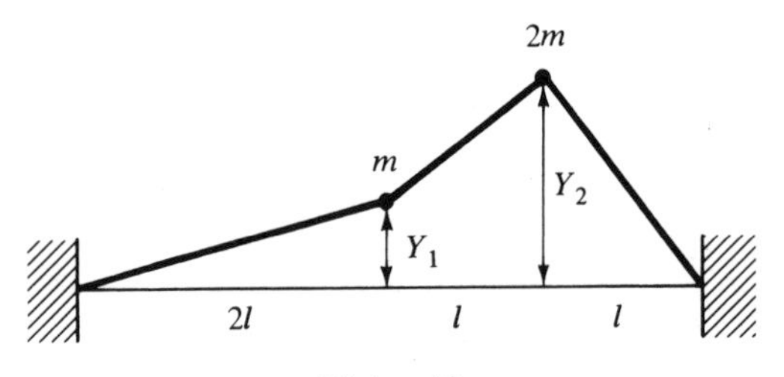

Figure 43

The kinetic energy of motion of the first bead of mass m is given as

$$T_1 = \tfrac{1}{2} m Y_1'(t)^2,$$

while the kinetic energy of motion of the second bead of mass $2m$ is

$$T_2 = \tfrac{1}{2}(2m) Y_2'(t)^2 = m Y_2'(t)^2.$$

We neglect the mass of the light string, so the total kinetic energy of the vibrating system will be the sum of T_1 and T_2,

$$T = m[\tfrac{1}{2}Y'_1(t)^2 + Y'_2(t)^2].$$

We assume that the potential energy of the system due to the stretching of the elastic string is proportional to the amount by which the string has been stretched. At time t the left half of the string is stretched by an amount equal to (see Figure 43)

$$\sqrt{(2l)^2 + Y_1(t)^2} - 2l = 2l\left(\sqrt{1 + \left(\frac{Y_1(t)}{2l}\right)^2} - 1\right).$$

We consider only small vibrations for which the displacements are small, and hence we can safely approximate the quantity $\sqrt{1 + (Y_1/2l)^2}$ by the quantity†

$$1 + \frac{1}{2}\left(\frac{Y_1}{2l}\right)^2. \tag{4.8.10}$$

Hence the stretch of the left half of the string can be given (approximately) as

$$\frac{Y_1(t)^2}{4l}.$$

Similarly, we find for the stretch of the portion of the string between the two beads

$$\frac{[Y_2(t) - Y_1(t)]^2}{2l},$$

and for the remaining piece of string

$$\frac{Y_2(t)^2}{2l}.$$

The total stretch is the sum of these three expressions. The potential energy, which is proportional to the stretch, is then given as

$$V = \tau\left\{\frac{Y_1(t)^2}{4l} + \frac{[Y_2(t) - Y_1(t)]^2}{2l} + \frac{Y_2(t)^2}{2l}\right\},$$

where the proportionality factor τ is a given positive constant which is known as the *tension* of the string.

†If we apply equation (3.7.27) to the function $f(x) = \sqrt{1 + x}$ and then set $x = u^2$ in the resulting equation, we find after a simple estimation that $0 \leq 1 + \frac{1}{2}u^2 - \sqrt{1 + u^2} \leq u^4/8$ for all numbers u. Hence if the displacements are small enough so that $u = |Y_1/2l| \leq .1$, then the two quantities $1 + \frac{1}{2}(Y_1/2l)^2$ and $\sqrt{1 + (Y_1/2l)^2}$ will differ by less than .000 012 5.

The *action* of the motion during a given time interval $t_0 \le t \le t_1$ is given by the integral

$$A = \int_{t_0}^{t_1} (T - V)\,dt$$
$$= \int_{t_0}^{t_1} [T(\mathbf{Y}'(t)) - V(\mathbf{Y}(t))]\,dt,$$

where $V = V(\mathbf{y})$ and $T = T(\mathbf{z})$ are defined in this case by

$$V(\mathbf{y}) = \tau\left[\frac{y_1^2}{4l} + \frac{(y_2 - y_1)^2}{2l} + \frac{y_2^2}{2l}\right]$$

and

$$T(\mathbf{z}) = m(\tfrac{1}{2}z_1^2 + z_2^2)$$

for any vectors $\mathbf{y} = (y_1, y_2)$ and $\mathbf{z} = (z_1, z_2)$ in $\mathcal{R}_2$. According to Hamilton's principle, the actual motion of the system will furnish a minimum (or at least a stationary) value to this action functional from among all possible motions which coincide with the actual motion at times $t = t_0$ and $t = t_1$. Hence we are led to the following Euler-Lagrange equations for the actual motion of the system,

$$-V_{Y_i}(\mathbf{Y}(t)) = \frac{d}{dt} T_{Y_i'}(\mathbf{Y}'(t)) \qquad \text{for } i = 1, 2,$$

where

$$V_{Y_1}(\mathbf{Y}(t)) = \frac{\partial}{\partial y_1} V(\mathbf{y})\Big|_{\mathbf{y}=\mathbf{Y}(t)} = \tau\left[\frac{Y_1(t)}{2l} + \frac{Y_1(t) - Y_2(t)}{l}\right]$$

$$V_{Y_2}(\mathbf{Y}(t)) = \frac{\partial}{\partial y_2} V(\mathbf{y})\Big|_{\mathbf{y}=\mathbf{Y}(t)} = \tau\left[\frac{Y_2(t) - Y_1(t)}{l} + \frac{Y_2(t)}{l}\right]$$

$$T_{Y_1'}(\mathbf{Y}'(t)) = \frac{\partial}{\partial z_1} T(\mathbf{z})\Big|_{\mathbf{z}=\mathbf{Y}'(t)} = mY_1'(t)$$

and

$$T_{Y_2'}(\mathbf{Y}'(t)) = \frac{\partial}{\partial z_2} T(\mathbf{z})\Big|_{\mathbf{z}=\mathbf{Y}'(t)} = 2mY_2'(t),$$

so that the differential equations of motion become

$$Y_1'' = \frac{\tau}{2lm}(-3Y_1 + 2Y_2)$$

$$Y_2'' = \frac{\tau}{2lm}(Y_1 - 2Y_2).$$

These equations can, of course, be derived alternatively directly from Newton's second law of motion.

We can most easily solve the present *coupled* system of differential equations for $Y_1 = Y_1(t)$ and $Y_2 = Y_2(t)$ by first introducing new unknown functions, say $U_1 = U_1(t)$ and $U_2 = U_2(t)$, in terms of which the equations become *uncoupled.* For example, if we multiply the second equation by a constant α and add the resulting equation to the first equation, we find that

$$\frac{d^2}{dt^2}[Y_1 + \alpha Y_2] = \frac{\tau(\alpha - 3)}{2lm}\left[Y_1 + \frac{2 - 2\alpha}{\alpha - 3}Y_2\right], \qquad (4.8.11)$$

which will be of the form

$$U'' = \text{constant} \cdot U$$

for $U = Y_1 + \alpha Y_2$ provided that we take α to satisfy the condition

$$\alpha = \frac{2 - 2\alpha}{\alpha - 3}.$$

This condition imples that $\alpha^2 - \alpha - 2 = 0$, which has the solutions $\alpha = 2$ and $\alpha = -1$. Taking first $\alpha = 2$ and then $\alpha = -1$ in (4.8.11), we find that

$$\begin{aligned} U_1'' &= -\omega^2 U_1 \\ U_2'' &= -4\omega^2 U_2 \end{aligned} \qquad (4.8.12)$$

with

$$\begin{aligned} U_1 &= Y_1 + 2Y_2 \\ U_2 &= Y_1 - Y_2 \end{aligned}$$

and $\omega^2 = \tau/2lm$. The displacements Y_1 and Y_2 can be obtained from U_1 and U_2 through

$$\begin{aligned} Y_1 &= \frac{U_1 + 2U_2}{3} \\ Y_2 &= \frac{U_1 - U_2}{3}, \end{aligned} \qquad (4.8.13)$$

as the reader can verify.

Each differential equation in (4.8.12) involves only *one* of the unknown functions U_1 or U_2, and these equations can easily be solved subject to any given initial condition. In fact, the most general solutions of these equations have the form†

$$\begin{aligned} U_1(t) &= a_1 \cos \omega(t - \theta_1) \\ U_2(t) &= a_2 \cos 2\omega(t - \theta_2) \end{aligned} \qquad (4.8.14)$$

†See, for example, Thomas, *op. cit.*, Chapter 20, or any elementary text on differential equations.

for arbitrary constants of integration a_1, a_2, θ_1, and θ_2, which are determined in any given situation by specifying the initial displacements and velocities at, say, $t = 0$.

For example, if initially the two beads are each pulled up a small distance b from the equilibrium position and released from rest at time $t = 0$, then initially we have

$$Y_1(0) = b, \qquad Y_1'(0) = 0$$
$$Y_2(0) = b, \qquad Y_2'(0) = 0,$$

with

$$U_1(0) = 3b, \qquad U_1'(0) = 0$$
$$U_2(0) = 0, \qquad U_2'(0) = 0.$$

If we impose these initial conditions on the solutions $U_1(t)$ and $U_2(t)$ in (4.8.14), we find for U_1 and U_2 that

$$U_1(t) = 3b \cos \omega t$$
$$U_2(t) = 0 \qquad \text{for } t \geq 0,$$

and then Y_1 and Y_2 become

$$Y_1(t) = b \cos \omega t$$
$$Y_2(t) = b \cos \omega t.$$

Hence in this case, and indeed in any case for which $a_2 = 0$ in (4.8.14), both particles oscillate with the same frequency ω, and $Y_1(t) = Y_2(t)$ holds for all time $t \geq 0$, as illustrated in Figure 44.

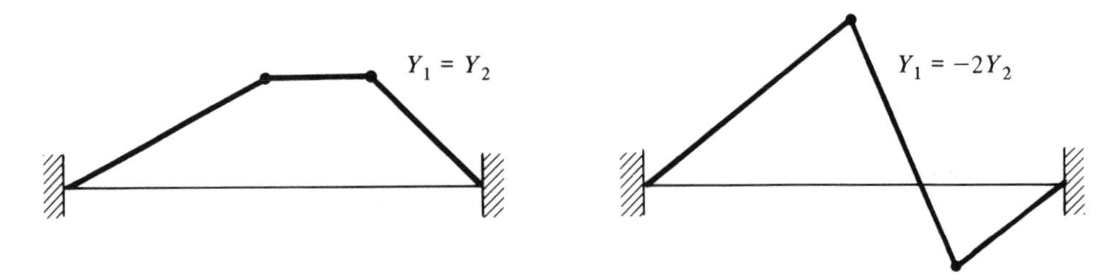

Fundamental modes of vibration

Figure 44

On the other hand, if initially the larger bead located at $3l$ is pulled *down* a small distance c while the smaller bead located at $2l$ is pulled *up* a distance $2c$ and if both beads are then released from rest at time $t = 0$, the initial conditions become

$$Y_1(0) = 2c, \qquad Y_1'(0) = 0$$
$$Y_2(0) = -c, \qquad Y_2'(0) = 0$$

with

$$U_1(0) = 0, \qquad U_1'(0) = 0$$
$$U_2(0) = 3c, \qquad U_2'(0) = 0.$$

The appropriate solutions in this case are found from (4.8.14) to be given as

$$U_1(t) = 0$$
$$U_2(t) = 3c \cos 2\omega t \qquad \text{for } t \geq 0,$$

with

$$Y_1(t) = 2c \cos 2\omega t$$
$$Y_2(t) = -c \cos 2\omega t.$$

In this case, as in any other case for which $a_1 = 0$ in (4.8.14), both particles oscillate with the same frequency 2ω, and $Y_1(t) = -2Y_2(t)$ holds for all $t \geq 0$, as illustrated in Figure 44.

These special vibrations for which both particles vibrate with the same frequency are known as *fundamental modes of vibration*. It is clear from equations (4.8.13) and (4.8.14) that the general motion of the system involves a superposition of the two fundamental modes. The phenomenon of fundamental modes of vibration for vibrating systems was discovered by Daniel Bernoulli (1700–1782), a son of John Bernoulli. (For the details of this discovery see Section 23 of C. Truesdell, "The Rational Mechanics of Flexible or Elastic Bodies, 1638–1788" which is cited at the beginning of this section.)

The reader who wishes a more thorough and systematic discussion of the general theory of small vibrations can consult Pars† or Goldstein.‡

Excerises

1. Find the displacement functions $Y_1 = Y_1(t)$ and $Y_2 = Y_2(t)$ which describe the subsequent motion of the beads shown in Figures 42 and 43 if initially the smaller bead located at $2l$ is pulled up a small distance $2a$ while the larger bead located at $3l$ is pulled up a distance a and the beads are then released from rest at time $t = 0$.

†L. A. Pars, *A Treatise on Analytical Dynamics* (New York: John Wiley & Sons, Inc., 1965), Chapter 9. This reference gives a clear discussion of the general theory along with several very interesting examples, including the example we have presented above.

‡Herbert Goldstein, *Classical Mechanics* (Reading, Mass.: Addison-Wesley Publishing Company, Inc., 1959), Chapter 10.

2. Assume that the tension τ satisfies the condition $\tau = 2lm$ in Exercise 1, and then sketch in the (t, y)-plane the graphs of the displacement functions $y = Y_1(t)$ and $y = Y_2(t)$ for $0 \leq t \leq 2\pi$.

3. Use the sketches of Exercise 2 to sketch the instantaneous configuration of the elastic string and beads in the (x, y)-plane at several different instants for t between $t = 0$ and $t = 2\pi$.

4.9. The McShane-Blankinship Curtain Rod Problem; Functionals Involving Higher-Order Derivatives

We first consider a problem in elasticity on the bending of a long thin elastic rod. To fix the ideas, we shall think of a long straight curtain rod of uniform cross section, density, and elasticity which is forced to pass through three given fixed points P_1, P_2, and P_3, as indicated in Figure 45. The rod is to

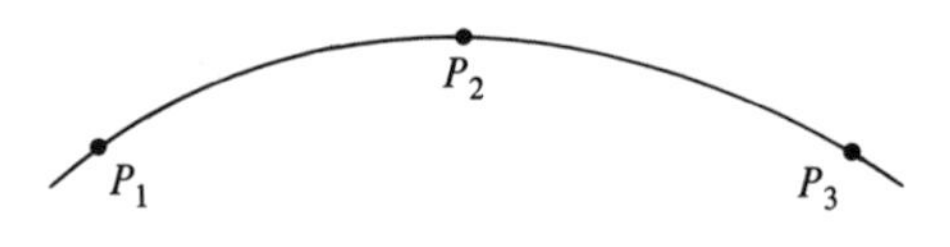

Figure 45

pass through three rings attached firmly to fixed supports located, respectively, at P_1, P_2, and P_3 so as to cause the rod to pass across a curved window. The outermost points P_1 and P_3 are assumed to be located near the ends of the rod, and we assume that the curved rod itself lies in a fixed plane. The supports located at P_1, P_2, and P_3 provide suitable forces $\mathbf{F}_1$, $\mathbf{F}_2$, and $\mathbf{F}_3$ which act at these points so as to hold the rod in a position of equilibrium, as shown in Figure 46. In this position the elastic forces of resistance of the bent rod are in equilibrium with the external forces, and each particle of the rod remains at rest.

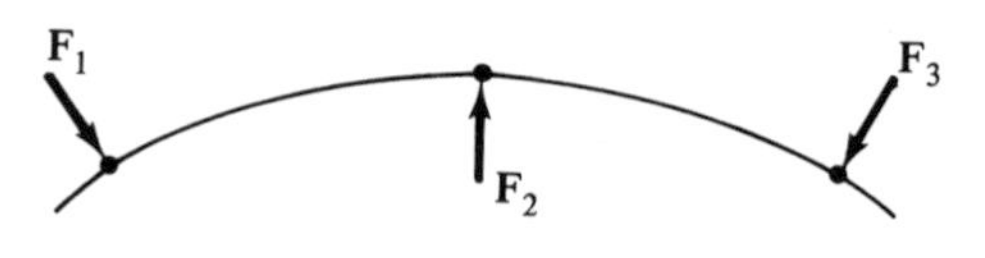

Figure 46

We place a Cartesian (x, y)-coordinate system in the plane of the rod so that the x-axis passes through the points P_1 and P_3 while the y-axis passes through P_2, as shown in Figure 47. We shall only consider the special case in which the three points are symmetrically located so that we may take $P_1 = (-a, 0)$, $P_2 = (0, b)$, and $P_3 = (a, 0)$ for suitable fixed positive num-

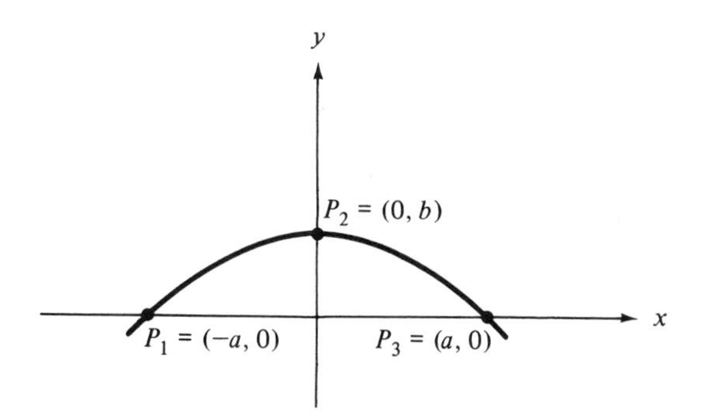

Figure 47

bers a and b. We replace the thin rod with the curve γ, which coincides with the central line of the rod, given as

$$\gamma\colon\ y = Y(x) \qquad \text{for } -a \leq x \leq a$$

for some suitable function $Y = Y(x)$. The rod actually extends slightly beyond the end points P_1 and P_3, but we shall only be interested in determining γ for $-a \leq x \leq a$. In fact, on physical grounds we expect γ to be symmetric about the y-axis, as shown in Figure 47, with

$$Y(-x) = Y(x),$$

and for this reason it will be enough to determine γ on the interval $0 \leq x \leq a$. Hence we shall consider curves γ given as

$$\gamma\colon\ y = Y(x) \qquad \text{for } 0 \leq x \leq a \tag{4.9.1}$$

subject to the constraints

$$Y(0) = b, \qquad Y(a) = 0, \qquad Y'(0) = 0. \tag{4.9.2}$$

The last constraint $Y'(0) = 0$ follows from the required symmetry of the elastic rod for $-a \leq x \leq a$, while the first two constraints ensure that the rod passes through the given fixed points.

During the period 1691–1705 James Bernoulli investigated the bending of elastic rods in terms of a certain theoretical model which was based on the assumption that the resistance of a bent rod is caused by the elongation and contraction of certain long fibres which compose the rod.† Daniel Bernoulli, a son of John, used the results of his uncle James to show in 1738 that the *work of deformation* done in bending a straight elastic rod is *proportional to the square of the curvature* of the resulting bent rod. Thus the work of defor-

†See Sections 12 and 13 of C. Truesdell, "The Rational Mechanics of Flexible or Elastic Bodies, 1638–1788."

mation V expended in the present case in bending the straight curtain rod into the curve γ can be given as the line integral

$$V = \alpha \int_{\text{rod}} \kappa^2 \, ds,$$

where α is a positive proportionality constant which depends on the material properties of the rod, κ is the *curvature* of the rod, which can be given as†

$$\kappa = \frac{Y''(x)}{[1 + Y'(x)^2]^{3/2}},$$

and ds is the differential element of arc length, which can be given as

$$ds = \sqrt{1 + Y'(x)^2} \, dx.$$

Hence the work of deformation can be written as

$$V = 2\alpha \int_0^a \frac{Y''(x)^2}{[1 + Y'(x)^2]^{5/2}} \, dx, \tag{4.9.3}$$

where we have included the factor of 2 to account for the symmetrical left half of the rod corresponding to the interval $-a \leq x \leq 0$. This work which is done in deforming the elastic rod is stored in the bent rod as **potential energy of strain**; it can be recovered by removing the supports which hold the rod at P_1, P_2, and P_3, thereby allowing the rod to snap back into its original straight configuration.‡

In 1738 Daniel Bernoulli suggested to Euler that from among all possible curves, an elastic rod subject to specified external forces of constraint will, when in stable equilibrium, actually bend along that particular curve which *minimizes the potential energy of strain* subject to the appropriate constraints. And, indeed, Euler showed that this principle of least potential energy of strain does lead to the same differential equation for the bent rod as is obtained directly from the principles of mechanics. Euler also classified the possible different types of solutions of the resulting differential equation for

†See Kline, *op. cit.*, Part One, p. 458.

‡The modern theory of elasticity is based on a somewhat different and more accurate model than that used by James and Daniel Bernoulli, and a somewhat different expression than (4.9.3) is obtained in the modern theory for the potential energy of strain. See, for example, I. S. Sokolnikoff, *Mathematical Theory of Elasticity* (New York: McGraw-Hill Book Company, 1946), p. 278, for a more accurate expression for the potential energy of strain. The approximate result (4.9.3) has nevertheless proved to be satisfactory in a wide range of engineering applications, and we shall use it here. See R. V. Southwell, *An Introduction to the Theory of Elasticity for Engineers and Physicists* (London: Oxford University Press, 1959), p. 46.

the case of an elastic rod of *fixed* length with both ends clamped.† We shall use Bernoulli's principle of least potential energy of strain to solve the curtain rod problem posed above. Thus we shall seek the curve γ which minimizes the potential energy functional of (4.9.3) among all curves given as (4.9.1) which satisfy the constraints of (4.9.2). Note that we do *not* require that γ have a constant length.‡

The functional $V = V(Y)$ of (4.9.3) is an example of a wider class of functionals which have the general form

$$J(Y) = \int_{x_0}^{x_1} F(x, Y(x), Y'(x), Y''(x))\,dx, \tag{4.9.4}$$

where in any particular case the function $F = F(x, y, z, w)$ is a specified given function defined for ali points $(x, y, z\ w)$ in some open set in four-dimensional Euclidean space $\mathcal{R}_4$. The potential energy functional given by formula (4.9.3) is of the form (4.9.4), with

$$F(x, y, z, w) = \frac{2\alpha w^2}{[1 + z^2]^{5/2}} \tag{4.9.5}$$

and $x_0 = 0$, $x_1 = a$. It is understood in formula (4.9.4) that $F(x, y, z, w)$ is evaluated at $y = Y(x)$, $z = Y'(x)$, and $w = Y''(x)$ for any x in the interval $[x_0, x_1]$, where $Y = Y(x)$ may be any suitable function of class $\mathcal{C}^2$ on $[x_0, x_1]$. We shall use the norm on the vector space $\mathcal{C}^2[x_0, x_1]$ given by formula (1.4.9) with $k = 2$, so that

$$||Y|| = \max_{x_0 \le x \le x_1} |Y(x)| + \max_{x_0 \le x \le x_1} |Y'(x)| + \max_{x_0 \le x \le x_1} |Y''(x)|$$

for any vector Y in $\mathcal{C}^2[x_0, x_1]$. We assume that the functional J is defined by (4.9.4) for all vectors Y in some open subset D of the normed vector space $\mathcal{C}^2[x_0, x_1]$. For example, if $F = F(x, y, z, w)$ is defined by (4.9.5), then we can take D to be the entire vector space $\mathcal{C}^2[x_0, x_1]$.

In general we may wish to find a local extremum vector Y in $D[K_i = k_i$ for $i = 1, 2, \ldots, m]$ for J, where $K_1, K_2, \ldots, K_m$ are given constraint functionals defined on D and where $k_1, k_2, \ldots, k_m$ are certain specified constants. We shall use the Euler-Lagrange multiplier theorem for this purpose, and therefore we shall need to calculate the variation of the functional J

†See Truesdell, "The Rational Mechanics of Flexible or Elastic Bodies, 1638–1788," 199–210.

‡This problem was solved in 1943 by W. A. Blankinship, who credited E. J. McShane with the statement of the problem. See W. A. Blankinship, "The Curtain Rod Problem," *American Mathematical Monthly*, **50** (1943), 186–189. Actually, the analysis used by Euler to solve a related problem for a rod of fixed length with both ends clamped can also be used to solve the present curtain rod problem.

given by (4.9.4). For this purpose we use (4.9.4) to calculate

$$J(Y + \epsilon\,\Delta Y) = \int_{x_0}^{x_1} F(x, Y(x) + \epsilon\,\Delta Y(x), Y'(x) + \epsilon\,\Delta Y'(x), Y''(x) + \epsilon\,\Delta Y''(x))\,dx$$

for any fixed vector Y in D, for any vector ΔY in $\mathfrak{C}^2[x_0, x_1]$, and for any small number ϵ. From this equation we find that [compare with (2.4.14)]

$$\frac{d}{d\epsilon} J(Y + \epsilon\,\Delta Y) = \int_{x_0}^{x_1} \frac{\partial}{\partial \epsilon} F(x, Y(x) + \epsilon\,\Delta Y(x), Y'(x) + \epsilon\,\Delta Y'(x), Y''(x) + \epsilon\,\Delta Y''(x))\,dx. \tag{4.9.6}$$

If we use the chain rule of differential calculus to calculate the derivative $\partial F/\partial \epsilon$ appearing in the last integral and then evaluate both sides of (4.9.6) at $\epsilon = 0$, we find that [compare with (2.4.17)]

$$\begin{aligned}\delta J(Y; \Delta Y) = \int_{x_0}^{x_1} [&F_Y(x, Y(x), Y'(x), Y''(x))\,\Delta Y(x) \\ &+ F_{Y'}(x, Y(x), Y'(x), Y''(x))\,\Delta Y'(x) \\ &+ F_{Y''}(x, Y(x), Y'(x), Y''(x))\,\Delta Y''(x)]\,dx\end{aligned} \tag{4.9.7}$$

for any vector $Y = Y(x)$ in the domain D of J and for any vector $\Delta Y = \Delta Y(x)$ in the vector space $\mathfrak{C}^2[x_0, x_1]$. Here the expressions F_Y, $F_{Y'}$, and $F_{Y''}$ are defined as

$$\begin{aligned} F_Y(x, Y(x), Y'(x), Y''(x)) &= \frac{\partial}{\partial y} F(x, y, z, w)\Big|_{\substack{y=Y(x)\\ z=Y'(x)\\ w=Y''(x)}} \\ F_{Y'}(x, Y(x), Y'(x), Y''(x)) &= \frac{\partial}{\partial z} F(x, y, z, w)\Big|_{\substack{y=Y(x)\\ z=Y'(x)\\ w=Y''(x)}} \\ F_{Y''}(x, Y(x), Y'(x), Y''(x)) &= \frac{\partial}{\partial w} F(x, y, z, w)\Big|_{\substack{y=Y(x)\\ z=Y'(x)\\ w=Y''(x)}}. \end{aligned} \tag{4.9.8}$$

We shall now consider the problem of minimizing (or maximizing) a functional J of the form (4.9.4) subject to the constraints

$$Y(x_0) = y_0, \qquad Y(x_1) = y_1, \qquad Y'(x_0) = m_0 \tag{4.9.9}$$

for given numbers y_0, y_1, and m_0. This problem clearly includes the previous curtain rod problem as a special case. If we define functionals K_1, K_2, and K_3 by

$$K_1(Y) = Y(x_0), \qquad K_2(Y) = Y(x_1), \qquad K_3(Y) = Y'(x_0) \tag{4.9.10}$$

for any function $Y = Y(x)$ in the vector space $\mathcal{C}^2[x_0, x_1]$, then the problem is to find extremum vectors in $D[K_1 = y_0, K_2 = y_1, K_3 = m_0]$ for the functional J of (4.9.4). Here we take D to be the entire normed vector space $\mathcal{C}^2[x_0, x_1]$ or some suitable given open subset of it. We easily find from (4.9.10) and (2.3.4) the variations

$$\begin{aligned} \delta K_1(Y; \Delta Y) &= \Delta Y(x_0) \\ \delta K_2(Y; \Delta Y) &= \Delta Y(x_1) \\ \delta K_3(Y; \Delta Y) &= \Delta Y'(x_0) \end{aligned} \tag{4.9.11}$$

for any function ΔY in $\mathcal{C}^2[x_0, x_1]$.

One easily checks that all the hypotheses of the Euler-Lagrange multiplier theorem of Section 3.6 are satisfied, at least provided that the given function F is "nice." Hence at least one of the two possibilities concluded by that theorem must hold for any local extremum vector Y in $D[K_1 = y_0, K_2 = y_1, K_3 = m_0]$ for J. We can eliminate the first possibility by observing with (4.9.11) that the determinant [see (3.6.2)]

$$\det \begin{vmatrix} \delta K_1(Y, \Delta Y_1) & \delta K_1(Y; \Delta Y_2) & \delta K_1(Y; \Delta Y_3) \\ \delta K_2(Y; \Delta Y_1) & \delta K_2(Y; \Delta Y_2) & \delta K_2(Y; \Delta Y_3) \\ \delta K_3(Y; \Delta Y_1) & \delta K_3(Y; \Delta Y_2) & \delta K_3(Y; \Delta Y_3) \end{vmatrix}$$

does *not* vanish identically for all functions ΔY_1, ΔY_2, and ΔY_3 in $\mathcal{C}^2[x_0, x_1]$. For example, if we take

$$\Delta Y_1(x) = 1 \qquad \text{for all } x$$

$$\Delta Y_2(x) = \frac{x - x_0}{x_1 - x_0} \qquad \text{for all } x$$

and

$$\Delta Y_3(x) = \left(\frac{x_1 - x}{x_1 - x_0}\right)^2 \qquad \text{for all } x,$$

then the corresponding value of the determinant is found to be $(x_0 - x_1)^{-1}$, which is negative. This eliminates the first possibility, and hence the second possibility of the Euler-Lagrange multiplier theorem must hold, so that if $Y = Y(x)$ is a local extremum vector in $D[K_1 = y_0, K_2 = y_1, K_3 = m_0]$ for J, there will be constants λ_1, λ_2, and λ_3 such that [see (3.6.3)]

$$\delta J(Y; \Delta Y) = \lambda_1 \, \delta K_1(Y; \Delta Y) + \lambda_2 \, \delta K_2(Y; \Delta Y) + \lambda_3 \, \delta K_3(Y; \Delta Y)$$

holds for *all* vectors ΔY in $\mathcal{C}^2[x_0, x_1]$. If we use (4.9.7) and (4.9.11), we may rewrite this condition as

$$\int_{x_0}^{x_1} [F_Y(x, Y(x), Y'(x), Y''(x))\, \Delta Y(x) + F_{Y'}(x, Y(x), Y'(x), Y''(x))\, \Delta Y'(x) \\ + F_{Y''}(x, Y(x), Y'(x), Y''(x))\, \Delta Y''(x)]\, dx \\ = \lambda_1\, \Delta Y(x_0) + \lambda_2\, \Delta Y(x_1) + \lambda_3\, \Delta Y'(x_0), \tag{4.9.12}$$

which must necessarily hold for *all* twice continuously differentiable functions $\Delta Y = \Delta Y(x)$.

As usual, we wish to eliminate the arbitrary function ΔY from (4.9.12) so as to obtain a simpler equation which will involve only the extremum vector and which may be solved to give $Y = Y(x)$. The earlier calculation leading to equation (4.1.8) (which amounts to a single integration by parts) can be used in the present case to give

$$\int_{x_0}^{x_1} F_{Y'}(x, Y(x), Y'(x), Y''(x))\, \Delta Y'(x)\, dx \\ = F_{Y'}(x_1, Y(x_1), Y'(x_1), Y''(x_1))\, \Delta Y(x_1) \\ - F_{Y'}(x_0, Y(x_0), Y'(x_0), Y''(x_0))\, \Delta Y(x_0) \\ - \int_{x_0}^{x_1} \left[\frac{d}{dx} F_{Y'}(x, Y(x), Y'(x), Y''(x))\right] \Delta Y(x)\, dx,$$

and a similar calculation (involving *two* integrations by parts) leads similarly to

$$\int_{x_0}^{x_1} F_{Y''}(x, Y(x), Y'(x), Y''(x))\, \Delta Y''(x)\, dx \\ = F_{Y''}(x_1, Y(x_1), Y'(x_1), Y''(x_1))\, \Delta Y'(x_1) \\ - F_{Y''}(x_0, Y(x_0), Y'(x_0), Y''(x_0))\, \Delta Y'(x_0) \\ - \left[\frac{d}{dx} F_{Y''}(x_1, Y(x_1), Y'(x_1), Y''(x_1))\right] \Delta Y(x_1) \\ + \left[\frac{d}{dx} F_{Y''}(x_0, Y(x_0), Y'(x_0), Y''(x_0))\right] \Delta Y(x_0) \\ + \int_{x_0}^{x_1} \left[\frac{d^2}{dx^2} F_{Y''}(x, Y(x), Y'(x), Y''(x))\right] \Delta Y(x)\, dx.$$

We leave the derivation of this equation as an exercise for the reader. We can now use the last two equations to eliminate the integrals involving $\Delta Y' = d\,\Delta Y(x)/dx$ and $\Delta Y'' = d^2\,\Delta Y(x)/dx^2$ in (4.9.12) so as to find [compare with (4.1.9)]

$$\int_{x_0}^{x_1}\Big[F_Y(x, Y(x), Y'(x), Y''(x)) - \frac{d}{dx}F_{Y'}(x, Y(x), Y'(x), Y''(x)) + \frac{d^2}{dx^2}F_{Y''}(x, Y(x), Y'(x), Y''(x))\Big]\Delta Y(x)\,dx$$
$$= \Big[\lambda_1 + F_{Y'}(x_0, Y(x_0), Y'(x_0), Y''(x_0)) - \frac{d}{dx}F_{Y''}(x_0, Y(x_0), Y'(x_0), Y''(x_0))\Big]\Delta Y(x_0)$$
$$+ \Big[\lambda_2 - F_{Y'}(x_1, Y(x_1), Y'(x_1), Y''(x_1)) + \frac{d}{dx}F_{Y''}(x_1, Y(x_1), Y'(x_1), Y''(x_1))\Big]\Delta Y(x_1)$$
$$+ [\lambda_3 + F_{Y''}(x_0, Y(x_0), Y'(x_0), Y''(x_0))]\Delta Y'(x_0)$$
$$- F_{Y''}(x_1, Y(x_1), Y'(x_1), Y''(x_1))\,\Delta Y'(x_1), \qquad (4.9.13)$$

which must then hold for *all* vectors ΔY in the vector space $\mathcal{C}^2[x_0, x_1]$. In particular, if we consider first only those functions ΔY of class $\mathcal{C}^2$ on $[x_0, x_1]$ which vanish at the end points along with their first derivatives, we find from (4.9.13) that

$$\int_{x_0}^{x_1}\Big[F_Y(x, Y(x), Y'(x), Y''(x)) - \frac{d}{dx}F_{Y'}(x, Y(x), Y'(x), Y''(x)) + \frac{d^2}{dx^2}F_{Y''}(x, Y(x), Y'(x), Y''(x))\Big]\Delta Y(x)\,dx = 0, \qquad (4.9.14)$$

which must hold for all functions ΔY *of class* $\mathcal{C}^2$ *on* $[x_0, x_1]$ *which satisfy the additional requirements* $\Delta Y(x_0) = 0$, $\Delta Y'(x_0) = 0$, $\Delta Y(x_1) = 0$, *and* $\Delta Y'(x_1) = 0$. It follows now from (4.9.14) and the lemma of Section A4 of the Appendix (with $n = 2$) that *the extremum function* $Y(x)$ *must satisfy*

$$F_Y(x, Y(x), Y'(x), Y''(x)) - \frac{d}{dx}F_{Y'}(x, Y(x), Y'(x), Y''(x)) + \frac{d^2}{dx^2}F_{Y''}(x, Y(x), Y'(x), Y''(x)) = 0 \qquad (4.9.15)$$

for all x *in the interval* $[x_0, x_1]$. This is the Euler-Lagrange equation obtained in the present case by eliminating the arbitrary function ΔY from (4.9.12).

This equation (4.9.15) is in general a *fourth*-order differential equation for the unknown extremum function $Y = Y(x)$. For example, if F is given by

(4.9.5), we find with (4.9.8) that

$$F_Y(x, Y(x), Y'(x), Y''(x)) = 0,$$

$$F_{Y'}(x, Y(x), Y'(x), Y''(x)) = -10\alpha \frac{Y'(x)Y''(x)^2}{[1 + Y'(x)^2]^{7/2}}$$

and

$$F_{Y''}(x, Y(x), Y'(x), Y''(x)) = 4\alpha \frac{Y''(x)}{[1 + Y'(x)^2]^{5/2}}, \tag{4.9.16}$$

and then in this case the Euler-Lagrange equation (4.9.15) becomes

$$5\frac{d}{dx}\left(\frac{Y'(x)Y''(x)^2}{[1 + Y'(x)^2]^{7/2}}\right) + 2\frac{d^2}{dx^2}\left(\frac{Y''(x)}{[1 + Y'(x)^2]^{5/2}}\right) = 0, \tag{4.9.17}$$

which will indeed involve derivatives of $Y = Y(x)$ of order up through and including fourth order. The most general solution of such a fourth-order differential equation will involve four arbitrary constants of integration which can be specified by imposing four auxiliary conditions or constraints. If, as in the present case, Y is an extremum vector in $D[K_1 = y_0, K_2 = y_1, K_3 = m_0]$ for J, then the given constraints of (4.9.9) will provide *three* such conditions, so that we need one additional condition in order to determine completely the extremum function as a suitable solution of (4.9.15). This remaining condition can be obtained from the Euler-Lagrange multiplier theorem as a natural boundary condition from (4.9.13). Indeed, if we use the Euler-Lagrange equation (4.9.15) along with the constraints (4.9.9) in (4.9.13), we find the requirement

$$\begin{aligned}\Big[\lambda_1 + F_{Y'}(x_0, y_0, m_0, Y''(x_0)) - \frac{d}{dx}F_{Y''}(x_0, y_0, m_0, Y''(x_0))\Big]\Delta Y(x_0)\\ + \Big[\lambda_2 - F_{Y'}(x_1, y_1, Y'(x_1), Y''(x_1))\\ + \frac{d}{dx}F_{Y''}(x_1, y_1, Y'(x_1), Y''(x_1))\Big]\Delta Y(x_1)\\ + [\lambda_3 + F_{Y''}(x_0, y_0, m_0, Y''(x_0))]\,\Delta Y'(x_0)\\ - F_{Y''}(x_1, y_1, Y'(x_1), Y''(x_1))\,\Delta Y'(x_1) = 0,\end{aligned} \tag{4.9.18}$$

which must still hold for *all* functions $\Delta Y = \Delta Y(x)$ of class $\mathcal{C}^2$ on $[x_0, x_1]$. If we take in particular ΔY to be a function such as the one illustrated in Figure 48 which vanishes along with its first derivative at $x = x_0$ and which, moreover, vanishes at $x = x_1$ but with $\Delta Y'(x_1) \neq 0$ (we show one way to construct such a function ΔY in Section A7 of the Appendix), we then find from (4.9.18) the condition

$$F_{Y''}(x_1, y_1, Y'(x_1), Y''(x_1)) = 0, \tag{4.9.19}$$

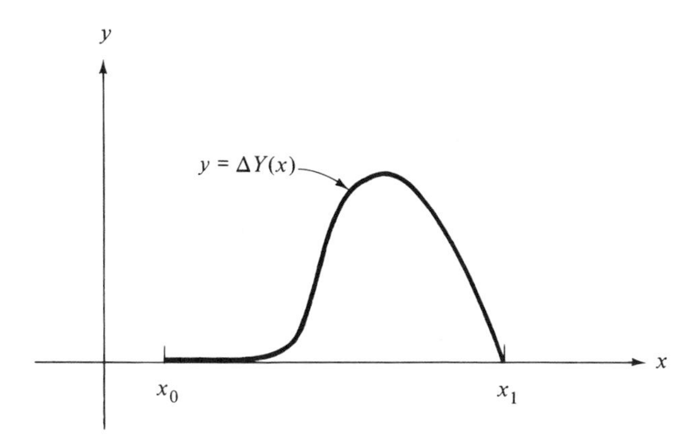

Figure 48

which supplies the fourth auxiliary condition needed in the present case to specify the extremum function $Y = Y(x)$. For example, if we take F to be given by (4.9.5), we find in this case from (4.9.16) and (4.9.19) the natural boundary condition [with $x_1 = a$ as in (4.9.1) and (4.9.2)]

$$Y''(a) = 0, \tag{4.9.20}$$

which must be satisfied by any local extremum vector Y in $D[K_1 = y_0, K_2 = y_1, K_3 = m_0]$ for J.

Thus in the general case any local extremum vector $Y = Y(x)$ in $D[K_1 = y_0, K_2 = y_1, K_3 = m_0]$ for J must satisfy the Euler-Lagrange equation (4.9.15) for $x_0 < x < x_1$ along with the three prescribed boundary conditions of (4.9.9) and the natural boundary condition (4.9.19). In the case of the previous curtain rod problem we can integrate the Euler-Lagrange equation (4.9.17) once to find that

$$\frac{d}{dx}\left(\frac{Y''(x)}{[1 + Y'(x)^2]^{5/2}}\right) + \frac{5}{2}\frac{Y'(x)Y''(x)^2}{[1 + Y'(x)^2]^{7/2}} = A$$

for some suitable constant of integration A. This equation can be rewritten as

$$Y'''(x) - \frac{5}{2}\frac{Y'(x)Y''(x)^2}{1 + Y'(x)^2} = A[1 + Y'(x)^2]^{5/2}, \tag{4.9.21}$$

and the solution of this *third*-order differential equation will involve three additional constants of integration, say B, C, and D. The four constants A, B, C, and D can be specified by imposing the four boundary conditions contained in (4.9.2) and (4.9.20). Equation (4.9.21) can actually be solved in (more or less) closed form for $Y(x)$ in terms of certain special functions studied by Carl Jacobi and known as *Jacobi elliptic functions*. At this point, however, we prefer to refer the interested reader to the paper of Blankinship† or the work of Euler‡ for an indication of the details. (See also Exercise 1.)

We might mention that it is possible to consider extremum problems for the functional $J = J(Y)$ of (4.9.4) subject to different prescribed conditions other than those given by (4.9.9). For example, we might wish to minimize or maximize J among all functions $Y = Y(x)$ of class $\mathcal{C}^2$ on $[x_0, x_1]$ which satisfy the four prescribed boundary conditions

$$Y(x_0) = y_0, \qquad Y(x_1) = y_1$$
$$Y'(x_0) = m_0, \qquad Y'(x_1) = m_1$$

for given constants y_0, y_1, m_0, and m_1. It is clear that any such local extremum function Y subject to these last conditions must again satisfy the same Euler-Lagrange equation (4.9.15) as before, but the natural boundary condition (4.9.19) no longer obtains. We leave the proofs of these assertions to the reader. Similarly, we can consider more general functionals of the form

$$J(Y) = \int_{x_0}^{x_1} F(x, Y(x), Y'(x), Y''(x), \ldots, Y^{(n)}(x))\, dx,$$

which depend on higher-order derivatives of the function $Y = Y(x)$. In this case the function $F = F(x, z_0, z_1, \ldots, z_n)$ is a given function of $n + 2$ variables which is evaluated in the integral at $z_0 = Y(x)$, $z_1 = Y'(x), \ldots,$ $z_n = Y^{(n)}(x)$. We leave these extensions of the theory to the interested reader. (See, for example, Exercise 6.) The great Euler considered problems of these and many other types in his book *Methodus Inveniendi Lineas Curvas* (1744), including applications of these variational methods to certain problems in elasticity similar to the curtain rod problem considered above.

Exercises

1. Let $a = 1$ and assume that b is a *small* positive number so that the curtain rod of Figure 47 is only *slightly* bent. In this case the slope of the rod will be small and the expression $1 + Y'(x)^2$ will be approximately equal to 1, so that the potential energy of strain of the rod given by (4.9.3) will be approximately equal to the quantity $J(Y) = 2\alpha \int_0^1 Y''(x)^2\, dx$, where we have put $a = 1$. Find the curve γ which minimizes this functional J among all curves given by (4.9.1) which satisfy the constraints of (4.9.2). Verify directly that a minimum is achieved.

†Blankinship, *op. cit.*,

‡See Truesdell, "The Rational Mechanics of Flexible or Elastic Bodies, 1638–1788," L. Euleri *Opera Omnia*, (2) **11**$_2$, 428 pp. (Zurich: Fussli, 1960), p. 203.

2. Find the Euler-Lagrange differential equation for the functional $J = \int_0^1 [x^4 + Y(x)^2 + 2Y'(x)^2 + Y''(x)^2]\,dx$. (The resulting differential equation has been considered in Section 4.6.)

3. Find the curve γ given as $\gamma: y = Y(x)$ for $0 \leq x \leq \pi/2$, for which the functional $J(Y) = \int_0^{\pi/2} [Y(x)^2 - Y''(x)^2]\,dx$ can have a local extreme value subject to the boundary conditions $Y(0) = 0$, $Y'(0) = 0$, $Y(\pi/2) = 0$, and $Y'(\pi/2) = 1$. *Hint:* The most general solution of the Euler-Lagrange equation in this case can be given as a sum of (any constants multiplied by) the functions e^x, e^{-x}, $\sin x$, and $\cos x$.

4. Suppose that the function $F = F(x, y, z, w)$ which appears in (4.9.4) does *not* depend on the independent variable x, as happens, for example, in the special case (4.9.5). In any such case with $F = F(y, z, w)$ independent of x, show that any possible solution of the Euler-Lagrange equation (4.9.15) must satisfy the third-order differential equation $F(Y(x), Y'(x), Y''(x)) - Y'(x)[F_{Y'}(Y(x), Y'(x), Y''(x)) - (d/dx)F_{Y''}(Y(x), Y'(x), Y''(x))] - Y''(x)F_{Y''}(Y(x), Y'(x), Y''(x)) = C$ for some constant of integration C.

5. Show that any possible solution of the Euler-Lagrange equation (4.9.15) must satisfy the third-order differential equation $F_{Y'}(x, Y'(x), Y''(x)) - (d/dx)\,F_{Y''}(x, Y'(x), Y''(x)) = C$ for some constant of integration C if the function $F = F(x, z, w)$ does *not* depend on the variable y.

6. Derive the Euler-Lagrange equation for the functional $J(Y) = \int_{x_0}^{x_1} F(x, Y(x), Y'(x), Y''(x), Y'''(x))\,dx$ subject to the constraints $Y(x_0) = y_0$, $Y'(x_0) = m_0$, $Y''(x_0) = l_0$, $Y(x_1) = y_1$, $Y'(x_1) = m_1$, and $Y''(x_1) = l_1$.

4.10. *Functionals Involving Several Independent Variables; the Minimal Surface Problem*

Certain extremum problems from physics and geometry which we shall want to consider lead to functionals whose domains of definition consist of certain vector spaces of functions of *several* independent variables. For simplicity we shall consider mainly the case of *two* such independent variables which we shall label as x and y, and we shall then consider functionals $J = J(Z)$ of the form

$$J(Z) = \iint_R F(x, y, Z(x, y), Z_x(x, y), Z_y(x, y))\,dx\,dy, \tag{4.10.1}$$

where R is some given fixed open region in the (x, y)-plane, and where $Z = Z(x, y)$ is a function defined for all points (x, y) in R. The function Z

is assumed to be of class $\mathcal{C}^1$ on R, which means that the partial derivatives

$$Z_x(x, y) = \frac{\partial}{\partial x} Z(x, y) = \lim_{h \to 0} \frac{Z(x + h, y) - Z(x, y)}{h}$$

$$Z_y(x, y) = \frac{\partial}{\partial y} Z(x, y) = \lim_{h \to 0} \frac{Z(x, y + h) - Z(x, y)}{h}$$

exist and are themselves continuous functions of (x, y) in R. The function $F = F(x, y, z, p, q)$ is a given function of the five variables x, y, z, p, and q, which in (4.10.1) is evaluated along $z = Z(x, y)$, $p = Z_x(x, y) = \partial Z(x, y)/\partial x$, and $q = Z_y(x, y) = \partial Z(x, y)/\partial y$ for (x, y) in R. For example, if we take F to be defined by

$$F(x, y, z, p, q) = \sqrt{1 + p^2 + q^2}, \tag{4.10.2}$$

then the functional given by (4.10.1) becomes

$$J(Z) = \iint_R \sqrt{1 + Z_x(x, y)^2 + Z_y(x, y)^2}\, dx\, dy, \tag{4.10.3}$$

which gives the *surface area*† of the graph of Z in 3-space. This graph of Z is the surface S in (x, y, z)-space given parametrically in terms of x and y as

$$S: \quad z = Z(x, y) \qquad \text{for all } (x, y) \text{ in } R, \tag{4.10.4}$$

as shown in Figure 49.

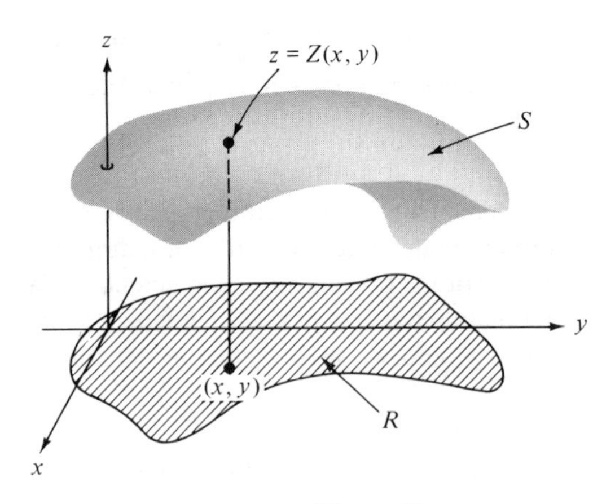

Figure 49

†See, for example, Lipman Bers, *Calculus* (New York: Holt, Rinehart and Winston, Inc., 1969), p. 889.

The problem of minimizing the surface area functional of (4.10.3) subject to certain specified boundary constraints goes back to Lagrange, who in 1760–1761 posed the problem of minimizing the surface area among all surfaces of the form (4.10.4) which have the same given boundary curve or perimeter in 3-space. The required fixed boundary curve is specified in this case by prescribing the values of $Z = Z(x, y)$ for all points (x, y) on the boundary of R. We assume that R is a bounded open region in the (x, y)-plane with a smooth boundary curve, which we denote as ∂R. This boundary curve ∂R may consist of one or several parts, as illustrated in Figure 50.

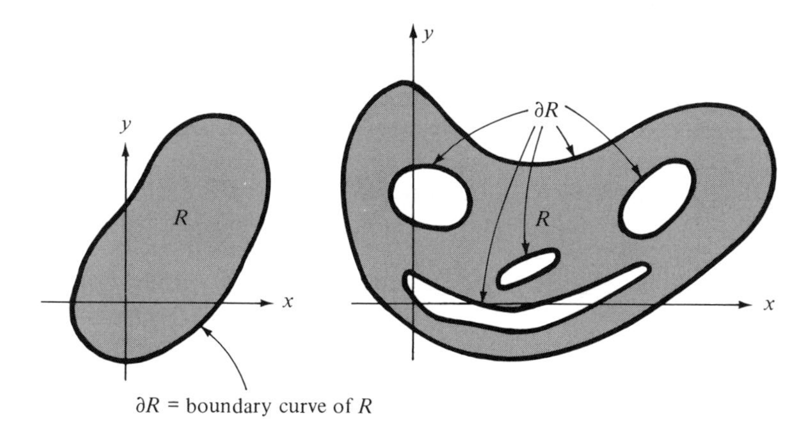

Figure 50

The specified boundary curve in 3-space is then prescribed by the condition

$$Z(x, y) = \phi(x, y) \qquad \text{for all } (x, y) \text{ on } \partial R, \tag{4.10.5}$$

where $\phi = \phi(x, y)$ is a *given* fixed continuous function defined on the boundary of R. Hence the problem posed by Lagrange is that of minimizing the surface area functional J of (4.10.3) among all functions Z which are continuously differentiable on the closed region $R + \partial R$ and which satisfy the boundary constraint given by (4.10.5). This boundary condition (4.10.5) simply requires that the graph of Z must span the fixed curve γ in 3-space given parametrically as

$$\gamma\colon \; z = \phi(x, y) \qquad \text{for all } (x, y) \text{ on } \partial R.$$

The graph of any such extremum function Z for the surface area functional (4.10.3) is called a **minimal surface**. It is a remarkable fact that such minimal surfaces are actually formed naturally by soap films spanning given closed wire contours in 3-space. Indeed, it can be shown to follow from

Hamilton's principle of stationary action that, from among all possible surfaces spanning a given wire contour, an actual soap film will form a surface which minimizes (or at least makes stationary) the potential energy due to surface tension in the film. The stated result then follows from the fact that this potential energy is proportional to the surface area (if we neglect the force of gravity). The general mathematical problem of the study of minimal surfaces is called **Plateau's problem** after the blind Belgian physicist J. Plateau (1801–1883), who determined several interesting properties of such surfaces through experimental studies of soap bubbles and soap films.

We shall now consider the general problem of minimizing or maximizing the functional $J = J(Z)$ given by (4.10.1) among all suitable functions Z defined on R which satisfy the boundary constraint of (4.10.5). This problem includes the above minimal surface problem along with several other extremum problems in physics of the type that we shall consider in Section 4.11. We always assume that the domain D of J is some given open subset of the normed vector space $\mathcal{C}^1(R + \partial R)$ which consists of all functions $Z = Z(x, y)$ of class $\mathcal{C}^1$ on the closed region $R + \partial R$, with norm given by

$$\| Z \| = \max_{R+\partial R} |Z(x, y)| + \max_{R+\partial R} \left| \frac{\partial Z(x, y)}{\partial x} \right| + \max_{R+\partial R} \left| \frac{\partial Z(x, y)}{\partial y} \right|$$

for any such function Z. (Other norms may also be used.) In calculating the norm here we take the maximum value of $|Z(x, y)|$, $|\partial Z(x, y)/\partial x|$, and $|\partial Z(x, y)/\partial y|$, respectively, for all points (x, y) in the closed region $R + \partial R$. Finally, we shall use the symbol

$$D[Z(x, y) = \phi(x, y) \text{ for all points } (x, y) \text{ on } \partial R]$$

to denote the subset of D consisting of all vectors Z in D which satisfy the boundary constraint (4.10.5) on ∂R. Here D is the domain of J. The extremum problem under consideration is to find local extremum vectors in $D[Z(x, y) = \phi(x, y)$ for all points (x, y) on $\partial R]$ for J.

We would like to use the Euler-Lagrange multiplier theorem to study the stated extremum problem. Unfortunately, however, our available version of this multiplier theorem applies only to extremum problems with a *finite* number of constraints of the type $K(Z) = k$, while our present extremum problem has in a certain sense *infinitely* many such constraints since there are infinitely many boundary points (x, y) on ∂R at each of which the constraint (4.10.5) must hold. There are several ways around this difficulty. We might consider the related problem of finding a local extremum vector for J in D subject to the m constraints

$$Z(x_i, y_i) = \phi(x_i, y_i) \qquad \text{for } i = 1, 2, \ldots, m,$$

where $(x_1, y_1), (x_2, y_2), \ldots, (x_m, y_m)$ are the first m points in an infinite sequence of points which is *dense* in ∂R.† Any local extremum vector in $D[Z(x_i, y_i) = \phi(x_i, y_i)$ for $i = 1, 2, \ldots, m]$ for J can be viewed as an approximate solution to our original problem, and in this way we may hope to obtain an exact solution of our original extremum problem by letting m tend toward infinity.

Alternatively, we can modify the original extremum problem slightly so as to account for the given boundary constraint (4.10.5) directly. To do this we proceed as follows. If Z^* is a local minimum vector in $D[Z(x, y) = \phi(x, y)$ for all points (x, y) on $\partial R]$ for J, then

$$J(Z^*) \leq J(Z^* + U) \tag{4.10.6}$$

must hold for all functions U which *vanish on the boundary* ∂R and which lie in some ball centered at the zero vector in the vector space $\mathcal{C}^1(R + \partial R)$. (The reader should convince himself of this assertion.) Hence we are led to consider the new vector space $\mathfrak{X}$ which is the *subspace* of $\mathcal{C}^1(R + \partial R)$ consisting of all functions of class $\mathcal{C}^1$ on $R + \partial R$ *which vanish on the boundary* ∂R, and the new functional $\mathcal{J} = \mathcal{J}(U)$ defined by

$$\mathcal{J}(U) = J(Z^* + U)$$

for all vectors U in some ball $B_\rho(0)$ centered at the zero vector in $\mathfrak{X}$. We take this ball as the domain of $\mathcal{J}$ in $\mathfrak{X}$ and denote it as $\mathfrak{D} = B_\rho(0)$. It now follows from (4.10.6) and the definition of $\mathcal{J}$ that

$$\mathcal{J}(0) \leq \mathcal{J}(U)$$

for all vectors U near the zero vector in $\mathfrak{D}$, so that the zero vector is a local minimum vector in $\mathfrak{D}$ for $\mathcal{J}$. The basic theorem on the vanishing of the variation at a local extremum vector then implies that

$$\delta \mathcal{J}(0; \Delta U) = 0$$

for all vectors ΔU in $\mathfrak{X}$. Alternatively, we can write this equation in the form [see (2.3.4)]

$$\operatorname*{limit}_{\epsilon \to 0} \frac{\mathcal{J}(\epsilon\, \Delta U) - \mathcal{J}(0)}{\epsilon} = 0$$

for all ΔU in $\mathfrak{X}$, or if we use the definition of $\mathcal{J}$, we have

$$\operatorname*{limit}_{\epsilon \to 0} \frac{J(Z^* + \epsilon\, \Delta Z) - J(Z^*)}{\epsilon} = 0 \tag{4.10.7}$$

†That is, the given infinite sequence contains for each point of ∂R a subsequence which converges to that point.

for all vectors ΔZ in $\mathfrak{X}$. Here we are using the symbol ΔZ rather than ΔU to denote an arbitrary function of class $\mathcal{C}^1$ on $R + \partial R$ which vanishes on the boundary of R. The same result (4.10.7) clearly holds also if Z^* is a local *maximum* vector in $D[Z(x, y) = \phi(x, y)$ for all points (x, y) on $\partial R)]$ for J. Note that (4.10.7) simply requires that the variation of J vanish at Z^*, not for all ΔZ in $\mathcal{C}^1(R + \partial R)$, but rather for all such ΔZ which vanish on ∂R; i.e.,

$$\delta J(Z^*; \Delta Z) = 0 \tag{4.10.8}$$

for all such ΔZ which vanish on the boundary of R.

To proceed, we must calculate the variation of J. From (4.10.1) we find that

$$J(Z + \epsilon\,\Delta Z) = \iint_R F(x, y, Z(x, y) + \epsilon\,\Delta Z(x, y), Z_x(x, y) + \epsilon\,\Delta Z_x(x, y), Z_y(x, y) + \epsilon\,\Delta Z_y(x, y))\,dx\,dy$$

for any fixed vector Z in D, for any vector ΔZ in $\mathcal{C}^1(R + \partial R)$, and for any small number ϵ. From this equation we find that [compare with (4.9.6)]

$$\frac{d}{d\epsilon}J(Z + \epsilon\,\Delta Z) = \iint_R \frac{\partial}{\partial\epsilon}F(x, y, Z(x, y) + \epsilon\,\Delta Z(x, y), Z_x(x, y) + \epsilon\,\Delta Z_x(x, y), Z_y(x, y) + \epsilon\,\Delta Z_y(x, y))\,dx\,dy.$$

If we use the chain rule of differentiation to calculate the derivative $\partial F/\partial\epsilon$ appearing in the last integral and then set $\epsilon = 0$, we find with (2.3.4) that

$$\begin{aligned}\delta J(Z; \Delta Z) = \iint_R [&F_Z(x, y, Z(x, y), Z_x(x, y), Z_y(x, y))\,\Delta Z(x, y)\\ &+ F_{Z_x}(x, y, Z(x, y), Z_x(x, y), Z_y(x, y))\,\Delta Z_x(x, y)\\ &+ F_{Z_y}(x, y, Z(x, y), Z_x(x, y), Z_y(x, y))\,\Delta Z_y(x, y)]\,dx\,dy\end{aligned} \tag{4.10.9}$$

for any vector $Z = Z(x, y)$ in the domain D of J and for any vector $\Delta Z = \Delta Z(x, y)$ in the vector space $\mathcal{C}^1(R + \partial R)$. Here $\Delta Z_x = \partial\,\Delta Z/\partial x$ and $\Delta Z_y = \partial\,\Delta Z/\partial y$, while the expressions F_Z, F_{Z_x}, and F_{Z_y} are defined by

$$\begin{aligned}F_Z(x, y, Z(x, y), Z_x(x, y), Z_y(x, y)) &= \frac{\partial}{\partial z}F(x, y, z, p, q)\Big|_{\substack{z=Z(x,y)\\ p=Z_x(x,y)\\ q=Z_y(x,y)}}\\ F_{Z_x}(x, y, Z(x, y), Z_x(x, y), Z_y(x, y)) &= \frac{\partial}{\partial p}F(x, y, z, p, q)\Big|_{\substack{z=Z(x,y)\\ p=Z_x(x,y)\\ q=Z_y(x,y)}}\\ F_{Z_y}(x, y, Z(x, y), Z_x(x, y), Z_y(x, y)) &= \frac{\partial}{\partial q}F(x, y, z, p, q)\Big|_{\substack{z=Z(x,y)\\ p=Z_x(x,y)\\ q=Z_y(x,y)}}.\end{aligned} \tag{4.10.10}$$

For later purposes it will be convenient to rewrite equation (4.10.9) in a form which involves only ΔZ and not the partial derivatives ΔZ_x and ΔZ_y. To accomplish this, we need an analogous result similar to the earlier equation (4.1.8), which was based on integration by parts. Such a result can be obtained from *Green's theorem*,† which imples that

$$\begin{aligned} \iint_R \frac{\partial}{\partial x} f(x, y)\, dx\, dy &= \oint_{\partial R} f(x, y) N_x\, ds \\ \iint_R \frac{\partial}{\partial y} f(x, y)\, dx\, dy &= \oint_{\partial R} f(x, y) N_y\, ds \end{aligned} \tag{4.10.11}$$

for any function f of class $\mathcal{C}^1$ on $R + \partial R$, where $\mathbf{N} = (N_x, N_y)$ denotes the exterior-directed unit normal vector on ∂R, as indicated in Figure 51,‡ and where *ds* is the differential element of arc length along the boundary curve ∂R. The integrals on the right-hand sides of equations (4.10.11) are line integrals.

If we take $f = F_{Z_x} \Delta Z$ in the first equation of (4.10.11), with

$$\frac{\partial}{\partial x} f = F_{Z_x} \Delta Z_x + \left(\frac{\partial}{\partial x} F_{Z_x}\right) \Delta Z,$$

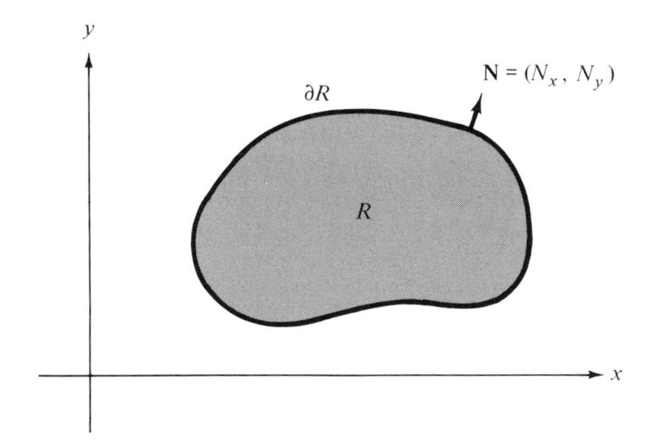

Figure 51

†See, for example, Richard E. Williamson, Richard H. Crowell, and Hale F. Trotter, *Calculus of Vector Functions* (Englewood Cliffs, N.J.: Prentice-Hall, Inc., 1968), pp. 346 and 381. George Green (1793–1841) discovered this theorem in the course of his studies on electricity and magnetism. It first appeared in 1828 in Green's memoir "Essay on the Application of Mathematical Analysis to the Theories of Electricity and Magnetism." Green's theorem allows one to reduce certain volume (or area) integrals to surface (or line) integrals and is closely related to the fundamental theorem of elementary calculus.

‡The subscripts x and y on N_x and N_y do *not* denote differentiation. Rather they simply label the x- and y-components of the normal vector $\mathbf{N}$. We require that $N_x^2 + N_y^2 = 1$ hold since $\mathbf{N}$ is a *unit* vector.

we find from (4.10.11) that

$$\iint_R F_{Z_x} \Delta Z_x \, dx \, dy = -\iint_R \left(\frac{\partial}{\partial x} F_{Z_x}\right) \Delta Z \, dx \, dy + \oint_{\partial R} F_{Z_x} \Delta Z N_x \, ds.$$

Similarly, we find from the second equation of (4.10.11) that

$$\iint_R F_{Z_y} \Delta Z_y \, dx \, dy = -\iint_R \left(\frac{\partial}{\partial y} F_{Z_y}\right) \Delta Z \, dx \, dy + \oint_{\partial R} F_{Z_y} \Delta Z N_y \, ds.$$

The last two equations are the required analogues of the earlier equation (4.1.8), where the present line integrals over the boundary ∂R replace the boundary terms in (4.1.8).

If we use the last equations in (4.10.9), we find for the variation of J that

$$\begin{aligned}\delta J(Z; \Delta Z) = \iint_R \Big[& F_Z(x, y, Z(x, y), Z_x(x, y), Z_y(x, y)) \\ & - \frac{\partial}{\partial x} F_{Z_x}(x, y, Z(x, y), Z_x(x, y), Z_y(x, y)) \\ & - \frac{\partial}{\partial y} F_{Z_y}(x, y, Z(x, y), Z_x(x, y), Z_y(x, y)) \Big] \Delta Z(x, y) \, dx \, dy \\ & + \oint_{\partial R} [F_{Z_x}(x, y, Z(x, y), Z_x(x, y), Z_y(x, y)) N_x \\ & + F_{Z_y}(x, y, Z(x, y), Z_x(x, y), Z_y(x, y)) N_y] \Delta Z(x, y) \, ds,\end{aligned} \tag{4.10.12}$$

where, of course, the components N_x and N_y of the normal vector $\mathbf{N}$ on ∂R will vary as functions of (x, y) along the boundary curve. We should perhaps also mention that the expression

$$\frac{\partial}{\partial x} F_{Z_x}(x, y, Z(x, y), Z_x(x, y), Z_y(x, y))$$

occurring in (4.10.12) denotes the derivative with respect to x of the function $g = g(x, y)$ defined as [see (4.10.10)]

$$g(x, y) = \frac{\partial}{\partial p} F(x, y, z, p, q) \Big|_{\substack{z = Z(x, y) \\ p = (\partial/\partial x) Z(x, y) \\ q = (\partial/\partial y) Z(x, y)}} .$$

A similar remark applies to the expression $(\partial/\partial y) F_{Z_y}$. For example, if we take F to be defined by formula (4.10.2), then

$$F_{Z_y} = \frac{Z_y(x, y)}{\sqrt{1 + Z_x(x, y)^2 + Z_y(x, y)^2}},$$

from which we find that

$$\frac{\partial}{\partial y} F_{Z_y} = \frac{[1 + (Z_x)^2]Z_{yy} - Z_x Z_y Z_{xy}}{[1 + (Z_x)^2 + (Z_y)^2]^{3/2}}. \tag{4.10.13}$$

Here $Z_{yy} = \partial^2 Z(x, y)/\partial y^2$ and $Z_{xy} = \partial^2 Z(x, y)/\partial x\, \partial y$.

Now that we have in hand the expression (4.10.12) for the variation of J, we can use equation (4.10.8) to seek a local extremum vector in $D[Z(x, y) = \phi(x, y)$ for all points (x, y) on $\partial R]$ for J. In this way we find for any such extremum vector $Z = Z(x, y)$ that

$$\begin{aligned}\iint_R \Big[& F_Z(x, y, Z(x, y), Z_x(x, y), Z_y(x, y)) \\ & - \frac{\partial}{\partial x} F_{Z_x}(x, y, Z(x, y), Z_x(x, y), Z_y(x, y)) \\ & - \frac{\partial}{\partial y} F_{Z_y}(x, y, Z(x, y), Z_x(x, y), Z_y(x, y)) \Big] \Delta Z(x, y)\, dx\, dy = 0,\end{aligned} \tag{4.10.14}$$

which must hold for all functions ΔZ of class $\mathfrak{C}^1$ on $R + \partial R$ *which vanish on the boundary of* R. Note that the line integrals over ∂R which occur in the variation of J do *not* appear in equation (4.10.14) since ΔZ is required here to vanish on the boundary. Equation (4.10.14) is analogous to the earlier equation (4.1.10), which led to the earlier Euler-Lagrange equation (4.1.11). We might expect in the present case to find from (4.10.14) the following Euler-Lagrange equation:

$$\begin{aligned} F_Z(x, y, Z(x, y), Z_x(x, y), Z_y(x, y)) & - \frac{\partial}{\partial x} F_{Z_x}(x, y, Z(x, y), Z_x(x, y), Z_y(x, y)) \\ & - \frac{\partial}{\partial y} F_{Z_y}(x, y, Z(x, y), Z_x(x, y), Z_y(x, y)) = 0 \end{aligned} \tag{4.10.15}$$

for all points (x, y) in the region R and for any extremum vector $Z = Z(x, y)$. And, indeed, the large degree of arbitrariness of ΔZ in (4.10.14) does allow us to conclude this result, as follows from a basic lemma of the calculus of variations which we prove in Section A8 of the Appendix. We leave the verification of this assertion as an exercise for the reader.

Hence any function $Z = Z(x, y)$ which minimizes or maximizes the functional J of (4.10.1) subject to the boundary constraint (4.10.5) must satisfy the *partial differential equation* (4.10.15) in R. For example, if we take F to be defined as in the surface area functional by formula (4.10.2), then (4.10.10) and calculations similar to those which led to (4.10.13) show that in this case the Euler-Lagrange equation becomes

$$\frac{\partial}{\partial x}\left(\frac{Z_x}{\sqrt{1 + Z_x^2 + Z_y^2}}\right) + \frac{\partial}{\partial y}\left(\frac{Z_y}{\sqrt{1 + Z_x^2 + Z_y^2}}\right) = 0,$$

or, after further differentiation and simplification,

$$(1 + Z_y^2)Z_{xx} - 2Z_xZ_yZ_{xy} + (1 + Z_x^2)Z_{yy} = 0. \qquad (4.10.16)$$

Hence any minimal surface S of the type (4.10.4) must be defined by a function $Z = Z(x, y)$ which satisfies this last differential equation. The general study of solutions of this *minimal surface equation* is very difficult indeed, and we shall not pursue it here.

A special (and simpler) case occurs when the given boundary curve spanned by the minimal surface in 3-space differs only slightly from a horizontally *plane* curve, as happens when the values of the given smooth function ϕ appearing in (4.10.5) are everywhere small. In this case we expect that *the minimal surface will differ only slightly from a plane surface*, and this means that the partial derivatives Z_x and Z_y will be everywhere small on R. Hence in this case the expression $\sqrt{1 + Z_x(x, y)^2 + Z_y(x, y)^2}$ will be approximately equal to†

$$1 + \tfrac{1}{2}[Z_x(x, y)^2 + Z_y(x, y)^2],$$

and the surface area functional J of (4.10.3) will be approximately equal to the functional

$$J_0(Z) = \iint_R \left\{1 + \frac{1}{2}[Z_x(x, y)^2 + Z_y(x, y)^2]\right\} dx\, dy. \qquad (4.10.17)$$

The Euler-Lagrange equation for J_0 is easily found to be given as

$$Z_{xx} + Z_{yy} = 0,$$

which is considerably easier to study than the minimal surface equation (4.10.16). The last equation is known as *Laplace's equation* or the *potential equation* and arises not only here in the minimal surface problem but also in problems on celestial mechanics, electrostatics, heat conduction, fluid dynamics, and many other problems in mathematical physics. We refer the reader to any elementary text on partial differential equations for the study of this important equation and its solutions.‡

†See the earlier footnote which refers to equation (4.8.10), p. 201.

‡See, for example, H. F. Weinberger, *A First Course in Partial Differential Equations* (Waltham, Mass.: Ginn/Blaisdell, 1965).

Exercises

1. Show how to derive the Euler-Lagrange equation (4.10.15) from (4.10.14).
2. Verify that Laplace's equation is the Euler-Lagrange equation for the functional J_0 given by (4.10.17).
3. Find the Euler-Lagrange partial differential equation for the functional

$$J(Z) = \iint_R [Z_x(x, y)^2 - Z_y(x, y)^2]\, dx\, dy.$$

4. Let the boundary of the plane region R be divided into two disjoint (distinct) parts $\partial_1 R$ and $\partial_2 R$, and let ϕ be a given function defined on $\partial_1 R$. Show that if $Z = Z(x, y)$ minimizes or maximizes the functional J of (4.10.1) among all functions satisfying the boundary condition $Z(x, y) = \phi(x, y)$ for all (x, y) on $\partial_1 R$, then on $\partial_2 R$ Z must satisfy the *natural boundary condition* $F_{Z_x}N_x + F_{Z_y}N_y = 0$ for all (x, y) on $\partial_2 R$, where $\mathbf{N} = (N_x, N_y)$ is the exterior-directed unit normal vector on ∂R. Show also in this case that any such extremum function Z must satisfy the same Euler-Lagrange equation (4.10.15) throughout R.
5. Let $\mathbf{x} = (x_1, x_2, x_3)$ and $\mathbf{p} = (p_1, p_2, p_3)$ denote arbitrary points in $\mathcal{R}_3$, and let $F = F(x_1, x_2, x_3, u, p_1, p_2, p_3) = F(\mathbf{x}, u, \mathbf{p})$ be a given function of the seven real variables $x_1, x_2, x_3, u, p_1, p_2, p_3$. Define a functional $J = J(U)$ by $J(U) = \int_R F(\mathbf{x}, U(\mathbf{x}), \nabla U(\mathbf{x}))\, d\mathbf{x}$ for any real-valued function $U = U(\mathbf{x})$ of class $\mathcal{C}^1$ on a given fixed open set R in $\mathcal{R}_3$, where ∇U is the gradient of U given as $\nabla U(\mathbf{x}) = (\partial U/\partial x_1, \partial U/\partial x_2, \partial U/\partial x_3) = (U_{x_1}, U_{x_2}, U_{x_3})$, and where the integral is a volume integral taken over the region R with $d\mathbf{x} = dx_1\, dx_2\, dx_3$. Show that any function U which furnishes a local extreme value to J among all functions of class $\mathcal{C}^1$ which coincide with U on the boundary of R must satisfy $F_U - \sum_{i=1}^{3} (\partial/\partial x_i) F_{U_{x_i}} = 0$ for all points $\mathbf{x}$ in R. *Hints:* You will need Green's theorem for regions in $\mathcal{R}_3$, according to which $\int_R (\partial/\partial x_i) f(\mathbf{x})\, d\mathbf{x} = \oint_{\partial R} f(\mathbf{x}) N_i\, d\sigma$ holds for $i = 1, 2, 3$ and for any function f of class $\mathcal{C}^1$ on $R + \partial R$, where ∂R denotes the boundary *surface* of R, $\mathbf{N} = (N_1, N_2, N_3)$ denotes the exterior-directed unit normal vector on the boundary surface ∂R, and $d\sigma$ is the differential element of surface area on ∂R. The integral on the right-hand side is a surface integral taken over the boundary surface ∂R.

4.11. The Vibrating String

In Section 4.8 we showed that Hamilton's principle of stationary action is valid for the motion of a single particle in a conservative force field, so that the particle actually moves so as to provide a *stationary* value to the following *action integral:*

$$A = \int_{t_0}^{t_1} (T - V)\, dt. \tag{4.11.1}$$

Here T and V denote the kinetic and potential energies of the motion. Hamilton's principle is actually valid in a much wider context, and we shall use the principle in this section to derive the differential equation that governs the small transverse vibrations of a continuous string. (In Section 7.1 we shall discuss one way to solve the resulting equation.)

We consider a vibrating elastic string of uniform cross section, such as a violin string or a guitar string. We let m denote the total *mass* of the string, and we let l be the *length* of the quiet string at rest in its equilibrium position. We take the quiet (stationary) string to coincide with the interval $0 \leq x \leq l$ along the x-axis, and we suppose that the vibrations occur in the (x, z)-plane. Hence at time t the string occupies a curve $\gamma = \gamma(t)$ which we assume can be given parametrically in terms of x as

$$\gamma: \quad z = Z(x, t) \qquad \text{for } 0 \leq x \leq l \tag{4.11.2}$$

for some suitable function $Z = Z(x, t)$, which gives the displacement of the string at the position x and at the time t. We shall only consider the case of a string with fixed end points, although various other cases involving variable end points can also be handled by these same methods. Hence we shall impose the constraints $Z(0, t) = 0$ and $Z(l, t) = 0$ for all t, so that the two end points of the string remain fixed throughout.

We wish to characterize the particular function $Z = Z(x, t)$ that actually corresponds to the motion of the string during some given time interval $t_0 \leq t \leq t_1$. For this purpose we shall use an extended version of Hamilton's principle which covers the present situation and which asserts that the actual motion will in a certain precise sense yield a stationary value to the action integral given by (4.11.1). Of course, we must first obtain suitable expressions for the kinetic and potential energies of the motion.

We shall assume that the *potential energy* V due to the stretching of the elastic string is *proportional to the amount by which the string has been stretched.* Since the length of the stationary string in equilibrium is l while the length of the vibrating string (at time t) is given by the integral

$$\int_0^l \sqrt{1 + Z_x(x, t)^2}\, dx = \text{length of } \gamma,$$

we find for V that

$$\begin{aligned} V &= \tau \left\{ \int_0^l \sqrt{1 + Z_x(x, t)^2}\, dx - \int_0^l dx \right\} \\ &= \tau \int_0^l \left(\sqrt{1 + Z_x(x, t)^2} - 1\right) dx \end{aligned} \tag{4.11.3}$$

for some proportionality factor τ which gives a measure of the *tension* in the string. We shall assume that τ is a given known positive constant. The expression $Z_x = Z_x(x, t)$ which appears in (4.11.3) is the partial derivative of the function Z with respect to x, $Z_x = \partial Z/\partial x$. Finally, we shall consider only the case of transversal motions in which each piece of the string vibrates up and down. In this case it can be shown that the kinetic energy T is given at time t by

$$T = \frac{m}{2l}\int_0^l Z_t(x, t)^2\, dx. \tag{4.11.4}$$

Here $Z_t = \partial Z/\partial t$; we refer the interested reader to Section A9 of the Appendix for a justification of this definition (4.11.4).

Hence we can now insert expressions (4.11.3) and (4.11.4) into (4.11.1) to obtain the *action* of the motion as a functional of $Z = Z(x, t)$. If we apply Hamilton's principle to the resulting action functional $A = A(Z)$, we obtain an Euler-Lagrange equation which will characterize the actual motion of the string. The resulting Euler-Lagrange equation is somewhat complicated, however, and we shall be content at this point to consider only the special (and much simpler) case in which the string undergoes *small* vibrations for which *the slope of γ remains everywhere small*. Hence we shall assume that the derivative $Z_x = \partial Z/\partial x$ is small. (This will ordinarily be the case for a vibrating violin or guitar string.) In particular, we shall assume that the slope Z_x is so small that the quantity $\sqrt{1 + Z_x(x, t)^2}$ can be safely approximated and replaced by the quantity† $1 + \frac{1}{2}Z_x(x, t)^2$. Hence we shall replace the expression V of (4.11.3) by

$$V_0 = \frac{\tau}{2}\int_0^l Z_x(x, t)^2\, dx.$$

We shall now apply Hamilton's principle to the model of a vibrating string in which the kinetic and potential energies of motion are given by the above expressions T and V_0. We let $A_0 = A_0(Z)$ be the action functional for this model, with

$$A_0 = \int_{t_0}^{t_1} (T - V_0)\, dt.$$

Hence we find that

$$A_0(Z) = \frac{1}{2}\int_{t_0}^{t_1}\int_0^l \left[\frac{m}{l} Z_t(x, t)^2 - \tau Z_x(x, t)^2\right] dx\, dt,$$

which can be rewritten in the form [compare with (4.10.1)]

$$A_0(Z) = \iint_R F(Z_x(x, t), Z_t(x, t))\, dx\, dt, \tag{4.11.5}$$

†See the earlier footnote which refers to equation (4.8.10), p. 201.

where the function $F = F(p, q)$ is defined by

$$F(p, q) = \frac{1}{2}\left[\frac{m}{l}q^2 - \tau p^2\right] \tag{4.11.6}$$

for any numbers p and q. The region of integration R appearing in (4.11.5) is the rectangular region in the (x, t)-plane given as

$$R = \{(x, t)\colon\ 0 < x < l,\, t_0 < t < t_1\}.$$

Hamilton's principle (as applied to our simplified version of the vibrating string problem) now asserts that the actual motion of the string is described by a function $Z = Z(x, t)$ which furnishes a minimum (or at least a stationary) value to the action functional of (4.11.5) from among all continuously differentiable functions which coincide with $Z(x, t)$ at $t = t_0$ and at $t = t_1$ (for $0 \leq x \leq l$) and which vanish at the end points $x = 0$ and $x = l$ (for $t_0 \leq t \leq t_1$). This means that Z will furnish a minimum (stationary) value to A_0 from among all functions of class $\mathfrak{C}^1$ on $R + \partial R$ which satisfy suitable *fixed boundary conditions* everywhere on the boundary of R. The results of Section 4.10 then imply that Z must satisfy the Euler-Lagrange equation [see (4.10.15)]

$$\frac{\partial}{\partial x}F_{Z_x}(Z_x(x, t), Z_t(x, t)) + \frac{\partial}{\partial t}F_{Z_t}(Z_x(x, t), Z_t(x, t)) = 0$$

since $F_Z = 0$ in the present case. Here the expressions F_{Z_x} and F_{Z_t} are found with (4.11.6) to be given as

$$F_{Z_x}(Z_x(x, t), Z_t(x, t)) = \frac{\partial}{\partial p}F(p, q)\Big|_{\substack{p=Z_x(x,t)\\ q=Z_t(x,t)}} = -\tau Z_x(x, t)$$

and

$$F_{Z_t}(Z_x(x, t), Z_t(x, t)) = \frac{\partial}{\partial q}F(p, q)\Big|_{\substack{p=Z_x(x,t)\\ q=Z_t(x,t)}} = \frac{m}{l}Z_t(x, t),$$

so that the present Euler-Lagrange equation can be written as

$$\rho Z_{tt} = \tau Z_{xx} \tag{4.11.7}$$

for $Z = Z(x, t)$ and with $\rho = m/l$. This important equation arises not only in the vibrating string problem but also in problems concerning the longitudinal vibrations of rods (where the displacement is along the rod), the propagation of sound in a gas, electrical vibrations in long conducting wires, and many other problems of mathematical physics. The equation was first discovered in 1746 by Jean le Rond D'Alembert (1717–1783) in his studies on vibrating strings using Newton's laws of motion. We refer the reader to Weinberger† for a fascinating study of this equation and its solutions.

The quantity $\rho = m/l$ appearing in equation (4.11.7) for the vibrating string gives the mass per unit length of the string and is known as the linear mass density. We have assumed that both ρ and τ are known positive *constants*. In some cases, however, the material properties of the string vary with position, so that $\rho = \rho(x)$ and $\tau = \tau(x)$ become nonnegative *functions* of x for $0 \leq x \leq l$. In this case a similar analysis as that which led to equation (4.11.7) shows that small vibrations of the string are governed by

$$\rho(x)\frac{\partial^2 Z}{\partial t^2} = \frac{\partial}{\partial x}\left[\tau(x)\frac{\partial Z}{\partial x}\right],$$

which reduces to the earlier equation if ρ and τ are constants. We leave the derivation of the last equation as an exercise for the reader.

The method which leads to the differential equation for the vibrating string can also be used to find the equation for a vibrating *membrane*, such as a drum head. (A membrane is a two-dimensional analogue of a string.) We assume that the surface S of the membrane in (x, y, z)-space can be given parametrically in terms of x and y as

$$S\colon \quad z = Z(x, y, t) \qquad \text{for all } (x, y) \text{ in } R$$

for some suitable function $Z = Z(x, y, t)$ which gives the displacement of the membrane at the position (x, y) and at the time t. The region R is a fixed region in the (x, y)-plane which might, for example, be identified with the equilibrium position of the membrane.‡ We assume that the potential energy of the stretched membrane is proportional to the amount by which the membrane has been stretched, and then if we consider only *small* vibrations we find for the potential energy the (approximate) expression

$$V = \frac{\tau}{2}\iint_R [Z_x(x, y, t)^2 + Z_y(x, y, t)^2]\, dx\, dy,$$

where the positive constant τ gives a measure of the tension in the membrane, assumed to be uniform. The kinetic energy becomes (for small vibrations)

$$T = \frac{m}{2A}\iint_R Z_t(x, y, t)^2\, dx\, dy,$$

where m is the total mass of the membrane and A is the area of the region R in the (x, y)-plane. We are assuming here that the material properties of

†Weinberger, *op. cit*. See also Section 7.1 in the present text.

‡In general, R will be the projection of the membrane surface onto the (x, y)-plane.

the membrane are everywhere uniform over its entire surface. The action functional now becomes

$$A(Z) = \int_{t_0}^{t_1} (T - V)\,dt$$
$$= \frac{1}{2}\int_{t_0}^{t_1}\iint_R \{\rho Z_t(x, y, t)^2 - \tau[Z_x(x, y, t)^2 + Z_y(x, y, t)^2]\}\,dx\,dy\,dt,$$

where we have let $\rho = m/A$ be the mass per unit area of the membrane. If we apply Hamilton's principle to this functional and use Exercise 5 of Section 4.10, we find that the actual motion of the membrane is described by a function $Z = Z(x, y, t)$ which satisfies the equation

$$\rho Z_{tt} = \tau(Z_{xx} + Z_{yy})$$

for all points (x, y) in R and for all t between t_0 and t_1. We leave the details of the derivation of this equation as an exercise for the reader.

Finally, if the material properties of the membrane vary from point to point, then the quantities $\rho = \rho(x, y)$ and $\tau = \tau(x, y)$ become nonnegative functions of x and y for (x, y) in R. In this case a similar analysis shows that small vibrations of the membrane are governed by

$$\rho(x, y)\frac{\partial^2 Z}{\partial t^2} = \frac{\partial}{\partial x}\left[\tau(x, y)\frac{\partial Z}{\partial x}\right] + \frac{\partial}{\partial y}\left[\tau(x, y)\frac{\partial Z}{\partial y}\right],$$

which reduces to the earlier equation if ρ and τ are constants.

5. Applications of the Euler-Lagrange Multiplier Theorem to Problems with Global Pointwise Inequality Constraints

In this chapter we shall consider certain extremum problems which involve *global pointwise* **inequality** *constraints* which are imposed along the entire extremum curve. It will be shown how the Euler-Lagrange multiplier theorem can be used in practice to solve such problems. The main ideas are given in Section 5.1, along with certain examples. The remaining sections present additional examples and should be considered as optional; they need not be covered during a first reading.

5.1. Slack Functions and Composite Curves

There often arise in practice various problems in which it is required to minimize or maximize a functional $J = J(Y)$ among all curves γ given as

$$\gamma: \quad y = Y(x), x_0 \leq x \leq x_1, \tag{5.1.1}$$

where the admissible functions $Y(x)$ are required to satisfy certain *global inequality constraints* of the form

$$Y(x) \geq \varphi(x) \qquad \text{for } x_0 \leq x \leq x_1 \tag{5.1.2}$$

or

$$Y(x) \leq \psi(x) \qquad \text{for } x_0 \leq x \leq x_1, \tag{5.1.3}$$

where in each case $\varphi(x)$ or $\psi(x)$ is a *given* known function of x. For example, the manufacturing company considered in Section 2.5 may wish to minimize the cost functional $C = C(P)$ of (1.3.24) but only among those production rates $P = P(t)$ which satisfy the inequality constraint

$$P(t) \leq P_{\max}(t), \tag{5.1.4}$$

where $P_{\max} = P_{\max}(t)$ is a known function of time which gives the *maximum possible production rate* that the manufacturing company is known to be capable of achieving. The company need only consider production rates $P(t)$ which satisfy (5.1.4) since any production rate which violates (5.1.4) could never be achieved in reality. In particular, the optimum production rate $P^* = P^*(t)$ given by (2.5.12) which was found in Section 2.5 without consideration of the constraint (5.1.4) may be *disqualified* when (5.1.4) is imposed since P^* may exceed the maximum production rate allowed by (5.1.4). For example, if the initial inventory level I_0 is *far enough below* the desired initial inventory level $\mathcal{I}_0$ so that (see the discussion at the end of Section 1.3 for the notation)

$$\frac{\beta^2(\mathcal{I}_0 - I_0)(e^{\gamma T} - e^{-\gamma T})}{(\gamma + \alpha)e^{\gamma T} + (\gamma - \alpha)e^{-\gamma T}} > P_{\max}(0) - \mathcal{P}(0) \tag{5.1.5}$$

holds, then it follows from (2.5.12) that P^* exceeds $P_{\max}$ on some initial time interval $0 \leq t \leq \bar{t}$, as shown schematically in Figure 1 (see also Figure 5 near the end of Section 2.5). In this case the analysis of Section 2.5 is no longer entirely relevant. *We must search for an optimum production rate among the functions $P = P(t)$ which satisfy* (5.1.4). Of course, in this case [when (5.1.5) holds] it seems reasonable to expect that the new optimum production rate among those satisfying (5.1.4) will be given by a **composite rate** which first *equals* the maximum rate $P_{\max}$ over some initial time period extending beyond $t = \bar{t}$ (so as to raise the inventory level up from its low initial level) and then *decreases* down to its terminal value, as shown in Figure 2. This is indeed the case, as we shall see later.

As another example, consider a brachistochrone problem in which a curve γ is sought down which a bead can slide in minimum time from the point $P_0 = (x_0, y_0)$ to the line $x = x_1$ $(x_1 > x_0)$, as shown in Figure 3, but

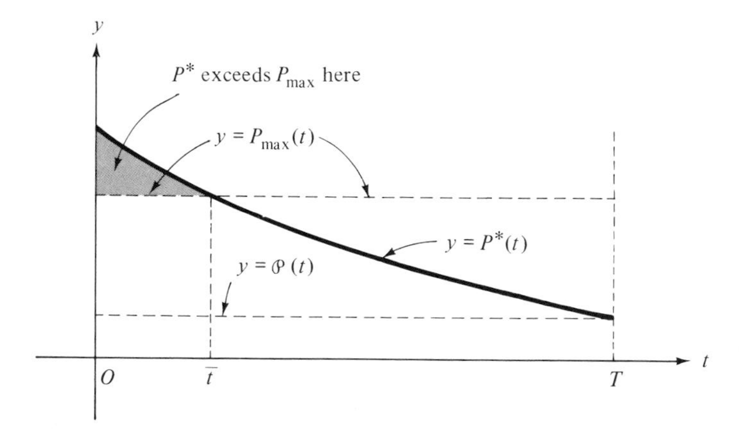

Figure 1

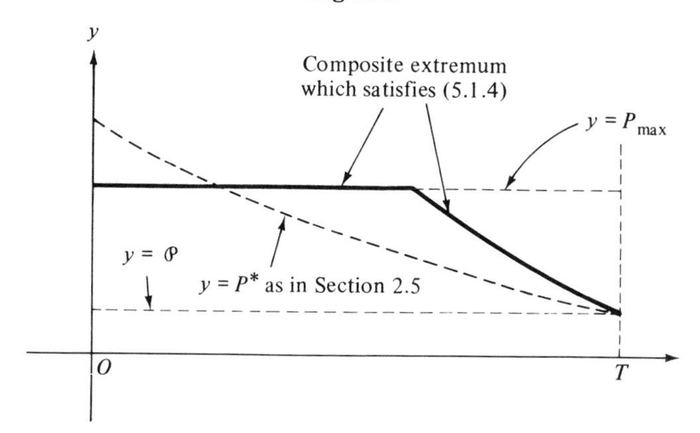

Figure 2

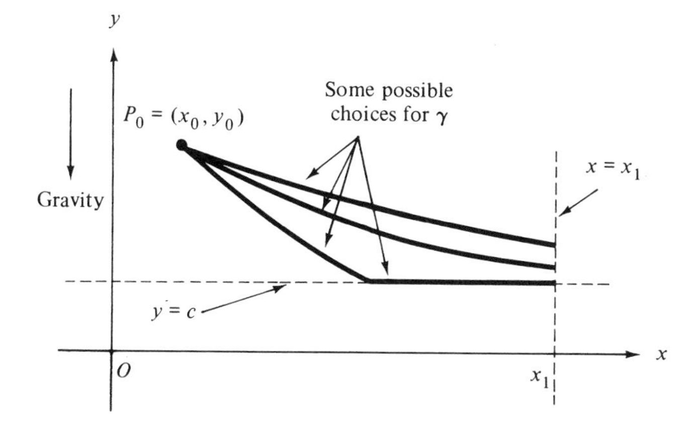

Figure 3

where γ is restricted to lie everywhere above a certain fixed level. Specifically, we might require γ to be given as in (5.1.1) for some function $Y(x)$ which is required to satisfy the inequality constraint

$$Y(x) \geq c \qquad \text{for } x_0 \leq x \leq x_1 \tag{5.1.6}$$

for some given fixed constant c which satisfies $c < y_0$ (see Figure 3). Of course, if the distance from P_0 down to the constraining line $y = c$ is *large* enough (i.e., if the difference $y_0 - c$ is large enough), then James Bernoulli's cycloid given by (4.4.40), (4.4.43) and (4.4.44), which furnishes the quickest descent in the *absence* of the inequality constraint (5.1.6), will automatically satisfy (5.1.6) (why?) and will then clearly give the quickest descent also when the constraint (5.1.6) is imposed. In this case we say that the constraint is **inactive** since it plays no role in the solution to the optimization problem. However, if the line $y = c$ is just slightly below P_0 (i.e., if the positive difference $y_0 - c$ is small), then the previous cycloid will be disqualified since it will violate (5.1.6). In this case we again expect a *composite* minimizing curve consisting of an initial arc dropping from P_0 down to (but not below) the constraining line $y = c$, joined with a segment of the constraining line extending over to $x = x_1$, as illustrated in Figure 4. We shall see later that this is indeed the case.

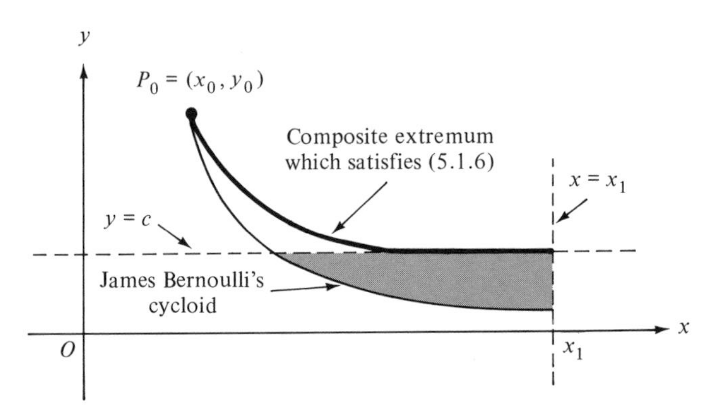

Figure 4

A General Example. By way of illustration we shall consider first the problem of minimizing or maximizing a functional J of the form

$$J(Y) = \int_{x_0}^{x_1} F(x, Y(x), Y'(x))\, dx \tag{5.1.7}$$

among all curves γ given by (5.1.1) which satisfy the global inequality constraint (5.1.2) along with any suitable prescribed fixed or variable end-point conditions. We assume that the given constraint function φ appearing in

(5.1.2) is continuously differentiable, and we assume that all specified end-point conditions are compatible with the constraint (5.1.2). For example, if we specify the end-point condition $Y(x_0) = y_0$, then we assume that the given number y_0 satisfies $y_0 \geq \varphi(x_0)$.

As mentioned earlier in connection with the previous examples (see Figures 2 and 4), we may *expect* in some cases that the extremum curve γ which furnishes a minimum or maximum value to J subject to (5.1.2) will be a composite curve consisting of one or more arcs along which $Y(x)$ agrees with the constraint (with $Y = \varphi$), joined with certain other appropriate arcs. Any such composite curve γ should be continuously differentiable *except possibly at a finite number of points*, where the various composite arcs join together and where γ may be allowed to have corners. In this case we say that the corresponding function $Y = Y(x)$ appearing in (5.1.1) is **piecewise continuously differentiable** on $[x_0, x_1]$, by which (in the present context) we shall mean that *$Y(x)$ is continuous everywhere on $[x_0, x_1]$ and continuously differentiable there with the possible exception of at most a finite number of points where the derivative Y' shall have well-defined limiting values both from the left and from the right.* [In Sections 5.2 and 5.3 we shall consider examples where $Y(x)$ itself may have points of discontinuity; the same methods will be seen to apply.] We let $\mathcal{PC}^1[x_0, x_1]$ denote the set of all such piecewise continuously differentiable functions on $[x_0, x_1]$. Note that $\mathcal{PC}^1[x_0, x_1]$ as defined here is a subspace of the vector space $\mathcal{C}^0[x_0, x_1]$ but is not a subspace of $\mathcal{C}^1[x_0, x_1]$. We shall equip the vector space $\mathcal{PC}^1[x_0, x_1]$ with the norm of $\mathcal{C}^1[x_0, x_1]$ given by (1.4.9) with $k = 1$.

Let D be the *subset* of $\mathcal{PC}^1[x_0, x_1]$ consisting of all piecewise continuously differentiable functions *Y which satisfy the inequality constraint* (5.1.2) everywhere on $[x_0, x_1]$. We then seek extremum vectors in D for J subject to certain specified fixed or variable end-point conditions.

It is tempting to try to use the Euler-Lagrange multiplier theorem to find the desired extremum vectors for J. However, *the multiplier theorem as stated in Section 3.6 need not apply if D is not an open set* in the given normed vector space $\mathfrak{X} = \mathcal{PC}^1[x_0, x_1]$. Indeed, the proof of the multiplier theorem given in Section 3.6 requires that any local extremum vector Y under consideration must be an *interior point* in D; i.e., *there must be an entire ball $B_\rho(Y)$ in $\mathfrak{X}$ centered at Y and contained in D.*† But suppose in the present case that the minimizing vector (function) $Y = Y(x)$ *agrees* with the constraining function $\varphi = \varphi(x)$ at certain points, with

$$Y(x) = \varphi(x) \tag{5.1.8}$$

for some numbers x in $[x_0, x_1]$. *Then we can obtain vectors* (functions) $Y + \Delta Y$

†Strictly speaking, it is only necessary that $Y + \epsilon \Delta Y$ should be in D for every fixed vector ΔY in $\mathfrak{X}$ and for all sufficiently small numbers ϵ. This will automatically be the case if D is an open set (relative to some norm).

arbitrarily near Y which violate (5.1.2) *and which are therefore not in D.* Indeed, we need only take $\Delta Y = \Delta Y(x)$ to be *negative* where (5.1.8) holds, and then the function $Y + \Delta Y$ will violate (5.1.2) at those points since

$$Y + \Delta Y = \varphi + \Delta Y < \varphi$$

if ΔY is negative and (5.1.8) holds. Hence the proof of the multiplier theorem breaks down for any such extremum vector Y which anywhere agrees with the constraining function φ. But we have already observed that we may in some cases *expect* the desired extremum function to be a composite function for which (5.1.8) may indeed hold along certain arcs. Hence *we cannot indiscriminately apply the Euler-Lagrange multiplier theorem in the present case.*

We can circumvent this difficulty [which is due to the global *inequality* constraint (5.1.2)] by rewriting the problem in terms of a **slack function**† $U = U(x)$, introduced by the relation

$$Y(x) = \varphi(x) + U(x)^2. \tag{5.1.9}$$

Clearly, the constraint (5.1.2) will automatically be satisfied by any function $Y(x)$ of the form (5.1.9) for *any* piecewise continuously differentiable function $U = U(x)$ in $\mathcal{PC}^1[x_0, x_1]$. From (5.1.9) we find that (except perhaps at a finite number of points)

$$Y'(x) = \varphi'(x) + 2U(x)U'(x),$$

so that the functional J of (5.1.7) can be rewritten as

$$J = \int_{x_0}^{x_1} F(x, \varphi(x) + U(x)^2, \varphi'(x) + 2U(x)U'(x))\, dx, \tag{5.1.10}$$

where now $U = U(x)$ may be *any* vector in $\mathcal{PC}^1[x_0, x_1]$, which need not satisfy (5.1.2). Hence the original problem of minimizing or maximizing the functional $J = J(Y)$ subject to the inequality constraint (5.1.2) can be recast as a problem of minimizing or maximizing a functional $J = \mathcal{J}(U)$ *without any such inequality constraint*, where we define

$$\mathcal{J}(U) = \int_{x_0}^{x_1} \mathfrak{F}(x, U(x), U'(x))\, dx \tag{5.1.11}$$

in terms of a new integrand function $\mathfrak{F}$ defined by [see (5.1.10) and (5.1.11)]

$$\mathfrak{F}(x, u, w) = F(x, \varphi(x) + u^2, \varphi'(x) + 2uw). \tag{5.1.12}$$

†The use of such slack functions was introduced by F. A. Valentine, "The Problem of Lagrange with Differential Inequalities as Added Side Conditions," in *Contributions to the Calculus of Variations 1933–1937* (Chicago: University of Chicago Press, 1937), pp. 407–447.

We emphasize that *the domain of the functional* $\mathcal{J}$ *of* (5.1.11) *is the entire normed vector space* $\mathcal{PC}^1[x_0, x_1]$ and is therefore automatically an open set in itself.

We can now observe that the problem of minimizing or maximizing the functional $\mathcal{J}$ of (5.1.11) on an open domain subject only to certain prescribed fixed or variable end-point conditions has already been considered in Sections 4.1 and 4.4, where it was shown that any local extremum function $U = U(x)$ must satisfy the Euler-Lagrange equation† [compare with (4.1.11)]

$$\mathfrak{F}_U(x, U(x), U'(x)) - \frac{d}{dx}\mathfrak{F}_{U'}(x, U(x), U'(x)) = 0, \tag{5.1.13}$$

where [see (2.4.16)]

$$\begin{aligned} \mathfrak{F}_U(x, U(x), U'(x)) &= \left.\frac{\partial \mathfrak{F}(x, u, w)}{\partial u}\right|_{\substack{u=U(x)\\ w=U'(x)}} \\ \mathfrak{F}_{U'}(x, U(x), U'(x)) &= \left.\frac{\partial \mathfrak{F}(x, u, w)}{\partial w}\right|_{\substack{u=U(x)\\ w=U'(x)}}. \end{aligned} \tag{5.1.14}$$

If we put $\varphi(x) + u^2 = y$ and $\varphi'(x) + 2uw = z$ in the right-hand side of (5.1.12) and use the chain rule of differential calculus there, we find that

$$\begin{aligned} \frac{\partial \mathfrak{F}(x, u, w)}{\partial u} &= \left.\left[\frac{\partial y}{\partial u}\frac{\partial F(x, y, z)}{\partial y} + \frac{\partial z}{\partial u}\frac{\partial F(x, y, z)}{\partial z}\right]\right|_{\substack{y=\varphi(x)+u^2\\ z=\varphi'(x)+2uw}} \\ &= \left. 2u\frac{\partial F(x, y, z)}{\partial y} + 2w\frac{\partial F(x, y, z)}{\partial z}\right|_{\substack{y=\varphi(x)+u^2\\ z=\varphi'(x)+2uw}} \end{aligned}$$

which with (5.1.9), (5.1.12), (5.1.14), and (2.4.16) gives

$$\begin{aligned} \mathfrak{F}_U(x, U(x), U'(x)) = 2U(x)F_Y(x, Y(x), Y'(x)) \\ + 2U'(x)F_{Y'}(x, Y(x), Y'(x)). \end{aligned}$$

Similarly, we find that

$$\mathfrak{F}_{U'}(x, U(x), U'(x)) = 2U(x)F_{Y'}(x, Y(x), Y'(x)),$$

so that

$$\begin{aligned} \frac{d}{dx}\mathfrak{F}_{U'}(x, U(x), U'(x)) = 2U(x)\frac{d}{dx}F_{Y'}(x, Y(x), Y'(x)) \\ + 2U'(x)F_{Y'}(x, Y(x), Y'(x)). \end{aligned}$$

If we now insert these results into the Euler-Lagrange equation (5.1.13), we

†The reader may wish to consider whether or not the possible discontinuities of $U'(x)$ can cause any difficulties in arriving at the necessary condition (5.1.13).

find that

$$U(x)\left[F_Y(x, Y(x), Y'(x)) - \frac{d}{dx}F_{Y'}(x, Y(x), Y'(x))\right] = 0,$$

which factors into

$$F_Y(x, Y(x), Y'(x)) - \frac{d}{dx}F_{Y'}(x, Y(x), Y'(x)) = 0 \tag{5.1.15}$$

and $U(x) = 0$. In view of (5.1.9), the latter equation $U(x) = 0$ can be given alternatively as

$$Y(x) = \varphi(x). \tag{5.1.16}$$

Hence any piecewise continuously differentiable extremum function $Y = Y(x)$ *for the functional* $J = J(Y)$ *of* (5.1.7) *subject to fixed or variable end-point conditions along with the global inequality constraint* (5.1.2) *must lead to a composite extremum curve* γ (see (5.1.1)) *consisting of pieces of arcs along which the usual Euler-Lagrange equation* (5.1.15) *holds and pieces of arcs* (or isolated points) *along which the equality constraint* (5.1.16) *holds.*

A Brachistochrone Example. As an example, we return to the brachistochrone problem mentioned earlier in this section in connection with Figure 3. In this case if

$$\frac{2(x_1 - x_0)}{\pi} \leq y_0 - c \tag{5.1.17}$$

holds, then clearly the quickest descent down from $P_0 = (x_0, y_0)$ to the line $x = x_1$ satisfying (5.1.6) will be along James Bernoulli's cycloid, shown in Figure 25 near the end of Section 4.4, since that cycloid remains everywhere above the line $y = c$ if (5.1.17) holds, and hence the inequality constraint (5.1.6) will be automatically satisfied in this case. Thus we need only consider the (more interesting) case where

$$\frac{2(x_1 - x_0)}{\pi} > y_0 - c \tag{5.1.18}$$

holds, in which case James Bernoulli's cycloid drops *below* the line $y = c$ and is therefore disqualified. In this case we expect that the quickest descent which satisfies (5.1.6) will be along a *composite* curve consisting first of a piece of some arc dropping down from the point P_0 to the line $y = c$ and along which the usual Euler-Lagrange equation (5.1.15) will hold, followed by a piece of the constraining line $y = c$ along which (5.1.16) will hold, as shown in Figure 5. If this is so, then *we need only consider such composite curves in the competition*; i.e., it will be enough to find the quickest descent among all such composite curves (why?).

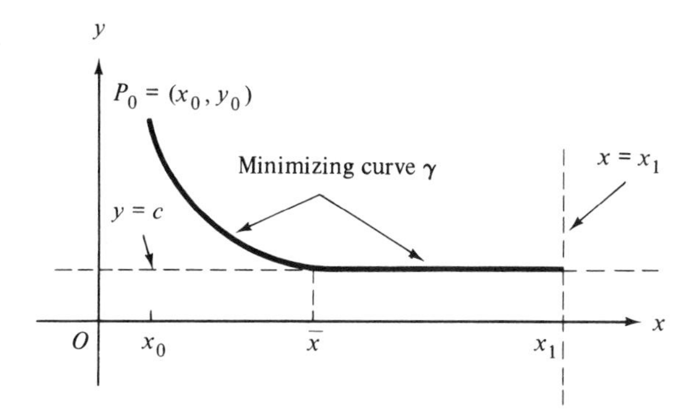

Figure 5

Hence let γ be any such composite curve as shown in Figure 5, and let $\bar{x}$ be the x-coordinate of the point where the two arcs of γ join, so that γ can be given as

$$\gamma: \quad y = \begin{cases} Y(x) & \text{for } x_0 \leq x \leq \bar{x} \\ c & \text{for } \bar{x} \leq x \leq x_1 \end{cases} \tag{5.1.19}$$

for some suitable function $Y = Y(x)$ which is to satisfy the constraint

$$Y(\bar{x}) = c \tag{5.1.20}$$

as well as the fixed end-point constraint

$$Y(x_0) = y_0. \tag{5.1.21}$$

The time T required for a bead to slide along γ is given by (1.3.3) and can be written as the sum of the corresponding times along each separate arc as

$$T = \int_{x_0}^{\bar{x}} \sqrt{\frac{1 + (dy/dx)^2}{2g(y_0 - y)}}\, dx + \int_{\bar{x}}^{x_1} \sqrt{\frac{1 + (dy/dx)^2}{2g(y_0 - y)}}\, dx,$$

or if we use (5.1.19), we find that (explain)

$$T(\bar{x}, Y) = \int_{x_0}^{\bar{x}} F(Y(x), Y'(x))\, dx + \frac{x_1 - \bar{x}}{\sqrt{2g(y_0 - c)}}, \tag{5.1.22}$$

where we have set

$$F(y, z) = \sqrt{\frac{1 + z^2}{2g(y_0 - y)}}. \tag{5.1.23}$$

We have written $T = T(\bar{x}, Y)$ in (5.1.22) to indicate that the time T depends explicitly on $\bar{x}$ as well as on the function $Y = Y(x)$ appearing in (5.1.19).

We take the underlying vector space $\mathfrak{X}$ corresponding to the functional T to be the set of all pairs $(\bar{x}, Y) = (\bar{x}, Y(x))$, where $\bar{x}$ may be any arbitrary number in $\mathfrak{R}$ and where $Y = Y(x)$ may be any arbitrary continuously differentiable function on the interval $[x_0, x_1]$. As in Section 4.5, there need not be any special relation between the number $\bar{x}$ and the function $Y = Y(x)$ occurring in the pair $(\bar{x}, Y)$. In particular, the relation (5.1.20) need *not* hold even if $\bar{x}$ happens to be in the interval $[x_0, x_1]$. If $(\bar{x}_1, Y_1)$ and $(\bar{x}_2, Y_2)$ are any two vectors in $\mathfrak{X}$, we define their sum by

$$(\bar{x}_1, Y_1) + (\bar{x}_2, Y_2) = (\bar{x}_1 + \bar{x}_2, Y_1 + Y_2),$$

which gives a vector in $\mathfrak{X}$. Similarly, we define the product $a(\bar{x}, Y)$ by

$$a(\bar{x}, Y) = (a\bar{x}, aY),$$

which gives a vector in $\mathfrak{X}$ for any number a in $\mathfrak{R}$ and any vector $(\bar{x}, Y)$. One easily checks that all the rules listed in Section 1.2 for vector spaces are satisfied by $\mathfrak{X}$. We equip $\mathfrak{X}$ with the norm $\|\cdot\|$ defined by

$$\|(\bar{x}, Y)\| = |\bar{x}| + \max_{x_0 \leq x \leq x_1} |Y(x)| + \max_{x_0 \leq x \leq x_1} |Y'(x)|$$

for any vector $(\bar{x}, Y)$ in $\mathfrak{X}$. Finally we take the domain D of the functional $T = T(\bar{x}, Y)$ of (5.1.22) to be the subset of $\mathfrak{X}$ consisting of all vectors $(\bar{x}, Y) = (\bar{x}, Y(x))$ in $\mathfrak{X}$ for which $x_0 < \bar{x} < x_1$. Clearly, D is an open set in $\mathfrak{X}$.

We now seek a vector $(\bar{x}, Y)$ in D that will minimize in D the functional T subject to the variable constraint (5.1.20) and the fixed end-point constraint (5.1.21). If we define functionals K_0 and K_1 on D by

$$K_0(\bar{x}, Y) = Y(x_0) \tag{5.1.24}$$

and

$$K_1(\bar{x}, Y) = Y(\bar{x}), \tag{5.1.25}$$

then *the problem is to find a minimum vector in* $D[K_0 = y_0, K_1 = c]$ *for the functional* T *of* (5.1.22). We shall see below that all the hypotheses of the Euler-Lagrange multiplier theorem are satisfied, and we shall be able to conclude that any minimum vector $(\bar{x}, Y)$ in $D[K_0 = y_0, K_1 = c]$ for T must satisfy the necessary condition [see equation (3.6.3)]

$$\delta T(\bar{x}, Y; \Delta\bar{x}, \Delta Y) = \lambda_0\, \delta K_0(\bar{x}, Y; \Delta\bar{x}, \Delta Y) + \lambda_1\, \delta K_1(\bar{x}, Y; \Delta\bar{x}, \Delta Y) \tag{5.1.26}$$

for suitable Euler-Lagrange multipliers λ_0 and λ_1, for all numbers $\Delta\bar{x}$, and for all continuously differentiable functions $\Delta Y = \Delta Y(x)$ on the interval $[x_0, x_1]$. Moreover, we shall see that this necessary condition (5.1.26) along with the constraints (5.1.20) and (5.1.21) will actually determine a minimum vector $(\bar{x}, Y)$ in $D[K_0 = y_0, K_1 = c]$ for T.

As usual, to proceed we must calculate the variations of the functionals involved. We easily find from (5.1.24) and (2.3.4) that

$$\delta K_0(\bar{x},\ Y;\ \Delta\bar{x},\ \Delta Y) = \Delta Y(x_0). \tag{5.1.27}$$

Also, comparing (4.4.11) and (5.1.25), it follows that the variation of K_1 can be given by (4.4.19) with

$$\Phi(x, y) = y;$$

hence we find that (explain)

$$\delta K_1(\bar{x},\ Y;\ \Delta\bar{x},\ \Delta Y) = Y'(\bar{x})\,\Delta\bar{x} + \Delta Y(\bar{x}) \tag{5.1.28}$$

for any vector $(\bar{x},\ Y)$ in D and any vector $(\Delta\bar{x},\ \Delta Y)$ in the vector space $\mathfrak{X}$. It only remains, then, to find the variation of T. From (5.1.22) we find that (explain)

$$T(\bar{x} + \epsilon\,\Delta\bar{x}, Y + \epsilon\,\Delta Y) = \int_{x_0}^{\bar{x}+\epsilon\Delta\bar{x}} F(Y(x) + \epsilon\,\Delta Y(x), Y'(x) + \epsilon\,\Delta Y'(x))\,dx + \frac{x_1 - \bar{x} - \epsilon\,\Delta\bar{x}}{\sqrt{2g(y_0 - c)}},$$

from which [see, for example, (4.4.14) and (4.4.15)]

$$\begin{aligned} &\frac{d}{d\epsilon} T(\bar{x} + \epsilon\,\Delta\bar{x}, Y + \epsilon\,\Delta Y)\bigg|_{\epsilon=0} \\ &= F(Y(\bar{x}), Y'(\bar{x}))\,\Delta\bar{x} + \int_{x_0}^{\bar{x}} [F_Y(Y(x), Y'(x))\,\Delta Y(x) \\ &+ F_{Y'}(Y(x), Y'(x))\,\Delta Y'(x)]\,dx - \frac{\Delta\bar{x}}{\sqrt{2g(y_0 - c)}} \end{aligned}$$

follows. This equation and (2.3.4) along with (4.1.8) gives for the variation of T

$$\begin{aligned} \delta T(\bar{x}, Y; \Delta\bar{x}, \Delta Y) = &\int_{x_0}^{\bar{x}} \left[F_Y(Y(x), Y'(x)) - \frac{d}{dx}F_{Y'}(Y(x), Y'(x))\right]\Delta Y(x)\,dx \\ &+ F_{Y'}(Y(\bar{x}), Y'(\bar{x}))\,\Delta Y(\bar{x}) - F_{Y'}(Y(x_0), Y'(x_0))\,\Delta Y(x_0) \\ &+ \left[F(Y(\bar{x}), Y'(\bar{x})) - \frac{1}{\sqrt{2g(y_0 - c)}}\right]\Delta\bar{x} \end{aligned} \tag{5.1.29}$$

for any vector $(\bar{x}, Y)$ in D and for any vector $(\Delta\bar{x}, \Delta Y)$ in the vector space $\mathfrak{X}$.

We can now use the Euler-Lagrange multiplier theorem to conclude that equation (5.1.26) must hold, at least provided that the following determinant [see (3.6.2)]

$$\det \begin{vmatrix} \delta K_0(\bar{x}, Y; \Delta\bar{x}_1, \Delta Y_1) & \delta K_0(\bar{x}, Y; \Delta\bar{x}_2, \Delta Y_2) \\ \delta K_1(\bar{x}, Y; \Delta\bar{x}_1, \Delta Y_1) & \delta K_1(\bar{x}, Y; \Delta\bar{x}_2, \Delta Y_2) \end{vmatrix}$$

is *different from zero* for some vectors $(\Delta\bar{x}_1, \Delta Y_1)$ and $(\Delta\bar{x}_2, \Delta Y_2)$ in $\mathfrak{X}$. But we may use (5.1.27) and (5.1.28) to calculate this determinant as

$$\begin{aligned} \text{determinant} &= \Delta Y_1(x_0)\,\Delta Y_2(\bar{x}) - \Delta Y_1(\bar{x})\,\Delta Y_2(x_0) \\ &\quad + Y'(\bar{x})[\Delta Y_1(x_0)\,\Delta\bar{x}_2 - \Delta Y_2(x_0)\,\Delta\bar{x}_1], \end{aligned}$$

which is seen to have the value 1 in the special case in which $\Delta\bar{x}_1 = \Delta\bar{x}_2 = 0$, with

$$\Delta Y_1(x) = 1 \qquad \text{for } x_0 \leq x \leq x_1$$

and

$$\Delta Y_2(x) = \frac{x - x_0}{\bar{x} - x_0} \qquad \text{for } x_0 \leq x \leq x_1.$$

Hence the second possibility of the multiplier theorem must hold [see (3.6.3)], and this verifies that (5.1.26) must hold in the present case.

If we now insert (5.1.27), (5.1.28), and (5.1.29) into (5.1.26) and use the constraints (5.1.20) and (5.1.21), we find for any local extremum vector $(\bar{x}, Y)$ in $D[K_0 = y_0, K_1 = c]$ for T the necessary condition

$$\begin{aligned} &\int_{x_0}^{\bar{x}} \left[F_Y(Y(x), Y'(x)) - \frac{d}{dx} F_{Y'}(Y(x), Y'(x))\right] \Delta Y(x)\,dx \\ &= [\lambda_0 + F_{Y'}(y_0, Y'(x_0))]\,\Delta Y(x_0) + [\lambda_1 - F_{Y'}(c, Y'(\bar{x}))]\,\Delta Y(\bar{x}) \\ &\quad + \left[\lambda_1 Y'(\bar{x}) - F(c, Y'(\bar{x})) + \frac{1}{\sqrt{2g(y_0 - c)}}\right] \Delta\bar{x}, \end{aligned} \tag{5.1.30}$$

which must hold for all numbers $\Delta\bar{x}$ in $\mathfrak{R}$ and for all continuously differentiable functions $\Delta Y = \Delta Y(x)$ on the interval $[x_0, \bar{x}]$. (Note that any such function ΔY on $[x_0, \bar{x}]$ can be extended so as to be defined and continuously differentiable on the larger interval $[x_0, x_1]$.) A standard argument based on the lemma of Section A4 of the Appendix and the arbitrariness of $\Delta\bar{x}$ and ΔY in (5.1.30) can now be used to obtain from (5.1.30) the necessary conditions

$$F_Y(Y(x), Y'(x)) - \frac{d}{dx} F_{Y'}(Y(x), Y'(x)) = 0 \qquad \text{for } x_0 < x < \bar{x} \tag{5.1.31}$$

$$\lambda_0 + F_{Y'}(y_0, Y'(x_0)) = 0 \tag{5.1.32}$$

$$\lambda_1 - F_{Y'}(c, Y'(\bar{x})) = 0 \tag{5.1.33}$$

and

$$\lambda_1 Y'(\bar{x}) - F(c, Y'(\bar{x})) + \frac{1}{\sqrt{2g(y_0 - c)}} = 0, \tag{5.1.34}$$

which must be satisfied by any local extremum vector $(\bar{x}, Y)$ in $D[K_0 = y_0, K_1 = c]$ for the functional $T = T(\bar{x}, Y)$. We leave the derivation of these conditions from (5.1.30) as an exercise for the reader.

The conditions (5.1.33) and (5.1.34) can be combined by eliminating the multiplier λ_1 to give the single (natural) condition

$$F(c, Y'(\bar{x})) - Y'(\bar{x})F_{Y'}(c, Y'(\bar{x})) = \frac{1}{\sqrt{2g(y_0 - c)}},$$

which must then hold at the variable point $x = \bar{x}$. Using (5.1.23), it follows that the last condition implies that (explain)

$$Y'(\bar{x}) = 0, \tag{5.1.35}$$

which must hold at $x = \bar{x}$. Hence this natural condition requires that any local extremum curve γ must be *tangent* to the constraining line $y = c$ at the point $x = \bar{x}$ as shown in Figure 6.

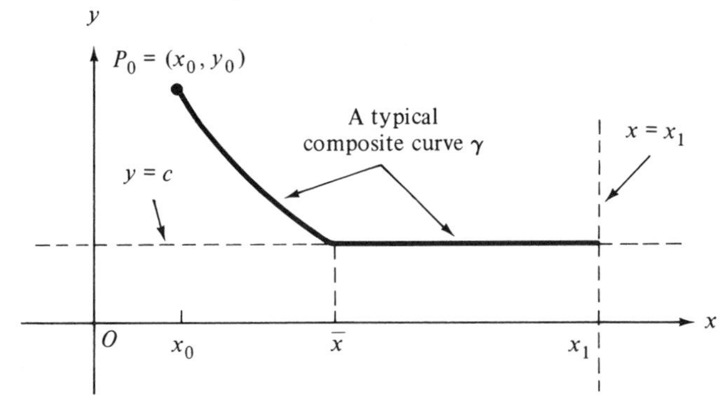

Figure 6

It now follows from (5.1.31) and (5.1.23) (as in Section 4.2) that any local extremum curve $y = Y(x)$ must represent a cycloid which can be given parametrically as [see (4.2.7)]

$$\begin{aligned} x &= x_0 + \frac{A}{2}(\theta - \sin\theta) \\ y &= y_0 - \frac{A}{2}(1 - \cos\theta) \qquad \text{for } 0 \leq \theta \leq \bar{\theta}, \end{aligned} \tag{5.1.36}$$

where we have imposed the constraint (5.1.21) on $y = Y(x)$ at $\theta = 0$ and where the two constants $\bar{\theta}$ and A must be chosen so that the remaining constraint (5.1.20) and the natural condition (5.1.35) are both satisfied at $x = \bar{x}$. [We take the parameter value $\theta = \bar{\theta}$ to correspond to the point $P = (x, y) = (\bar{x}, c)$.]

The natural condition (5.1.35) is seen to imply, as in Section 4.4 [see (4.4.43)], that $\bar{\theta} = \pi$, which with (5.1.20) and (5.1.36) implies also that $A = y_0 - c$. Inserting these choices for $\bar{\theta}$ and A into (5.1.36), we find the extremum cycloid

$$\begin{aligned} x &= x_0 + \frac{y_0 - c}{2}(\theta - \sin\theta) \\ y &= y_0 - \frac{y_0 - c}{2}(1 - \cos\theta) \qquad \text{for } 0 \leq \theta \leq \pi. \end{aligned} \tag{5.1.37}$$

The corresponding value of $\bar{x}$ is obtained by setting $\theta = \bar{\theta} = \pi$ in the first equation of (5.1.37) to find

$$\bar{x} = x_0 + \frac{\pi}{2}(y_0 - c). \tag{5.1.38}$$

One easily checks that the condition (5.1.18) implies with (5.1.38) that $x_0 < \bar{x} < x_1$, so that the number $\bar{x}$ given by (5.1.38) is in the open interval (x_0, x_1). The situation is as shown in Figure 6, with the extremum curve γ consisting of the cycloid given by (5.1.37) for $x_0 \leq x \leq \bar{x}$ joined with the line segment [see (5.1.19)]

$$y = c \qquad \text{for } \bar{x} \leq x \leq x_1, \tag{5.1.39}$$

with $\bar{x}$ given by (5.1.38).

Hence we have shown that formulas (5.1.37), (5.1.38), and (5.1.39) define a composite curve γ as in (5.1.19) *which satisfies all the conditions obtained from the Euler-Lagrange multiplier theorem*. These conditions are, of course, necessary conditions which must be satisfied by the desired minimizing curve, and therefore the composite curve γ just found is a *candidate* which *may* give the quickest descent of the bead in the present case. Strictly speaking, however, we have only shown that *if* a minimizing curve γ exists for the present problem in the form of a composite curve as shown in Figure 5, then that minimizing curve must be given by the *particular* composite curve defined by formulas (5.1.37), (5.1.38), and (5.1.39). It can actually be shown that this particular composite curve γ does give the quickest descent. However, the proof of this assertion would take us outside the intended scope of this book, and we shall omit it here.†

Finally, we remark that the value of the shortest time of descent of the bead subject to the global constraint (5.1.6) when (5.1.18) holds can be found

†The interested reader may refer to Gilbert Ames Bliss, *Calculus of Variations* (Chicago: The Open Court Publishing Company, 1925), where such questions are considered for certain brachistochrone problems.

from (5.1.22), (5.1.23), (5.1.37), (5.1.38), and (4.4.42) to be

$$T_{\text{minimum}} = \int_0^\pi \sqrt{\frac{(dx/d\theta)^2 + (dy/d\theta)^2}{2g(y_0 - y)}}\, d\theta + \frac{x_1 - \bar{x}}{\sqrt{2g(y_0 - c)}}$$
$$= \frac{\pi}{2}\sqrt{\frac{y_0 - c}{2g}} + \frac{x_1 - x_0}{\sqrt{2g(y_0 - c)}}.$$

On the other hand, if we *omit* the global inequality constraint (5.1.6), we find, as in Section 4.4, the minimum time [see (4.4.45)]

$$T_{\text{minimum}} = \sqrt{\frac{\pi(x_1 - x_0)}{g}}.$$

Clearly, the unconstrained minimum time must be *less* than the constrained minimum time. Hence

$$\sqrt{\frac{\pi(x_1 - x_0)}{g}} < \frac{\pi}{2}\sqrt{\frac{y_0 - c}{2g}} + \frac{x_1 - x_0}{\sqrt{2g(y_0 - c)}} \tag{5.1.40}$$

must hold whenever (5.1.18) holds. One can easily show that (5.1.40) is indeed valid using (5.1.18) and the simple inequality

$$\sqrt{1 + 2a^2} < 1 + a^2,$$

which holds for any positive number a (why?). We leave the verification of this assertion as an exercise for the reader.

Exercises

1. Give a careful derivation of the conditions (5.1.31), (5.1.32), (5.1.33), and (5.1.34) from the necessary condition (5.1.30).
2. Show that the number $\bar{x}$ given by (5.1.38) lies in the open interval (x_0, x_1) when (5.1.18) holds.
3. Verify that (5.1.40) holds when (5.1.18) is satisfied.

5.2. An Optimum Consumption Policy with Terminal Savings Constraint Without Extreme Hardship†

We return again to the problem in investment planning already discussed in Section 3.7, except now we shall impose the global inequality consumption constraint

$$C(t) \geq C_0, \tag{5.2.1}$$

†The remaining sections in this chapter provide additional examples and should be considered as optional.

which is required to hold for all t in the given time interval $0 \leq t \leq T$. The quantity C_0 is a given positive constant

$$C_0 > 0$$

which is specified by the individual investor as the *lowest acceptable consumption level* that he wishes to experience. Hence every *admissible* consumption rate $C = C(t)$ must now equal or exceed the constant rate C_0, which the investor considers to be the *hardship level* below which he prefers his consumption rate *not* to fall. As in Section 3.7, we assume also that the individual requires that a certain specified level of savings be accumulated at the end of T years, which leads to the terminal constraint

$$S(t) = S_T \qquad \text{at } t = T, \tag{5.2.2}$$

where S_T is a given nonnegative constant and where the savings level $S = S(t)$ is given by equation (3.7.4) as

$$S(t) = e^{\alpha t}\left(S_0 + \int_0^t e^{-\alpha\tau}[I(\tau) - C(\tau)]\,d\tau\right).$$

If we evaluate this expression at $t = T$ and use (5.2.2), we find the requirement

$$S_0 + \int_0^T e^{-\alpha t} I(t)\,dt = e^{-\alpha T} S_T + \int_0^T e^{-\alpha t} C(t)\,dt, \tag{5.2.3}$$

which must be satisfied by any admissible consumption rate $C = C(t)$. The initial savings level S_0 and the investment return rate α which appear in (5.2.3) are assumed to be known, as is the income function $I = I(t)$ for $0 \leq t \leq T$. The notation is the same as in Section 3.7.

The individual wishes to find the particular consumption rate $C = C(t)$ that will maximize the satisfaction functional $\mathcal{S} = \mathcal{S}(C)$ of (3.7.11) among all possible consumption rates which satisfy the two conditions (5.2.1) and (5.2.3). However, there are available to the individual only certain limited resources represented by the given initial savings level S_0 and the known income function $I = I(t)$, and therefore he must be careful to set realistic goals. (He may not eat his cake and then later have it too.) In fact, the global inequality constraint (5.2.1) leads to the inequality (explain)

$$\int_0^T e^{-\alpha t} C(t)\,dt \geq C_0 \int_0^T e^{-\alpha t}\,dt = \frac{1 - e^{-\alpha T}}{\alpha} C_0$$

for any (integrable) function $C(t)$ which satisfies (5.2.1), and this result along with (5.2.3) implies the condition (why?)

$$S_0 + \int_0^T e^{-\alpha t} I(t)\,dt \geq e^{-\alpha T} S_T + \left(\frac{1 - e^{-\alpha T}}{\alpha}\right) C_0, \tag{5.2.4}$$

which must be satisfied by the given data. If (5.2.4) is violated, then the constraints (5.2.1) and (5.2.3) are together incompatible and cannot simultaneously be achieved by any consumption rate $C = C(t)$ (see Exercise 1 at the end of this section). Hence we assume that the goals represented by the specified constants S_T and C_0 are chosen so that (5.2.4) is satisfied. This means that the nonnegative numbers S_T and C_0 are restricted to lie in the shaded region of Figure 7 (explain), where we have denoted in Figure 7 the positive quantity $S_0 + \int_0^T e^{-\alpha t} I(t)\,dt$ as

$$\gamma = S_0 + \int_0^T e^{-\alpha t} I(t)\,dt. \tag{5.2.5}$$

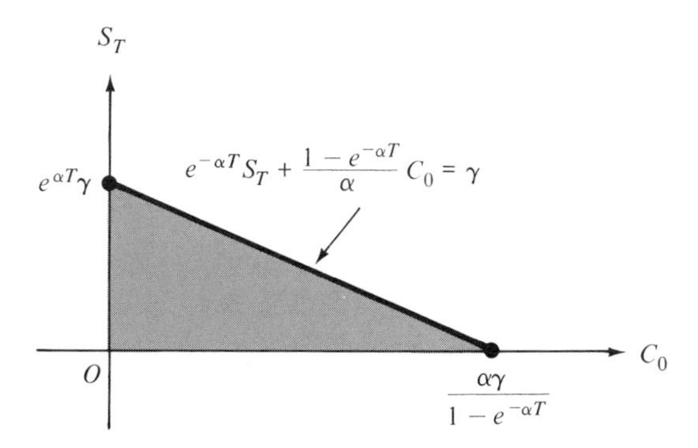

Figure 7

This known quantity γ gives a measure of the *resources* expected to be available to the individual.

We consider now in detail only the case in which the investment return rate α dominates the discount (inflation) rate β with

$$\alpha > \beta. \tag{5.2.6}$$

The other case in which $\alpha \leq \beta$ can be handled similarly with the same methods, and we leave that case as an exercise for the reader.

Now if (5.2.6) holds, *we might just as well assume further that the specified constants S_T and C_0 satisfy the additional condition*

$$S_0 + \int_0^T e^{-\alpha t} I(t)\,dt + \frac{1 - e^{-\alpha T}}{\alpha} < e^{-\alpha T} S_T + \left(\frac{1 - e^{-\beta T}}{\beta}\right)(1 + C_0) \tag{5.2.7}$$

since otherwise the optimum consumption rate $C^*(t)$ found in Section 3.7 will automatically satisfy the inequality constraint (5.2.1), as the reader can verify, and then the present optimization problem is already solved by the solution

C^* found in Section 3.7. Note that (5.2.7) restricts the nonnegative numbers S_T and C_0 to lie in the shaded region of Figure 8, where γ is given by (5.2.5) and δ is defined by

$$\delta = \frac{1 - e^{-\alpha T}}{\alpha} - \frac{1 - e^{-\beta T}}{\beta}. \tag{5.2.8}$$

It can be easily shown that the shaded regions of Figures 7 and 8 *overlap* somewhat in a common region, as shown in Figure 9 (see Exercise 3), and then both conditions (5.2.4) and (5.2.7) will simultaneously hold throughout that common region.

In particular, then, we assume that the individual has set realistic goals and that both (5.2.4) and (5.2.7) hold, and we then seek a new consumption rate $C = C(t)$ that will *maximize* the satisfaction functional $\mathcal{S}$ given by formula (3.7.11) subject to the constraints (5.2.1) and (5.2.3). The optimum rate $C^*(t)$ found earlier in Section 3.7 is disqualified in the present case since the

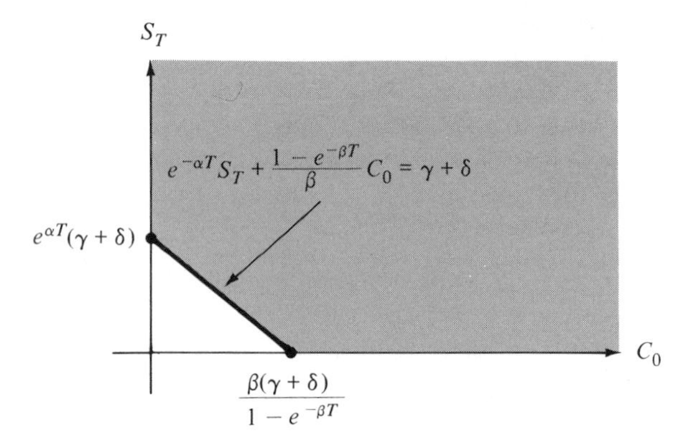

Figure 8

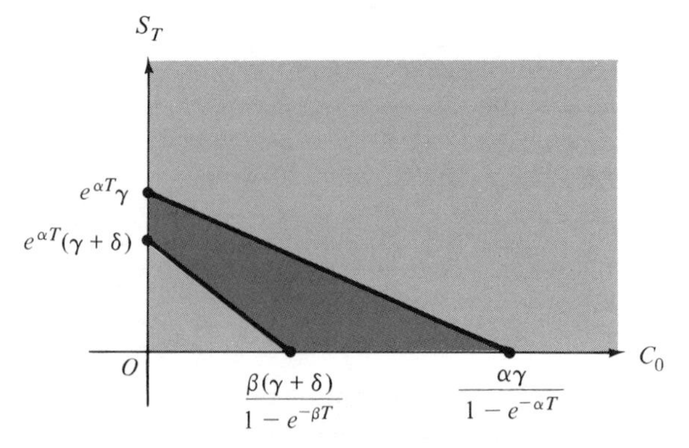

Figure 9

condition (5.2.7) implies that C^* will be initially somewhat *lower* than the *hardship level* C_0 and will therefore violate the inequality constraint (5.2.1), as shown in Figure 10. In this case the analysis of Section 3.7 is no longer entirely relevant, and we must search for a new optimum consumption rate among those functions $C = C(t)$ which satisfy both (5.2.1) and (5.2.3).

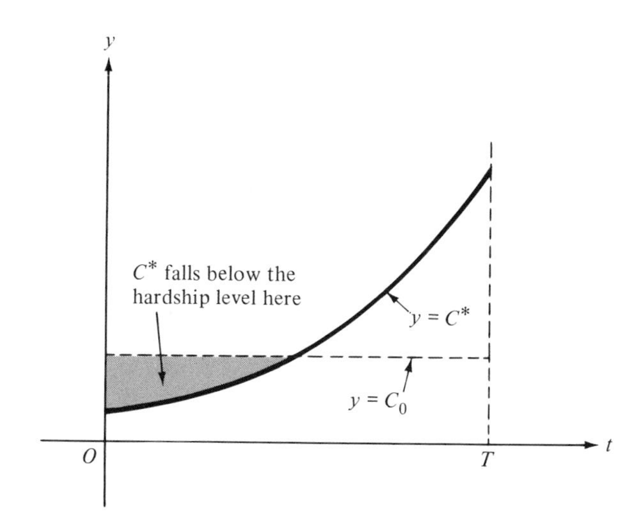

Figure 10

Actually the results of Section 3.7 may furnish us with certain valuable insight in the present case even though the optimum solution C^* found in Section 3.7 violates the inequality constraint (5.2.1). In fact, precisely because C^* falls initially *below* C_0, as shown in Figure 10, it seems natural to expect in the present case that *the new optimum consumption rate should be initially as low as possible subject to* (5.2.1). Hence as a consequence of the results of Section 3.7, we may expect the new optimum consumption rate to be a *composite rate* (as shown in Figure 11) which is initially equal to the lowest admissible rate C_0 so as to allow the savings to grow initially, and which then increases so as to give the maximum possible satisfaction from consumption subject to the terminal savings constraint (5.2.3). We shall see that this is indeed the case.

A Reformulation of the Problem. Hence we are led more or less naturally to consider composite consumption rates $C = C(t)$ as shown, for example, in Figure 11, with

$$C(t) = \begin{cases} C_0 & \text{for } 0 \leq t \leq \bar{t} \\ \mathcal{C}(t) & \text{for } \bar{t} \leq t \leq T \end{cases} \tag{5.2.9}$$

for some suitable time value $\bar{t}$ and some suitable function $\mathcal{C}(t)$ *both of which*

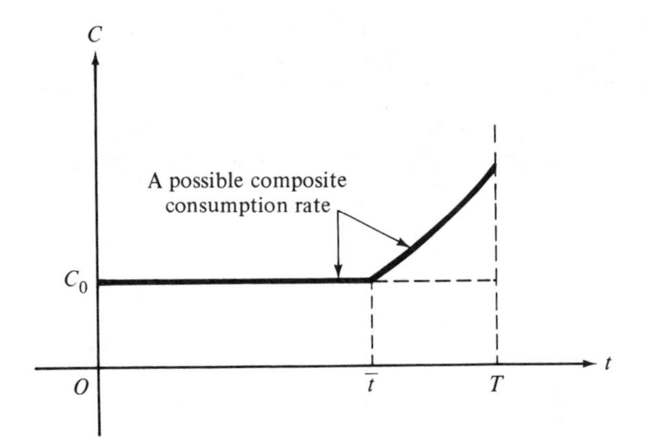

Figure 11

may depend on the particular composite rate considered. The satisfaction $\mathcal{S}$ obtained from any such composite rate is found from (3.7.11) to be given as

$$\begin{aligned}\mathcal{S}(C) &= \int_0^{\bar{t}} e^{-\beta t} \log (1 + C_0)\, dt + \int_{\bar{t}}^{T} e^{-\beta t} \log [1 + \mathcal{C}(t)]\, dt \\ &= \frac{1 - e^{-\beta \bar{t}}}{\beta} \log (1 + C_0) + \int_{\bar{t}}^{T} e^{-\beta t} \log [1 + \mathcal{C}(t)]\, dt,\end{aligned}$$

where C is given by (5.2.9). We shall want to consider the last expression as a function of both $\bar{t}$ and $\mathcal{C}$, and so we define a new functional $\mathcal{S}_0$ by

$$\mathcal{S}_0(\bar{t}, \mathcal{C}) = \frac{1 - e^{-\beta \bar{t}}}{\beta} \log (1 + C_0) + \int_{\bar{t}}^{T} e^{-\beta t} \log [1 + \mathcal{C}(t)]\, dt \qquad (5.2.10)$$

for any number $\bar{t}$ satisfying $0 < \bar{t} < T$ and for any suitable function $\mathcal{C} = \mathcal{C}(t)$. Of course, we have the identity

$$\mathcal{S}_0(\bar{t}, \mathcal{C}) = \mathcal{S}(C)$$

which is valid whenever C is given by (5.2.9).

Similarly, we can rewrite the terminal savings constraint (5.2.3) for any such composite rate C satisfying (5.2.9) as (explain)

$$S_0 + \int_0^T e^{-\alpha t} I(t)\, dt = e^{-\alpha T} S_T + \frac{1 - e^{-\alpha \bar{t}}}{\alpha} C_0 + \int_{\bar{t}}^{T} e^{-\alpha t} \mathcal{C}(t)\, dt,$$

or if we define a new functional $K_0 = K_0(\bar{t}, \mathcal{C})$ by [compare with (3.7.7)]

$$K_0(\bar{t}, \mathcal{C}) = \frac{1 - e^{-\alpha \bar{t}}}{\alpha} C_0 + \int_{\bar{t}}^{T} e^{-\alpha t} \mathcal{C}(t)\, dt, \qquad (5.2.11)$$

we then have the constraint [compare with (3.7.8)]

$$K_0(\bar{t}, \mathcal{C}) = S_0 - e^{-\alpha T}S_T + \int_0^T e^{-\alpha t}I(t)\,dt. \tag{5.2.12}$$

Note that K_0 as defined by (5.2.11) satisfies the identity $K_0(\bar{t}, \mathcal{C}) = K(C)$ for any C which satisfies (5.2.9), where here $K = K(C)$ is defined as in Section 3.7 by formula (3.7.7).

It is natural to use as the underlying vector space $\mathfrak{X}$ for the functionals $\mathcal{S}_0$ and K_0 the collection of all pairs $(\bar{t}, \mathcal{C}) = (\bar{t}, \mathcal{C}(t))$, where $\bar{t}$ may be *any arbitrary number* in $\mathcal{R}$ and where $\mathcal{C} = \mathcal{C}(t)$ may be *any arbitrary continuous function* defined on the fixed interval $[0, T]$. There need not be any special relation between the number $\bar{t}$ and the function $\mathcal{C} = \mathcal{C}(t)$ occurring in the pair $(\bar{t}, \mathcal{C})$. If $(\bar{t}_1, \mathcal{C}_1)$ and $(\bar{t}_2, \mathcal{C}_2)$ are any two vectors in $\mathfrak{X}$, we define their sum by

$$(\bar{t}_1, \mathcal{C}_1) + (\bar{t}_2, \mathcal{C}_2) = (\bar{t}_1 + \bar{t}_2, \mathcal{C}_1 + \mathcal{C}_2),$$

which gives a vector in $\mathfrak{X}$. Similarly, we define the product $a(\bar{t}, \mathcal{C})$ by

$$a(\bar{t}, \mathcal{C}) = (a\bar{t}, a\mathcal{C}),$$

which gives a vector in $\mathfrak{X}$ for any number a in $\mathcal{R}$ and any vector $(\bar{t}, \mathcal{C})$. It is easy to show that all the rules for vector spaces as listed in Section 1.2 are satisfied by $\mathfrak{X}$. Finally we furnish $\mathfrak{X}$ with the norm $||\cdot||$ defined by

$$||(\bar{t}, \mathcal{C})|| = |\bar{t}| + \max_{0 \le t \le T} |\mathcal{C}(t)|$$

for any vector $(\bar{t}, \mathcal{C})$ in $\mathfrak{X}$. We then take the domain D of the functionals $\mathcal{S}_0$ and K_0 to be the open subset D of $\mathfrak{X}$ consisting of all vectors $(\bar{t}, \mathcal{C})$ in $\mathfrak{X}$ for which $0 < \bar{t} < T$ and

$$1 + \mathcal{C}(t) > 0 \qquad \text{for } 0 \le t \le T. \tag{5.2.13}$$

The latter inequality plays no direct role in the present optimization problem other than serving to specify a fixed open subset D of $\mathfrak{X}$ which can be used as the domain of the functionals $\mathcal{S}_0$ and K_0. [The constraint (5.2.13) is *inactive* insofar as the optimization problem is concerned.]

We shall now use the Euler-Lagrange multiplier theorem to find a vector $(\bar{t}, \mathcal{C}) = (\bar{t}, \mathcal{C}(t))$ in D that will maximize the satisfaction functional $\mathcal{S}_0$ of (5.2.10) subject to the single constraint

$$K_0(\bar{t}, \mathcal{C}) = k_0, \tag{5.2.14}$$

where K_0 is defined by (5.2.11) and where the constant k_0 is given as in Section 3.7 by formula (3.7.13) [see (5.2.12)]. Hence we shall maximize the functional $\mathcal{S}_0$ in the set $D[K_0 = k_0]$ consisting of all vectors in D which satisfy (5.2.14). Note that *the inequality constraint* (5.2.1) *no longer appears in the statement of the latter problem.* Rather we have used our previous knowledge of the optimum solution C^* (which was found in Section 3.7 without any consideration of the present inequality constraint) to reformulate the present optimization problem in terms of suitable composite consumption rates in such a way as to avoid the need for any further consideration of the global inequality constraint in the reformulated problem. We shall see that once we have the desired extremum vector $(\bar{t}, \mathcal{C})$ in $D[K_0 = k_0]$ for $\mathcal{S}_0$, it will then be an easy matter to show directly that the resulting consumption rate $C = C(t)$ given by (5.2.9) actually maximizes the satisfaction $\mathcal{S}$ of (3.7.11) among all possible rates which satisfy the global inequality constraint (5.2.1) along with the terminal savings constraint (5.2.3).

The Required Variations. To proceed, we must calculate the variations of the functionals $\mathcal{S}_0$ and K_0. From (5.2.11) we find that (explain)

$$K_0(\bar{t} + \epsilon\,\Delta\bar{t}, \mathcal{C} + \epsilon\,\Delta\mathcal{C}) = \frac{1 - e^{-\alpha(\bar{t}+\epsilon\Delta\bar{t})}}{\alpha}C_0 + \int_{\bar{t}+\epsilon\Delta\bar{t}}^{T} e^{-\alpha t}[\mathcal{C}(t) + \epsilon\,\Delta\mathcal{C}(t)]\,dt$$

for any vector $(\Delta\bar{t}, \Delta\mathcal{C})$ in $\mathcal{X}$ and for any small number ϵ, from which [explain; see, for example, (4.4.14)]

$$\begin{aligned}\frac{d}{d\epsilon}K_0(\bar{t} + \epsilon\,\Delta\bar{t}, \mathcal{C} + \epsilon\,\Delta\mathcal{C}) = {} & \Delta\bar{t}e^{-\alpha(\bar{t}+\epsilon\Delta\bar{t})}C_0 \\ & - \Delta\bar{t}e^{-\alpha(\bar{t}+\epsilon\Delta\bar{t})}[\mathcal{C}(\bar{t} + \epsilon\,\Delta\bar{t}) + \epsilon\,\Delta\mathcal{C}(\bar{t} + \epsilon\,\Delta\bar{t})] \\ & + \int_{\bar{t}+\epsilon\Delta\bar{t}}^{T} e^{-\alpha t}\,\Delta\mathcal{C}(t)\,dt\end{aligned}$$

follows. If we evaluate this result at $\epsilon = 0$, we find for the variation of K_0 that

$$\delta K_0(\bar{t}, \mathcal{C}; \Delta\bar{t}, \Delta\mathcal{C}) = \Delta\bar{t}e^{-\alpha\bar{t}}[C_0 - \mathcal{C}(\bar{t})] + \int_{\bar{t}}^{T} e^{-\alpha t}\,\Delta\mathcal{C}(t)\,dt \tag{5.2.15}$$

for any vector $(\bar{t}, \mathcal{C})$ in D and for any vector $(\Delta\bar{t}, \Delta\mathcal{C})$ in the vector space $\mathcal{X}$. Similarly, we find from (5.2.10) that

$$\begin{aligned}\delta\mathcal{S}_0(\bar{t}, \mathcal{C}; \Delta\bar{t}, \Delta\mathcal{C}) = {} & \Delta\bar{t}e^{-\beta\bar{t}}[\log(1 + C_0) - \log(1 + \mathcal{C}(\bar{t}))] \\ & + \int_{\bar{t}}^{T} e^{-\beta t}\frac{\Delta\mathcal{C}(t)}{1 + \mathcal{C}(t)}dt\end{aligned} \tag{5.2.16}$$

for any vector $(\bar{t}, \mathcal{C})$ in D and for any vector $(\Delta\bar{t}, \Delta\mathcal{C})$ in the vector space $\mathcal{X}$.

The Extremum Conditions. It is now clear from (5.2.15) and (5.2.16) that the functionals $\mathcal{S}_0$ and K_0 satisfy all the hypotheses of the Euler-Lagrange multiplier theorem as stated in Section 3.3. Moreover,

$$\delta K_0(\bar{t}, \mathcal{C}; \Delta\bar{t}, \Delta\mathcal{C}) > 0$$

will hold for any vector $(\bar{t}, \mathcal{C})$ in D *when* $\Delta\bar{t} = 0$ *and* $\Delta\mathcal{C}(t) = 1$ *for all* t. Hence the first possibility (3.3.1) of the multiplier theorem *cannot hold*; so if $(\bar{t}, \mathcal{C})$ is any local extremum vector in $D[K_0 = k_0]$ for $\mathcal{S}_0$, it follows from (3.3.2) that there is a constant λ such that

$$\delta\mathcal{S}_0(\bar{t}, \mathcal{C}; \Delta\bar{t}, \Delta\mathcal{C}) = \lambda\, \delta K_0(\bar{t}, \mathcal{C}; \Delta\bar{t}, \Delta\mathcal{C})$$

holds for *all* numbers $\Delta\bar{t}$ and for *all* continuous functions $\Delta\mathcal{C} = \Delta\mathcal{C}(t)$ on $[0, T]$. We can use (5.2.15) and (5.2.16) to rewrite this equation as (explain)

$$\Delta\bar{t}\left[e^{-\beta\bar{t}} \log\frac{1+C_0}{1+\mathcal{C}(\bar{t})} - \lambda e^{-\alpha\bar{t}}(C_0 - \mathcal{C}(\bar{t}))\right] + \int_{\bar{t}}^{T}\left[\frac{e^{-\beta t}}{1+\mathcal{C}(t)} - \lambda e^{-\alpha t}\right]\Delta\mathcal{C}(t)\,dt = 0, \tag{5.2.17}$$

which must then hold for *all numbers* $\Delta\bar{t}$ and *all continuous functions* $\Delta\mathcal{C}$ if $(\bar{t}, \mathcal{C})$ is a local extremum vector in $D[K_0 = k_0]$ for $\mathcal{S}_0$. It follows now by a standard argument from (5.2.17) that *any such extremum vector* $(\bar{t}, \mathcal{C})$ *must satisfy the necessary conditions* (explain)

$$e^{-\beta\bar{t}} \log\frac{1+C_0}{1+\mathcal{C}(\bar{t})} = \lambda e^{-\alpha\bar{t}}[C_0 - \mathcal{C}(\bar{t})] \tag{5.2.18}$$

and

$$\frac{e^{-\beta t}}{1+\mathcal{C}(t)} = \lambda e^{-\alpha t} \qquad \text{for } \bar{t} \le t \le T. \tag{5.2.19}$$

We leave the derivation of these last two relations from (5.2.17) as an exercise for the reader (see Exercise 5).

The Optimum Consumption Rate. We can now show that the conditions (5.2.18) and (5.2.19) along with the constraint (5.2.12) actually determine a maximum vector $(\bar{t}, \mathcal{C})$ in $D[K_0 = k_0]$ for $\mathcal{S}_0$. As a first step in this direction we can evaluate equation (5.2.19) at $t = \bar{t}$ and use the resulting equation to eliminate the multiplier λ in (5.2.18); in this way we find the condition (explain)

$$\log\frac{1+C_0}{1+\mathcal{C}(\bar{t})} = \frac{C_0 - \mathcal{C}(\bar{t})}{1+\mathcal{C}(\bar{t})}, \tag{5.2.20}$$

which must hold at the variable point $t = \bar{t}$. Relation (5.2.20) will be satisfied if

$$\mathcal{C}(\bar{t}) = C_0, \tag{5.2.21}$$

and, in fact, we can show that (5.2.21) follows automatically as a *natural condition* from (5.2.20). Indeed, we can rewrite (5.2.20) as

$$\log(1 + x) = x,$$

with $x = (C_0 - \mathcal{C}(\bar{t}))/(1 + \mathcal{C}(\bar{t}))$, and this equation implies that $x = 0$ (see, for example, Figure 8 of Section 3.7) from which the stated result follows. Hence even though we have allowed into the competition certain *discontinuous* consumption rates $C = C(t)$, as shown, for example, in Figure 12

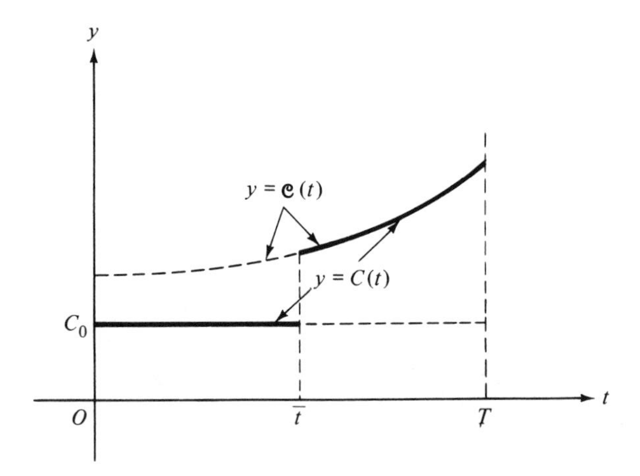

Figure 12

[see also equation (5.2.9)], it follows from the Euler-Lagrange multiplier theorem that the *optimum* consumption rate must automatically satisfy (5.2.21) and must therefore actually be continuous, as shown, for example, in Figure 11. Note also now that *the condition* (5.2.18) *need not be considered further* since it holds identically for all λ as a consequence of (5.2.21).

We can now use (5.2.21) to eliminate the Euler-Lagrange multiplier λ appearing in (5.2.19). Indeed, if we evaluate (5.2.19) at $t = \bar{t}$ and use (5.2.21), we find that (explain)

$$\lambda = \frac{e^{(\alpha-\beta)\bar{t}}}{1 + C_0},$$

and this equation can be used to rewrite (5.2.19) in the form

$$\mathcal{C}(t) = -1 + (1 + C_0)e^{(\alpha-\beta)(t-\bar{t})} \qquad \text{for } \bar{t} \leq t \leq T, \tag{5.2.22}$$

where, of course, the time value $\bar{t}$ appearing here is still unknown.

Finally, we shall use (5.2.22) and the constraint (5.2.12) to determine $\bar{t}$. [The optimum consumption rate will then be given by (5.2.9) and (5.2.22) *once $\bar{t}$ is known.*] If we insert (5.2.22) into the constraint (5.2.12), we find with (5.2.11) that (the reader should carry through the details)

$$\begin{aligned} S_0 - e^{-\alpha T}S_T + \frac{1 - e^{-\alpha T}}{\alpha} + \int_0^T e^{-\alpha t}I(t)\,dt \\ = (1 + C_0)\left[\frac{1 - e^{-\alpha \bar{t}}}{\alpha} + e^{-(\alpha-\beta)\bar{t}}\left(\frac{e^{-\beta \bar{t}} - e^{-\beta T}}{\beta}\right)\right], \end{aligned} \tag{5.2.23}$$

which gives a single equation which relates the unknown time value $\bar{t}$ to the known data of the problem.

Hence finally we need only show that (5.2.23) determines a unique value of $\bar{t}$ in the interval $0 < \bar{t} < T$. That this is the case will be seen to follow from the two conditions (5.2.4) and (5.2.7). Indeed, if we evaluate separately both sides of (5.2.23) at $\bar{t} = 0$, we find with (5.2.7) that (explain)

$$\text{left-hand side of (5.2.23)} < \text{right-hand side of (5.2.23)} \qquad \text{at } \bar{t} = 0. \tag{5.2.24}$$

On the other hand, if we evaluate both sides of (5.2.23) at $\bar{t} = T$, we find with (5.2.4) that

$$\text{left-hand side of (5.2.23)} \geq \text{right-hand side of (5.2.23)} \qquad \text{at } \bar{t} = T. \tag{5.2.25}$$

Since both sides of (5.2.23) are continuous functions of $\bar{t}$ on the interval $0 \leq \bar{t} \leq T$ (indeed, the left-hand side is a constant independent of $\bar{t}$), it follows now from (5.2.24) and (5.2.25) that the graph of the left-hand side of (5.2.23) (considered as a function of $\bar{t}$) must actually intersect the graph of the right-hand side in at least one point (why?). *Hence equation* (5.2.23) *has at least one solution $\bar{t}$ with $0 \leq \bar{t} \leq T$.* Moreover, one easily shows that the right-hand side of (5.2.23) is a *decreasing* function of $\bar{t}$ for $0 \leq \bar{t} \leq T$ (see Exercise 6), from which one concludes that *equation* (5.2.23) *has at most one solution $\bar{t}$ with $0 \leq \bar{t} \leq T$* (explain). Hence (5.2.23) has precisely one solution $\bar{t}$ in the interval $[0, T]$. In practice one would use a suitable numerical algorithm to solve equation (5.2.23) for $\bar{t}$. For example, Newton's method would be appropriate, or Picard's method of successive approximation.†

Verification of the Optimum Rate. Summarizing, we have shown that if an optimum consumption rate exists for the given optimization problem in the form of a composite rate such as (5.2.9), then $\bar{t}$ must be the unique solution of (5.2.23) and the function $\mathfrak{C}$ must be given by (5.2.22) for $\bar{t} \leq t \leq T$. In fact, *it is now easy to prove that this particular composite consumption rate $C =$*

†See, for example, Peter Henrici, *Elements of Numerical Analysis* (New York: John Wiley & Sons, Inc., 1964), where such methods are discussed.

$C(t)$ given by (5.2.9), (5.2.22), *and* (5.2.23) *does actually maximize the satisfaction functional* $\mathcal{S} = \mathcal{S}(C)$ *of* (3.7.11) *among all possible competing rates which satisfy the global inequality constraint* (5.2.1) *and the terminal savings constraint* (5.2.3).

Indeed, if $C = C(t)$ is determined by (5.2.9), (5.2.22), and (5.2.23), and if $C + \Delta C = C(t) + \Delta C(t)$ is any other *admissible* competing consumption rate, then

$$\Delta C(t) \geq 0 \qquad \text{for } 0 \leq t \leq \bar{t} \tag{5.2.26}$$

and

$$\int_0^T e^{-\alpha t} \Delta C(t)\, dt = 0 \tag{5.2.27}$$

must hold (why?). Note that we are using implicitly here the fact that $\bar{t}$ satisfies (5.2.23) so that C satisfies (5.2.3). Then $C + \Delta C$ will satisfy both (5.2.1) and (5.2.3) provided that ΔC satisfies (5.2.26) and (5.2.27). We can now use (3.7.11) to calculate

$$\begin{aligned} \mathcal{S}(C + \Delta C) - \mathcal{S}(C) &= \int_0^T e^{-\beta t} \log\,[1 + C(t) + \Delta C(t)]\, dt \\ &\quad - \int_0^T e^{-\beta t} \log\,[1 + C(t)]\, dt \\ &= \int_0^T e^{-\beta t} \log\left[1 + \frac{\Delta C(t)}{1 + C(t)}\right] dt, \end{aligned}$$

which with (5.2.9) becomes

$$\begin{aligned} \mathcal{S}(C + \Delta C) - \mathcal{S}(C) &= \int_0^{\bar{t}} e^{-\beta t} \log\left[1 + \frac{\Delta C(t)}{1 + C_0}\right] dt \\ &\quad + \int_{\bar{t}}^T e^{-\beta t} \log\left[1 + \frac{\Delta C(t)}{1 + \mathcal{C}(t)}\right] dt. \end{aligned}$$

Since $\log\,(1 + x) \leq x$ (see Figure 8 in Section 3.7), we find from this equation that

$$\mathcal{S}(C + \Delta C) - \mathcal{S}(C) \leq \int_0^{\bar{t}} e^{-\beta t} \frac{\Delta C(t)}{1 + C_0} dt + \int_{\bar{t}}^T e^{-\beta t} \frac{\Delta C(t)}{1 + \mathcal{C}(t)}\, dt,$$

which with (5.2.22) becomes

$$\mathcal{S}(C + \Delta C) - \mathcal{S}(C) \leq \int_0^{\bar{t}} e^{-\beta t} \frac{\Delta C(t)}{1 + C_0} dt + \frac{e^{(\alpha - \beta)\bar{t}}}{1 + C_0} \int_{\bar{t}}^T e^{-\alpha t}\, \Delta C(t)\, dt. \tag{5.2.28}$$

On the other hand, (5.2.27) gives

$$\int_{\bar{t}}^T e^{-\alpha t}\, \Delta C(t)\, dt = -\int_0^{\bar{t}} e^{-\alpha t}\, \Delta C(t)\, dt,$$

so that (5.2.28) becomes (explain)

$$\mathcal{S}(C + \Delta C) - \mathcal{S}(C) \leq \frac{e^{-\beta \bar{t}}}{1 + C_0} \int_0^{\bar{t}} \Delta C(t)[e^{\beta(\bar{t}-t)} - e^{\alpha(\bar{t}-t)}]\, dt. \tag{5.2.29}$$

The exponential function is an increasing function of its argument, and we are assuming that $\alpha > \beta$ [see (5.2.6)], so that

$$e^{\beta(\bar{t}-t)} \leq e^{\alpha(\bar{t}-t)}$$

for $0 \leq t \leq \bar{t}$, and this result along with (5.2.26) and (5.2.29) yields the desired result

$$\mathcal{S}(C + \Delta C) - \mathcal{S}(C) \leq 0$$

for any admissible competing consumption rate $C + \Delta C$. *This proves that the composite rate $C = C(t)$ determined by* (5.2.9), (5.2.22), *and* (5.2.23) *actually furnishes a maximum value to the satisfaction functional among all possible competing rates which satisfy the global inequality constraint* (5.2.1) *and the terminal savings constraint* (5.2.3). This result justifies our "composite approach" to the original problem, which involved reformulating that problem in terms of suitable composite consumption rates so as to avoid the need for any further direct consideration of the global inequality constraint in the reformulated problem.

In closing this section, it may be of interest to compare the optimum consumption rates C and C^* found, respectively, in this section and in Section 3.7. The graphs of these two optimum rates *corresponding to the same terminal savings constraint* are shown superimposed in Figure 13 (for the case

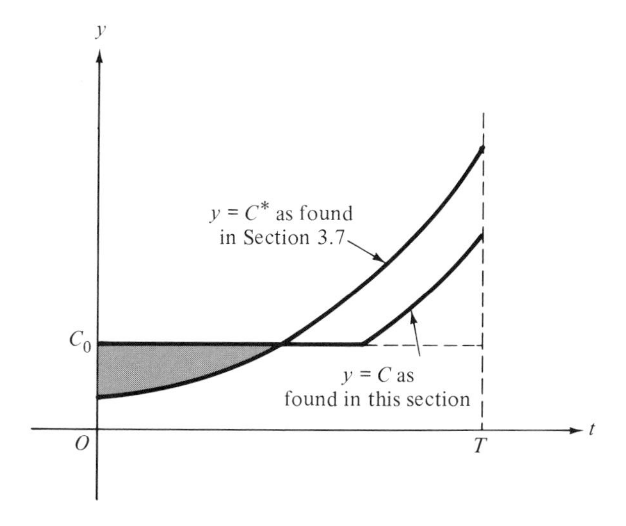

Figure 13

$\alpha > \beta$ considered here). Both rates achieve the same terminal savings constraint, while the earlier optimum rate $C^* = C^*(t)$ furnishes to the individual a *higher total satisfaction* in this case as measured by the functional $\mathcal{S}$ of (3.7.11). Even so, the individual may prefer to accept the *smaller* total satisfaction furnished by the present optimum rate C in order to avoid the initial hardship that would occur with C^* in this case where $C^*(t)$ lies initially below the level C_0. Hence the individual may purposely choose to forego part of his potential total satisfaction that could be achieved with C^* in order to avoid any temporary undue hardship.

Exercises

1. Show that if the terminal savings requirement S_T and the minimum consumption rate C_0 are chosen to be so large that $e^{-\alpha T}S_T + [(1 - e^{-\alpha T})/\alpha]\, C_0 > S_0 + \int_0^T e^{-\alpha t}I(t)\, dt$ holds, then every consumption rate $C = C(t)$ which satisfies (5.2.1) will automatically violate (5.2.3). [This shows that if (5.2.4) is violated, then the constraints (5.2.1) and (5.2.3) are incompatible.]

2. If the investment return rate α dominates the discount (inflation) rate β with $\alpha > \beta$, show that the earlier optimum consumption rate $C^*(t)$ found in Section 3.7 will furnish the solution to the present optimization problem subject to the global inequality constraint (5.2.1) and the terminal savings constraint (5.2.3) *provided that* (5.2.7) *is violated. Hint:* The function $C^*(t)$ is an increasing function of t in this case, and so the inequality (5.2.1) will hold for $C^*(t)$ for all t $(0 \leq t \leq T)$ if it already holds at $t = 0$.

3. Assume that $\alpha > \beta$ and let γ and δ be defined by (5.2.5) and (5.2.8). Show in this case that the shaded regions of Figures 7 and 8 overlap somewhat, as shown in Figure 9. *Hint:* The function $(1 - e^{-x})/x$ is monotonic.

4. Give a careful derivation of the result (5.2.16) for the variation of the functional $\mathcal{S}_0$ of (5.2.10).

5. Show how to obtain the two conditions (5.2.18) and (5.2.19) from the necessary condition (5.2.17).

6. Show that the right-hand side of (5.2.23) is a decreasing function of the variable $\bar{t}$ for $0 \leq \bar{t} \leq T$ when $0 < \beta < \alpha$.

7. Suppose that the discount (inflation) rate β dominates the investment return rate α with $\beta \geq \alpha > 0$. Show in this case that the earlier optimum consumption rate C^* found in Section 3.7 will furnish the solution to the present optimization problem subject to the global inequality constraint (5.2.1) and the terminal savings constraint (5.2.3) *provided that*

$$\left(\frac{e^{\beta T} - 1}{\beta}\right)(1 + C_0) + S_T \leq \frac{e^{\alpha T} - 1}{\alpha} + e^{\alpha T}\left(S_0 + \int_0^T e^{-\alpha t}I(t)\, dt\right) \tag{5.2.30}$$

holds. Hint: The function C^* is monotonic.

8. Let $\beta \geq \alpha > 0$ as in Exercise 7, but assume now that (5.2.30) *fails* to hold. What do you expect the graph of the new optimum rate $C = C(t)$ to look like in this case? Explain. (The interested reader should be able to solve this problem by reformulating it in terms of suitable composite rates, but we do not ask here that this be done.)

5.3. A Problem in Production Planning with Inequality Constraints

We return again to the problem in production planning already discussed in Section 2.5, except we now add the inequality constraints

$$0 \leq P(t) \tag{5.3.1}$$

and

$$P(t) \leq P_m(t), \tag{5.3.2}$$

which are now required to hold for all t in the interval $0 \leq t \leq T$. Hence every admissible production rate $P = P(t)$ must now be *nonnegative* and, moreover, *bounded above* by the quantity $P_m = P_m(t)$, which gives the maximum possible production rate that the manufacturing company is known to be capable of achieving. We assume that the known *desired* production rate $\mathcal{P} = \mathcal{P}(t)$ given by (1.3.20) satisfies both (5.3.1) and (5.3.2), with

$$0 \leq \mathcal{P}(t) \leq P_m(t) \tag{5.3.3}$$

for all $0 \leq t \leq T$.

There are several cases to consider concerning the initial inventory level I_0. (The notation is the same as that introduced at the end of Section 1.3.) If the given initial inventory level I_0 *agrees* with the desired initial level $\mathcal{I}_0$ with $I_0 = \mathcal{I}_0$, then there is *no initial disturbance* to be corrected and the optimum new production rate $P(t)$ will clearly agree with the known desired rate $\mathcal{P}(t)$, so that

$$P(t) = \mathcal{P}(t) \qquad \text{for } 0 \leq t \leq T.$$

We need not consider this case further. Hence we need only consider the two remaining cases:

$$I_0 > \mathcal{I}_0 \tag{5.3.4}$$

and

$$I_0 < \mathcal{I}_0. \tag{5.3.5}$$

If (5.3.4) holds, then the given initial inventory level I_0 *exceeds* the desired initial level $\mathcal{I}_0$, and the company will want to *lower* its inventory by choosing a new production rate $P = P(t)$ somewhat *lower* than the otherwise desired rate $\mathcal{P} = \mathcal{P}(t)$ so as to minimize the cost functional $C = C(P)$ of (1.3.24). Any such new production rate P (which is lower than $\mathcal{P}$) will automatically satisfy the upper constraint (5.3.2) [see (5.3.3)], so that the constraint (5.3.2) is *inactive* in this case and plays no role in the solution of the optimization problem. On the other hand, if (5.3.5) holds with the initial inventory level I_0 *lower* than the desired level $\mathcal{I}_0$, then the company will want to *raise* its inventory by choosing a new production rate $P = P(t)$ somewhat *higher* than the rate $\mathcal{P} = \mathcal{P}(t)$. In this case the lower constraint (5.3.1) plays no role (why?).

For brevity we shall consider in detail only the latter case in which (5.3.5) holds. It will be evident that the same methods can also be used to handle the other case (5.3.4). Hence we assume that (5.3.5) holds, so that the given initial inventory level I_0 is *low* as compared with $\mathcal{I}_0$. *We need only consider the upper constraint* (5.3.2) *in this case* since the other constraint (5.3.1) is inactive. Moreover, *we shall only consider the case where the initial disturbance* (as measured by the difference $\mathcal{I}_0 - I_0$) *is large enough so that*

$$\frac{\beta^2(e^{\gamma T} - e^{-\gamma T})(\mathcal{I}_0 - I_0)}{(\gamma + \alpha)e^{\gamma T} + (\gamma - \alpha)e^{-\gamma T}} > P_m(0) - \mathcal{P}(0) \tag{5.3.6}$$

holds [see (5.1.5)]. Indeed, if this condition holds, then the previous optimum production rate P^* found in Section 2.5 will be *disqualified* since P^* will be initially higher than the upper constraint P_m, and therefore P^* could not in this case be achieved in practice. Note that (5.3.5) holds automatically as a consequence of (5.3.6) and (5.3.3).

Hence we assume that (5.3.6) holds, and *we then seek a new production rate $P = P(t)$ that will minimize the cost functional given by* (1.3.24) *subject to the active inequality constraint* (5.3.2). Since (5.3.6) implies that the initial inventory level I_0 is substantially lower than the desired initial level $\mathcal{I}_0$, it is natural to expect that the company should choose a new production rate $P = P(t)$ that is initially *as high as possible*, subject to the constraint (5.3.2). Hence we may expect in this case that the new optimum production rate will be a *composite rate* which first equals the maximum permitted rate $P_m = P_m(t)$ so as to raise the inventory level and which later decreases down to its terminal value in such a way as to minimize the total cost as measured by (1.3.24). We shall see that this is indeed the case.

Hence we are led more or less naturally to consider composite production rates $P = P(t)$, as shown, for example, in Figure 14, with

$$P(t) = \begin{cases} P_m(t) & \text{for } 0 \leq t \leq \bar{t} \\ p(t) & \text{for } \bar{t} \leq t \leq T \end{cases} \tag{5.3.7}$$

for some suitable time value $\bar{t}$ and some suitable function $p(t)$, both of which may depend on the particular composite rate considered. In general there may be *several* time intervals during which such a composite production rate

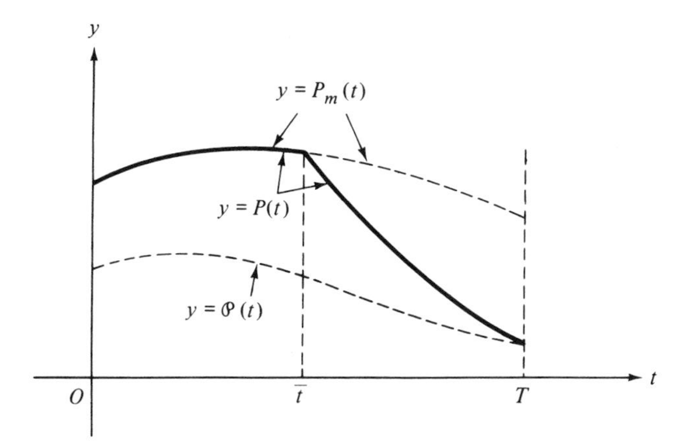

Figure 14

$P(t)$ agrees with the known maximum rate, as shown, for example, in Figure 15. [In this case the function $p(t)$ occurring in (5.3.7) will itself agree with $P_m(t)$ over certain time periods.] For simplicity, however, *we shall assume that the known functions* P_m *and* $\mathcal{P}$ *satisfy the condition*

$$\frac{d}{dt}[P_m(t) - \mathcal{P}(t)] \geq 0 \qquad \text{for } 0 < t < T \tag{5.3.8}$$

(i.e., the difference $P_m - \mathcal{P}$ *never decreases*), in which case it can be shown that the new optimum production rate $P(t)$ will agree with $P_m(t)$ *only* for $0 \leq t \leq \bar{t}$ for some suitable number $\bar{t}$, as we shall see below. It will be clear that the same methods can also be used to handle situations where (5.3.8) fails to hold, as shown, for example, in Figure 15.

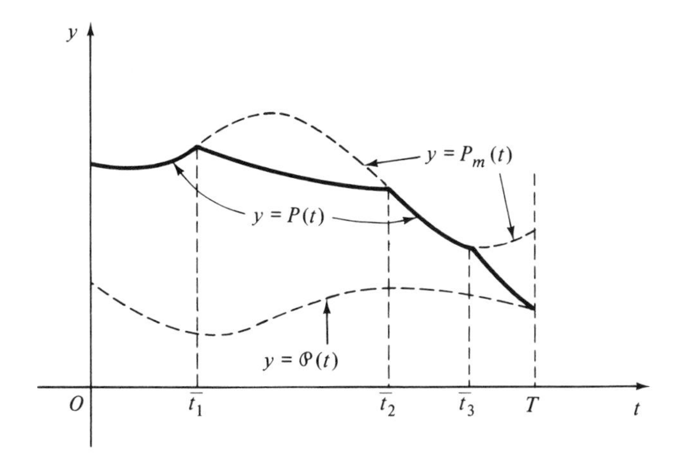

Figure 15

A Reformulation of the Problem. We now use (1.3.24) to calculate the value of the cost functional $C = C(P)$ for any such composite production rate $P = P(t)$ as given by (5.3.7); we find that

$$C(P) = \int_0^{\bar{t}} \{\beta^2[I_{P_m}(t) - \mathcal{I}(t)]^2 + [P_m(t) - \mathcal{P}(t)]^2\}\, dt$$
$$+ \int_{\bar{t}}^{T} \{\beta^2[I_P(t) - \mathcal{I}(t)]^2 + [p(t) - \mathcal{P}(t)]^2\}\, dt, \tag{5.3.9}$$

where P is given by (5.3.7). The function I_{P_m} appearing in the first integral on the right-hand side of (5.3.9) is known and can be calculated from (1.3.23) as

$$I_{P_m}(t) = e^{-\alpha t}\left(I_0 + \int_0^t e^{\alpha\tau}[P_m(\tau) - \mathcal{S}(\tau)]\, d\tau\right) \tag{5.3.10}$$

for $0 \leq t \leq \bar{t}$. Similarly, we find [see (1.3.20)]

$$\mathcal{I}(t) = e^{-\alpha t}\left(\mathcal{I}_0 + \int_0^t e^{\alpha\tau}[\mathcal{P}(\tau) - \mathcal{S}(\tau)]\, d\tau\right), \tag{5.3.11}$$

which is known, and an analogous formula for $I_P(t)$ given in terms of the production rate P. Actually, since P is a composite rate as given by (5.3.7), it follows from (1.3.23) that I_P will depend explicitly on the number $\bar{t}$ and on the function $p = p(t)$. Hence we shall change notation somewhat in this case and write

$$I_P(t) = I(\bar{t}, p; t)$$

whenever P is given by (5.3.7). Using this notation, then, we find from (1.3.23) that

$$I_P(t) = I(\bar{t}, p; t) = e^{-\alpha t}\left(I_0 + \int_0^{\bar{t}} e^{\alpha\tau}[P_m(\tau) - \mathcal{S}(\tau)]\, d\tau\right.$$
$$\left. + \int_{\bar{t}}^{t} e^{\alpha\tau}[p(\tau) - \mathcal{S}(\tau)]\, d\tau\right) \qquad \text{for } \bar{t} \leq t \leq T \tag{5.3.12}$$

whenever P is given by (5.3.7). Finally, then, inserting all these results into (5.3.9), we find that

$$C(P) = \int_0^{\bar{t}} f(t)\, dt + \int_{\bar{t}}^{T} \{\beta^2[I(\bar{t}, p; t) - \mathcal{I}(t)]^2 + [p(t) - \mathcal{P}(t)]^2\}\, dt$$

whenever P is given by (5.3.7), where $I(\bar{t}, p; t)$ is given by (5.3.12) and where the known function $f = f(t)$ is defined by

$$f(t) = \beta^2[I_{P_m}(t) - \mathcal{I}(t)]^2 + [P_m(t) - \mathcal{P}(t)]^2. \tag{5.3.13}$$

We shall want to consider $C(P)$ as a function of both $\bar{t}$ and p, and so we define a new functional C_0 by

$$C_0(\bar{t}, p) = \int_0^{\bar{t}} f(t)\, dt + \int_{\bar{t}}^{T} \{\beta^2[I(\bar{t}, p; t) - \mathcal{I}(t)]^2 + [p(t) - \mathcal{P}(t)]^2\}\, dt \tag{5.3.14}$$

for any number $\bar{t}$ in the interval $0 < \bar{t} < T$ and for any continuous function $p = p(t)$. Of course, we have $C_0(\bar{t}, p) = C(P)$ whenever P is given by (5.3.7).

It is natural to take as our basic vector space $\mathfrak{X}$ for the functional C_0 the collection of all pairs $(\bar{t}, p) = (\bar{t}, p(t))$, where $\bar{t}$ may be any arbitrary number in $\mathcal{R}$ and where $p = p(t)$ may be any arbitrary continuous function defined on the fixed interval $[0, T]$. The definitions of addition of vectors and multiplication of vectors by numbers should be clear to the reader. Similarly, we let the reader define a suitable norm on the vector space $\mathfrak{X}$. We take the domain D of the functional C_0 to be the open subset D of $\mathfrak{X}$ consisting of all vectors $(\bar{t}, p)$ in $\mathfrak{X}$ for which $0 < \bar{t} < T$.

We shall now use the theorem of Section 2.2 on the vanishing of the variation at a local extremum vector to find a vector $(\bar{t}, p)$ in D that will minimize in D the functional $C_0 = C_0(\bar{t}, p)$ of (5.3.14). The active inequality constraint (5.3.2) no longer appears in the statement of the latter problem. Rather we have reformulated the optimization problem in terms of suitable composite production rates so as to avoid any need for further consideration of the inequality constraint in the reformulated problem. Once we have the desired extremum vector $(\bar{t}, p)$ for the reformulated problem, it will be an easy matter to show directly that the resulting production rate P of (5.3.7) actually minimizes the cost functional C of (1.3.24) among all production rates which satisfy the inequality constraints (5.3.1) and (5.3.2).

The Variation of C_0.† To proceed, we must calculate the variation of the functional $C_0 = C_0(\bar{t}, p)$ at an arbitrary vector $(\bar{t}, p)$ in D. From (5.3.14) we find that

$$C_0(\bar{t} + \epsilon\,\Delta\bar{t}, p + \epsilon\,\Delta p) = \int_0^{\bar{t}+\epsilon\Delta\bar{t}} f(t)\,dt + \int_{\bar{t}+\epsilon\Delta\bar{t}}^{T} \{\beta^2[I(\bar{t} + \epsilon\,\Delta\bar{t}, p + \epsilon\,\Delta p; t) - \mathcal{I}(t)]^2 + [p(t) + \epsilon\,\Delta p(t) - \mathcal{P}(t)]^2\}\,dt$$

for any vector $(\Delta\bar{t}, \Delta p)$ in $\mathfrak{X}$ and for any small number ϵ. If we use (4.4.14) several times now, we obtain from the last equation

$$\begin{aligned}\frac{d}{d\epsilon}C_0(\bar{t} + \epsilon\,\Delta\bar{t}, p + \epsilon\,\Delta p) &= \Delta\bar{t}f(\bar{t} + \epsilon\,\Delta\bar{t}) - \Delta\bar{t}\{\beta^2[I(\bar{t} + \epsilon\,\Delta\bar{t}, p + \epsilon\,\Delta p; \bar{t} + \epsilon\,\Delta\bar{t}) - \mathcal{I}(\bar{t} + \epsilon\,\Delta\bar{t})]^2 \\ &+ [p(\bar{t} + \epsilon\,\Delta\bar{t}) + \epsilon\,\Delta p(\bar{t} + \epsilon\,\Delta\bar{t}) - \mathcal{P}(\bar{t} + \epsilon\,\Delta\bar{t})]^2\} \\ &+ 2\int_{\bar{t}+\epsilon\Delta\bar{t}}^{T} \Big\{\beta^2[I(\bar{t} + \epsilon\,\Delta\bar{t}, p + \epsilon\,\Delta p; t) - \mathcal{I}(t)]\frac{\partial I}{\partial\epsilon} \\ &+ [p(t) + \epsilon\,\Delta p(t) - \mathcal{P}(t)]\,\Delta p(t)\Big\}\,dt,\end{aligned}$$

†The details of the following calculation may safely be skipped over.

where the quantity $\partial I/\partial\epsilon = (\partial/\partial\epsilon)I(\bar{t} + \epsilon\,\Delta\bar{t}, p + \epsilon\,\Delta p; t)$ appearing in the last integral is found from (5.3.12) to be given as

$$\frac{\partial I}{\partial \epsilon} = e^{-\alpha t}\Big\{\Delta\bar{t}e^{\alpha(\bar{t}+\epsilon\Delta\bar{t})}[P_m(\bar{t} + \epsilon\,\Delta\bar{t}) - \mathcal{S}(\bar{t} + \epsilon\,\Delta\bar{t})]$$
$$- \Delta\bar{t}e^{\alpha(\bar{t}+\epsilon\Delta\bar{t})}[p(\bar{t} + \epsilon\,\Delta\bar{t}) + \epsilon\,\Delta p(\bar{t} + \epsilon\,\Delta\bar{t}) - \mathcal{S}(\bar{t} + \epsilon\,\Delta\bar{t})]$$
$$+ \int_{\bar{t}+\epsilon\Delta\bar{t}}^{t} e^{\alpha\tau}\,\Delta p(\tau)\,dt\Big\}.$$

If we evaluate the last two equations at $\epsilon = 0$ and use (2.3.4), we find for the variation of C_0 that

$$\delta C_0(\bar{t}, p; \Delta\bar{t}, \Delta p) = \Delta\bar{t}\Big\{f(\bar{t}) - \beta^2[I(\bar{t}, p; \bar{t}) - \mathcal{I}(\bar{t})]^2 - [p(\bar{t}) - \mathcal{P}(\bar{t})]^2$$
$$+ 2\beta^2 e^{\alpha\bar{t}}[P_m(\bar{t}) - p(\bar{t})]\int_{\bar{t}}^{T} e^{-\alpha t}[I(\bar{t}, p; t) - \mathcal{I}(t)]\,dt\Big\}$$
$$+ 2\int_{\bar{t}}^{T}\Big\{\beta^2[I(\bar{t}, p; t) - \mathcal{I}(t)]e^{-\alpha t}\int_{\bar{t}}^{t} e^{\alpha\tau}\,\Delta p(\tau)\,d\tau$$
$$+ [p(t) - \mathcal{P}(t)]\,\Delta p(t)\Big\}\,dt.$$

Finally, if we use (5.3.10), (5.3.11), (5.3.12), and (5.3.13) with the last result, we find after a brief calculation that [note that $I_{P_m}(\bar{t}) = I(\bar{t}, p; \bar{t})$]

$$\delta C_0(\bar{t}, p; \Delta\bar{t}, \Delta p) = \Delta\bar{t}[P_m(\bar{t}) - p(\bar{t})]\Big\{P_m(\bar{t}) + p(\bar{t}) - 2\mathcal{P}(\bar{t})$$
$$+ 2\beta^2 e^{\alpha\bar{t}}\int_{\bar{t}}^{T} e^{-\alpha t}[I(\bar{t}, p; t) - \mathcal{I}(t)]\,dt\Big\}$$
$$+ 2\int_{\bar{t}}^{T}\Big\{\beta^2 e^{-\alpha t}[I(\bar{t}, p; t) - \mathcal{I}(t)]\int_{\bar{t}}^{t} e^{\alpha\tau}\,\Delta p(\tau)\,d\tau$$
$$+ [p(t) - \mathcal{P}(t)]\,\Delta p(t)\Big\}\,dt. \tag{5.3.15}$$

The last integral can be put into a slightly more usable form by using the identity

$$\int_{\bar{t}}^{T} e^{-\alpha t}[I(\bar{t}, p; t) - \mathcal{I}(t)]\int_{\bar{t}}^{t} e^{\alpha\tau}\,\Delta p(\tau)\,d\tau\,dt$$
$$= \int_{\bar{t}}^{T} \Delta p(t)e^{\alpha t}\int_{t}^{T} e^{-\alpha\tau}[I(\bar{t}, p; \tau) - \mathcal{I}(\tau)]\,d\tau\,dt,$$

which is similar to the earlier identity (2.5.2). Using the last identity in (5.3.15), we have finally

$$\delta C_0(\bar{t}, p; \Delta\bar{t}, \Delta p) = \Delta\bar{t}[P_m(\bar{t}) - p(\bar{t})]\Big\{P_m(\bar{t}) + p(\bar{t}) - 2\mathcal{P}(\bar{t})$$
$$+ 2\beta^2 e^{\alpha\bar{t}} \int_{\bar{t}}^{T} e^{-\alpha t}[I(\bar{t}, p; t) - \mathcal{S}(t)]\, dt\Big\}$$
$$+ 2\int_{\bar{t}}^{T} \Delta p(t)\Big\{p(t) - \mathcal{P}(t)$$
$$+ \beta^2 e^{\alpha t} \int_{t}^{T} e^{-\alpha\tau}[I(\bar{t}, p; \tau) - \mathcal{S}(\tau)]\, d\tau\Big\}\, dt \qquad (5.3.16)$$

for any vector $(\bar{t}, p)$ in D and for any vector $(\Delta\bar{t}, \Delta p)$ in the vector space $\mathfrak{X}$.

The Extremum Conditions. We have seen that the functional C_0 has a variation at each vector $(\bar{t}, p)$ in the open set D, and this variation may be given by formula (5.3.16). Hence by the theorem of Section 2.2

$$\delta C_0(\bar{t}, p; \Delta\bar{t}, \Delta p) = 0$$

must hold at any local minimum vector $(\bar{t}, p)$ in D for all vectors $(\Delta\bar{t}, \Delta p)$ in the vector space $\mathfrak{X}$. We can use (5.3.16) to write the last condition as

$$\Delta\bar{t}[P_m(\bar{t}) - p(\bar{t})]\Big\{P_m(\bar{t}) + p(\bar{t}) - 2\mathcal{P}(\bar{t}) + 2\beta^2 e^{\alpha\bar{t}} \int_{\bar{t}}^{T} e^{-\alpha t}[I(\bar{t}, p; t) - \mathcal{S}(t)]\, dt\Big\}$$
$$+ 2\int_{\bar{t}}^{T} \Delta p(t)\Big\{p(t) - \mathcal{P}(t) + \beta^2 e^{\alpha t} \int_{t}^{T} e^{-\alpha\tau}[I(\bar{t}, p; \tau) - \mathcal{S}(\tau)]\, d\tau\Big\}\, dt = 0, \qquad (5.3.17)$$

which must then hold for *all numbers* $\Delta\bar{t}$ and for *all continuous functions* Δp if $(\bar{t}, p)$ is a local extremum vector in D for C_0. It follows now from (5.3.17) by a standard argument that *any such local extremum vector* $(\bar{t}, p)$ *must satisfy the necessary conditions*

$$[P_m(\bar{t}) - p(\bar{t})]\Big\{P_m(\bar{t}) + p(\bar{t}) - 2\mathcal{P}(\bar{t}) + 2\beta^2 e^{\alpha\bar{t}} \int_{\bar{t}}^{T} e^{-\alpha t}[I(\bar{t}, p; t) - \mathcal{S}(t)]\, dt\Big\} = 0 \qquad (5.3.18)$$

and

$$p(t) - \mathcal{P}(t) + \beta^2 e^{\alpha t} \int_{t}^{T} e^{-\alpha\tau}[I(\bar{t}, p; \tau) - \mathcal{S}(\tau)]\, d\tau = 0 \qquad \text{for } \bar{t} \le t \le T. \qquad (5.3.19)$$

We leave the derivation of the last two conditions from (5.3.17) as an exercise for the reader.

The Optimum Production Rate. We can now use the conditions (5.3.18) and (5.3.19) to determine the desired minimum vector $(\bar{t}, p)$ for C_0. As a first step in this direction we note that if we evaluate (5.3.19) at $t = \bar{t}$ and use the resulting equation in (5.3.18), we find the *natural condition*

$$p(t) = P_m(t) \qquad \text{at } t = \bar{t}, \tag{5.3.20}$$

which must be satisfied by the extremum vector $(\bar{t}, p)$. Hence even though we have allowed the composite production rate P of (5.3.7) to be discontinuous at $t = \bar{t}$, it follows now from (5.3.20) that the *optimum* production rate will automatically be continuous (explain). Note also that the earlier condition (5.3.18) need not be considered further provided that we impose (5.3.20). Finally, we can get a second natural condition by evaluating (5.3.19) at $t = T$; we find that

$$p(t) = \mathcal{P}(t) \qquad \text{at } t = T. \tag{5.3.21}$$

We turn now directly to equation (5.3.19). If we argue as we did in Section 2.5 when going from (2.5.3) to (2.5.8), we find in the present case from (5.3.19) that

$$p(t) - \mathcal{P}(t) = Ae^{\gamma t} + Be^{-\gamma t} \qquad \text{for } \bar{t} \leq t \leq T, \text{ with } \gamma = \sqrt{\alpha^2 + \beta^2}, \tag{5.3.22}$$

which must hold for some suitable constants A and B. If we require that (5.3.22) satisfy both the natural conditions (5.3.20) and (5.3.21), we find necessarily that the constants A and B must take on the values

$$A = -\frac{[P_m(\bar{t}) - \mathcal{P}(\bar{t})]e^{-\gamma T}}{e^{\gamma(T-\bar{t})} - e^{-\gamma(T-\bar{t})}}, \qquad B = \frac{[P_m(\bar{t}) - \mathcal{P}(\bar{t})]e^{\gamma T}}{e^{\gamma(T-\bar{t})} - e^{-\gamma(T-\bar{t})}},$$

so that (5.3.22) becomes

$$p(t) - \mathcal{P}(t) = \frac{[P_m(\bar{t}) - \mathcal{P}(\bar{t})][e^{\gamma(T-t)} - e^{-\gamma(T-t)}]}{e^{\gamma(T-\bar{t})} - e^{-\gamma(T-\bar{t})}} \qquad \text{for } \bar{t} \leq t \leq T, \tag{5.3.23}$$

where, of course, the time value $\bar{t}$ appearing here is still unknown. Finally, $\bar{t}$ may be determined by inserting (5.3.23) into equation (5.3.19) and evaluating the resulting equation at $t = \bar{t}$ to find the condition

$$\beta^2 e^{-\alpha\bar{t}}(\mathcal{I}_0 - I_0) = \beta^2 e^{-\alpha\bar{t}} \int_0^{\bar{t}} e^{\alpha\tau}[P_m(\tau) - \mathcal{P}(\tau)]\,d\tau + [P_m(\bar{t}) - \mathcal{P}(\bar{t})]\left\{\gamma + \alpha + \frac{2\gamma e^{-\gamma(T-\bar{t})}}{e^{\gamma(T-\bar{t})} - e^{-\gamma(T-\bar{t})}}\right\}, \tag{5.3.24}$$

which must be satisfied by the number $\bar{t}$. Note that all quantities which appear in equation (5.3.24) are known as part of the data of the problem except for the number $\bar{t}$. Equation (5.3.24) can be obtained alternatively with (5.3.23) from the condition

$$\frac{d}{dt}[p(t) - \mathcal{P}(t)] = \alpha[p(t) - \mathcal{P}(t)] + \beta^2[I(\bar{t}, p; t) - \mathcal{I}(t)] \qquad \text{at } t = \bar{t},$$

which follows from (5.3.19) [compare with (2.5.10)].

Hence finally we need only show that equation (5.3.24) determines a unique value for $\bar{t}$ in the interval $0 < \bar{t} < T$. This is indeed the case, as will now be seen to follow from (5.3.6) and (5.3.8). Indeed, if we evaluate separately both sides of (5.3.24) at $\bar{t} = 0$, we find with (5.3.6) that

$$\text{left-hand side of (5.3.24)} > \text{right-hand side of (5.3.24)} \qquad \text{at } \bar{t} = 0.$$

On the other hand, the right-hand side of (5.3.24) becomes unbounded as $\bar{t}$ approaches $\bar{t} = T$, so that

$$\text{left-hand side of (5.3.24)} < \text{right-hand side of (5.3.24)} \qquad \text{as } \bar{t} \text{ approaches } T$$

holds. [There might be some difficulty in obtaining the last result if $P_m(T) - \mathcal{P}(T) = 0$, but in view of (5.3.8) we need not consider this case; why?] Since both sides of (5.3.24) are continuous functions of $\bar{t}$ on the interval $0 \leq \bar{t} < T$, it follows now from these observations that the graph of the left-hand side of (5.3.24) must actually intersect the graph of the right-hand side in at least one point, so that *equation* (5.3.24) *has at least one solution* $\bar{t}$ *with* $0 < \bar{t} < T$. In fact, there is *precisely one* such solution as a consequence of (5.3.8). Indeed, it follows from (5.3.8) that the right-hand side of (5.3.24) is an *increasing* function of $\bar{t}$ (see Exercises 3 and 4), while clearly the left-hand side of (5.3.24) is a *decreasing* function of $\bar{t}$. These observations prove that equation (5.3.24) has precisely one solution for $\bar{t}$. In practice one would use a suitable numerical method to solve (5.3.24) for $\bar{t}$. For example, successive bisection of the interval would work.

Verification of the Optimum Rate. Summarizing, we have shown that the only possible candidate for an optimum production rate in the form (5.3.7) is obtained in the present case with the function $p = p(t)$ given by (5.3.23) and with the number $\bar{t}$ given as the unique solution of (5.3.24). We can now show that this particular composite production rate $P = P(t)$ given by (5.3.7), (5.3.23), and (5.3.24) *does actually minimize* the cost functional $C = C(P)$ of (1.3.24) among all possible production rates which satisfy the global inequality constraints (5.3.1) and (5.3.2).†

Indeed, if $P = P(t)$ is determined by (5.3.7), (5.3.23), and (5.3.24), and if $P + \Delta P = P(t) + \Delta P(t)$ is any other admissible production rate satisfying

†The following calculation can safely be skipped during a first reading.

the inequality constraint (5.3.2) with

$$P(t) + \Delta P(t) \leq P_m(t),$$

then necessarily

$$\Delta P(t) \leq 0 \qquad \text{for } 0 \leq t \leq \bar{t} \tag{5.3.25}$$

will hold (why?). On the other hand, the calculation following (2.4.1) implies that

$$\begin{aligned} C(P + \Delta P) - C(P) = 2 \int_0^T \Big\{\beta^2[I_P(t) - \mathcal{I}(t)]\, \Delta I(t) + [P(t) - \mathcal{P}(t)]\, \Delta P(t)\Big\}\, dt \\ + \int_0^T \{\beta^2\, \Delta I(t)^2 + \Delta P(t)^2\}\, dt, \end{aligned} \tag{5.3.26}$$

where $I_P(t)$ is defined by (1.3.23) and where we have defined $\Delta I(t)$ by

$$\Delta I(t) = e^{-\alpha t} \int_0^t e^{\alpha \tau}\, \Delta P(\tau)\, d\tau. \tag{5.3.27}$$

If we insert (5.3.27) into the first term on the right-hand side of (5.3.26) and interchange the order of the repeated integrals in the appropriate place, we find that

$$\begin{aligned} C(P + \Delta P) - C(P) = 2 \int_0^T \Big\{P(t) - \mathcal{P}(t) \\ + \beta^2 e^{\alpha t} \int_t^T e^{-\alpha \tau}[I_P(\tau) - \mathcal{I}(\tau)]\, d\tau\Big\}\, \Delta P(t)\, dt \\ + \int_0^T \{\beta^2\, \Delta I(t)^2 + \Delta P(t)^2\}\, dt. \end{aligned} \tag{5.3.28}$$

Now for $\bar{t} \leq t \leq T$ we have by the construction of $P(t)$ [see (5.3.7) and (5.3.23)]

$$P(t) - \mathcal{P}(t) + \beta^2 e^{\alpha t} \int_t^T e^{-\alpha \tau}[I_P(\tau) - \mathcal{I}(\tau)]\, d\tau = 0 \qquad \text{for } \bar{t} \leq t \leq T, \tag{5.3.29}$$

so that (5.3.28) becomes

$$\begin{aligned} C(P + \Delta P) - C(P) = 2 \int_0^{\bar{t}} \Big\{P_m(t) - \mathcal{P}(t) \\ + \beta^2 e^{\alpha t} \int_t^T e^{-\alpha \tau}[I_P(\tau) - \mathcal{I}(\tau)]\, d\tau\Big\}\, \Delta P(t)\, dt \\ + \int_0^T \{\beta^2\, \Delta I(t)^2 + \Delta P(t)^2\}\, dt \end{aligned} \tag{5.3.30}$$

since $P(t) = P_m(t)$ for $0 \leq t \leq \bar{t}$. On the other hand, (5.3.29) evaluated at

$t = \bar{t}$ gives with (5.3.7) and (5.3.23)

$$\beta^2 e^{\alpha \bar{t}} \int_{\bar{t}}^{T} e^{-\alpha\tau}[I_P(\tau) - \mathcal{I}(\tau)]\, d\tau = -P_m(\bar{t}) + \mathcal{P}(\bar{t}),$$

so that for $0 \leq t \leq \bar{t}$ we have (explain)

$$\beta^2 \int_t^T e^{-\alpha\tau}[I_P(\tau) - \mathcal{I}(\tau)]\, d\tau = \beta^2 \int_t^{\bar{t}} e^{-\alpha\tau}[I_{P_m}(\tau) - \mathcal{I}(\tau)]\, d\tau - e^{-\alpha\bar{t}}[P_m(\bar{t}) - \mathcal{P}(\bar{t})].$$

Hence (5.3.30) can be written as

$$\begin{aligned} C(P + \Delta P) - C(P) = 2 \int_0^{\bar{t}} e^{\alpha t}\Big\{ & e^{-\alpha t}[P_m(t) - \mathcal{P}(t)] - e^{-\alpha\bar{t}}[P_m(\bar{t}) - \mathcal{P}(\bar{t})] \\ & + \beta^2 \int_t^{\bar{t}} e^{-\alpha\tau}[I_{P_m}(\tau) - \mathcal{I}(\tau)]\, d\tau \Big\} \Delta P(t)\, dt \\ & + \int_0^T \{\beta^2\, \Delta I(t)^2 + \Delta P(t)^2\}\, dt. \end{aligned}$$

Since $\Delta P(t) \leq 0$ for $0 \leq t \leq \bar{t}$ [see (5.3.25)], *we shall be able to conclude from the last equation the desired result*

$$C(P + \Delta P) - C(P) \geq 0 \tag{5.3.31}$$

provided that

$$e^{-\alpha t}[P_m(t) - \mathcal{P}(t)] + \beta^2 \int_t^{\bar{t}} e^{-\alpha\tau}[I_{P_m}(\tau) - \mathcal{I}(\tau)]\, d\tau \leq e^{-\alpha\bar{t}}[P_m(\bar{t}) - \mathcal{P}(\bar{t})] \qquad \text{for } 0 \leq t \leq \bar{t}. \tag{5.3.32}$$

Hence we finally need only prove this last inequality, to which we now turn.

From (5.3.10) and (5.3.11)

$$I_{P_m}(\tau) - \mathcal{I}(\tau) = e^{-\alpha\tau}\Big(I_0 - \mathcal{I}_0 + \int_0^{\tau} e^{\alpha\sigma}[P_m(\sigma) - \mathcal{P}(\sigma)]\, d\sigma\Big)$$

follows for $0 \leq \tau \leq \bar{t}$, and so (5.3.32) holds if and only if

$$e^{-\alpha t}[P_m(t) - \mathcal{P}(t)] + \beta^2 \int_t^{\bar{t}} e^{-2\alpha\tau} \int_0^{\tau} e^{\alpha\sigma}[P_m(\sigma) - \mathcal{P}(\sigma)]\, d\sigma\, d\tau \overset{?}{\leq} \beta^2 \int_t^{\bar{t}} e^{-2\alpha\tau}(\mathcal{I}_0 - I_0)\, d\tau + e^{-\alpha\bar{t}}[P_m(\bar{t}) - \mathcal{P}(\bar{t})] \qquad \text{for } 0 \leq t \leq \bar{t}. \tag{5.3.33}$$

On the other hand, the choice of $\bar{t}$ implies with (5.3.24) that

$$\beta^2 e^{-\alpha\bar{t}} \int_0^{\bar{t}} e^{\alpha\sigma}[P_m(\sigma) - \mathcal{P}(\sigma)]\, d\sigma$$
$$= \beta^2 e^{-\alpha\bar{t}}(\mathcal{I}_0 - I_0) - [P_m(\bar{t}) - \mathcal{P}(\bar{t})]\left\{\gamma + \alpha + 2\gamma \frac{e^{-\gamma(T-\bar{t})}}{e^{\gamma(T-\bar{t})} - e^{-\gamma(T-\bar{t})}}\right\},$$

while $\int_0^\tau = \int_0^{\bar{t}} - \int_\tau^{\bar{t}}$, and so (5.3.33) will hold if and only if

$$e^{-\alpha t}[P_m(t) - \mathcal{P}(t)]$$
$$\overset{?}{\leq} \beta^2 \int_t^{\bar{t}} e^{-2\alpha\tau} \int_\tau^{\bar{t}} e^{\alpha\sigma}[P_m(\sigma) - \mathcal{P}(\sigma)]\, d\sigma\, d\tau$$
$$+ e^{-\alpha\bar{t}}[P_m(\bar{t}) - \mathcal{P}(\bar{t})]\left\{1 + \frac{e^{2\alpha(\bar{t}-t)} - 1}{2\alpha}\left[\gamma + \alpha + \frac{2\gamma e^{-\gamma(T-\bar{t})}}{e^{\gamma(T-\bar{t})} - e^{-\gamma(T-\bar{t})}}\right]\right\}$$

holds. Finally, since [why?; see, for example, (2.5.17)]

$$\int_t^{\bar{t}} e^{-2\alpha\tau} \int_\tau^{\bar{t}} e^{\alpha\sigma}[P_m(\sigma) - \mathcal{P}(\sigma)]\, d\sigma\, d\tau$$
$$= \int_t^{\bar{t}} e^{\alpha\sigma}[P_m(\sigma) - \mathcal{P}(\sigma)]\, d\sigma \int_t^\sigma e^{-2\alpha\tau}\, d\tau$$
$$= e^{-\alpha t} \int_t^{\bar{t}} [P_m(\sigma) - \mathcal{P}(\sigma)]\left\{\frac{e^{\alpha(\sigma-t)} - e^{-\alpha(\sigma-t)}}{2\alpha}\right\} d\sigma,$$

we can rewrite the previous inequality again as

$$P_m(t) - \mathcal{P}(t) \overset{?}{\leq} \beta^2 \int_t^{\bar{t}} \frac{e^{\alpha(\tau-t)} - e^{-\alpha(\tau-t)}}{2\alpha}[P_m(\tau) - P(\tau)]\, d\tau$$
$$+ \frac{(P_m(\bar{t}) - \mathcal{P}(\bar{t}))}{2\alpha}\Big\{2\alpha e^{-\alpha(\bar{t}-t)} + (e^{\alpha(\bar{t}-t)} - e^{-\alpha(\bar{t}-t)})\Big[\gamma + \alpha$$
$$+ \frac{2\gamma e^{-\gamma(T-\bar{t})}}{e^{\gamma(T-\bar{t})} - e^{-\gamma(T-\bar{t})}}\Big]\Big\} \qquad \text{for } 0 \leq t \leq \bar{t}. \tag{5.3.34}$$

Now the left-hand side of (5.3.34) *increases* with increasing t as a consequence of the assumption (5.3.8), while the right-hand side of (5.3.34) *decreases* (we leave the proof of this last assertion as an exercise for the reader; see Exercises 5 and 6). Hence the desired inequality (5.3.34) will certainly hold for all $0 \leq t \leq \bar{t}$ provided that the inequality is valid at $t = \bar{t}$ (explain). But at $t = \bar{t}$, equality holds! This, then, completes the proof of the desired inequality (5.3.31) and shows that *the production rate* $P = P(t)$ *determined by* (5.3.7), (5.3.23), *and* (5.3.24) *does indeed furnish a minimum value to the cost functional C among all possible competing production rates which satisfy the global inequality constraints* (5.3.1) *and* (5.3.2). [Note that P automatically satisfies (5.3.1) as a consequence of (5.3.3).]

Exercises

1. Show how to obtain the two conditions (5.3.18) and (5.3.19) from the necessary condition (5.3.17).

2. Give a careful derivation of the natural condition (5.3.20), which must hold for any local extremum vector $(\bar{t}, p)$.

3. [This exercise and Exercise 4 are designed to show that the right-hand side of (5.3.24) is an *increasing* function of $\bar{t}$.] Define a function $g = g(t)$ by $g(t) = e^{-\alpha t} \int_0^t e^{\alpha\tau}[P_m(\tau) - \mathcal{P}(\tau)]\, d\tau$ for $0 \leq t \leq T$. Show that g satisfies the differential equation $g'(t) = -\alpha g(t) + P_m(t) - \mathcal{P}(t)$. On the other hand, estimate the integral appearing in $\alpha g(t) = \alpha e^{-\alpha t} \int_0^t e^{\alpha\tau}[P_m(\tau) - \mathcal{P}(\tau)]\, d\tau$ and obtain the result $\alpha g(t) \leq P_m(t) - \mathcal{P}(t)$ provided that (5.3.8) holds. Can you conclude from these results that $g(t)$ is an increasing function of t?

4. Define a function $h(t)$ by $h(t) = e^{-\gamma(T-t)}/(e^{\gamma(T-t)} - e^{-\gamma(T-t)})$ for $0 \leq t < T$. Show directly that $h'(t) \geq 0$. Use this result and the result of Exercise 3 to show that the right-hand side of (5.3.24) is an increasing function of $\bar{t}$.

5. [The purpose of this exercise and Exercise 6 is to show that the right-hand side of (5.3.34) is a decreasing function of t for $0 \leq t \leq \bar{t}$.] Let $k(t)$ be defined by $k(t) = \int_t^{\bar{t}} [(e^{\alpha(\tau-t)} - e^{-\alpha(\tau-t)})/2\alpha][P_m(\tau) - \mathcal{P}(\tau)]\, d\tau$. Show directly that $k'(t) \leq 0$ if (5.3.3) holds.

6. Define the function $f = f(t)$ by $f(t) = 2\alpha e^{-\alpha(\bar{t}-t)} + (\gamma + \alpha)(e^{\alpha(\bar{t}-t)} - e^{-\alpha(\bar{t}-t)})$ for $0 \leq t \leq \bar{t}$. Show directly that $f'(t) \leq 0$. (Recall that $\gamma = \sqrt{\alpha^2 + \beta^2}$.) Show now that the right-hand side of (5.3.34) decreases with increasing $t (0 \leq t \leq \bar{t})$.

7. Explain why it is sufficient to check the inequality (5.3.34) at the single time $t = \bar{t}$.

6. Applications of the Euler-Lagrange Multiplier Theorem in Elementary Control Theory

In this chapter we shall show how to use Euler-Lagrange multipliers to solve certain classes of problems in elementary control theory. We shall consider several examples, including certain problems in the control of manufacturing systems and several problems in vehicular control. We shall show how to solve certain *bang-bang* control problems with the aid of certain auxiliary variables related to the slack functions of Chapter 5.

6.1. Introduction

It often happens in various problems in engineering, rocket control, economics, and business management that one wishes to minimize or maximize a functional J of the form

$$J = \int_{t_0}^{t_1} F(t, X_1(t), X_2(t), \ldots, X_n(t), U_1(t), U_2(t), \ldots, U_m(t))\, dt \tag{6.1.1}$$

over all functions $X_1(t), X_2(t), \ldots, X_n(t)$ satisfying various constraints *which include certain differential equations* such as

$$\frac{d}{dt} X_i(t) = G_i(t, X_1(t), X_2(t), \ldots, X_n(t), U_1(t), U_2(t), \ldots, U_m(t))$$
$$\text{for } i = 1, 2, \ldots, n \text{ and for } t > t_0. \tag{6.1.2}$$

The specified functions $F = F(t, x_1, x_2, \ldots, x_n, u_1, \ldots, u_m)$ and $G_i = G_i(t, x_1, \ldots, x_n, u_1, \ldots, u_m)$ (for $i = 1, 2, \ldots, n$) are given real-valued functions of the $n + m + 1$ variables $t, x_1, \ldots, x_n, u_1, \ldots, u_m$ which are evaluated along $x_i = X_i(t)$ (for $i = 1, 2, \ldots, n$) and $u_j = U_j(t)$ (for $j = 1, 2, \ldots, m$) in (6.1.1) and (6.1.2). In many applications the independent variable is *time*, and we have labeled it as t. The numbers t_0 and t_1 are the initial and final times.

We shall see that many extremum problems of many different types can be encompassed by this single formulation since the integers n and m can be given different values and the functions $F, G_1, \ldots, G_n$ can be chosen in many different ways. For example, if we take $m = n$ and define the functions $G_1, \ldots, G_n$ by the special formula

$$G_i(t, x_1, \ldots, x_n, u_1, \ldots, u_n) = u_i \qquad \text{for } i = 1, 2, \ldots, n,$$

then J can be written with (6.1.2) in the form

$$J = \int_{t_0}^{t_1} F(t, X_1(t), \ldots, X_n(t), X_1'(t), \ldots, X_n'(t))\, dt$$

since $X_i'(t) = dX_i(t)/dt = G_i = U_i(t)$. This special case has already been considered in Section 4.6.

Another special problem that has already been considered can be obtained by taking $m = n = 1$ with $x_1 = x$, $u_1 = u$, $G_1 = G$, and

$$F(t, x, u) = \beta^2[x - \mathcal{I}(t)]^2 + [u - \mathcal{P}(t)]^2$$
$$G(t, x, u) = -\alpha x + u - \mathcal{S}(t) \qquad \text{for all numbers } t, x, \text{ and } u,$$

where $\mathcal{I} = \mathcal{I}(t)$, $\mathcal{P} = \mathcal{P}(t)$, and $\mathcal{S} = \mathcal{S}(t)$ are prescribed known functions of t, while α and β are given constants. If we put $t_0 = 0$ and $t_1 = T$ in (6.1.1), then in this case the functional J becomes

$$J = \int_0^T \{\beta^2[X(t) - \mathcal{I}(t)]^2 + [U(t) - \mathcal{P}(t)]^2\}\, dt,$$

while the (single) differential equation of (6.1.2) becomes

$$\frac{d}{dt}X(t) = -\alpha X(t) + U(t) - \mathcal{S}(t).$$

If we identify the function X with the *inventory level* $I = X$ and the function U with the *production rate* $P = U$, then the problem of minimizing this functional J subject to this last differential equation is the same optimization problem in production planning already considered in Section 2.5. In this problem the inventory function $I = X(t)$ can be considered to describe the *state* of the manufacturing system as a function of time, and the production rate $P = U(t)$ can be varied by the manufacturer so as to *control* the resulting state of the system.

In the general case the functions $X_1(t), \ldots, X_n(t)$ are called *state functions*, while the functions $U_1(t), \ldots, U_m(t)$ are called *control functions*. The differential equations (6.1.2) along with any other specified constraints serve to govern the behavior of the state functions once the control functions have been specified. The given functional J usually represents a cost or a payoff which is to be minimized or maximized subject to the given constraints. In practice the business manager or the system engineer can vary the values of the control functions so as to control the resulting state of the system as described by the state functions $X_1, \ldots, X_n$. The object is to choose the control functions so as to minimize or maximize J subject to the given constraints.

For notational purposes it is often convenient to combine the n state functions $X_1, X_2, \ldots, X_n$ into a single *vector* state function $\mathbf{X}$ given as

$$\mathbf{X} = \mathbf{X}(t) = (X_1(t), X_2(t), \ldots, X_n(t)),$$

and, similarly, it is convenient to combine the m control functions $U_1, U_2, \ldots, U_m$ into a single *vector* control function $\mathbf{U}$ given as

$$\mathbf{U} = \mathbf{U}(t) = (U_1(t), U_2(t), \ldots, U_m(t)).$$

Then the previous functional J can be written more briefly as

$$J = \int_{t_0}^{t_1} F(t, \mathbf{X}(t), \mathbf{U}(t))\,dt,$$

while the differential equations of (6.1.2) can be written as a single vector equation as

$$\frac{d}{dt}\mathbf{X}(t) = \mathbf{G}(t, \mathbf{X}(t), \mathbf{U}(t)),$$

where the vector function $\mathbf{G}$ is given as $\mathbf{G} = (G_1, G_2 \ldots, G_n)$. It is important to remember that the state vector $\mathbf{X}$ has n components, while the control vector $\mathbf{U}$ has m components.

6.2. A Rocket Control Problem: Minimum Time

We shall consider a simplified version of a rocket control problem in which a rocket is propelled vertically upward from the earth's surface so as to reach a prescribed height above the earth's surface in *minimum time* while using a given fixed quantity of fuel. The problem is to find the optimum *thrust rate* at each instant of time so that the rocket will consume the given quantity of fuel at the proper rate in order to minimize the total flight time required to reach the specified altitude.

We let $X(t)$ be the vertical distance or altitude of the rocket above the earth's surface at time t. From Newton's second law of motion we have

$$mX''(t) = F(t),$$

where m is the mass of the rocket and $F(t)$ is the total (vertical) force acting on the rocket at time t. We consider only the earth's gravitational force and the thrust force of the rocket, so that we have

$$F(t) = -mg + mU(t).$$

Here $U(t)$ is the acceleration provided by the thrust of the rocket engine at time t, while g is a representative value of the earth's gravitational acceleration taken in the flight region of the rocket. We are assuming that g is a constant, although it would be possible to allow for the known variation of g with altitude. The effect of the earth's gravitational force is to pull the rocket back toward the surface of the earth, while the effect of the rocket thrust force is to lift the rocket away from the earth. We are neglecting all other forces, such as the force of resistance of the earth's atmosphere on the rocket. We have also neglected the variation of the mass of the rocket during the burning of the rocket fuel.† We are letting m be a constant that is taken to be representative of the rocket mass during the entire motion of the rocket.

Therefore we are considering a simplified model of vertical rocket motion in which the rocket height $X = X(t)$ satisfies

$$X''(t) = -g + U(t) \qquad \text{for } t > 0, \tag{6.2.1}$$

where $t_0 = 0$ is the initial launch time. The rocket begins from rest with zero initial height and speed, and so we impose the initial conditions

$$X(0) = 0, \qquad X'(0) = 0. \tag{6.2.2}$$

†We have represented the thrust force of the rocket as $mU = mU(t)$. A more realistic model of the thrust force would involve both the rocket speed $X'(t)$ and the rate of change of the rocket mass. See, for example, J. M. A. Danby, *Fundamentals of Celestial Mechanics* (New York: The Macmillan Company, 1962), pp. 52–53.

Finally, we impose the terminal constraint

$$X(t_1) = h, \tag{6.2.3}$$

which is required to hold at the final time $t = t_1$, where h is a given positive constant representing the specified height which must be achieved by the rocket.

We seek to find the particular thrust function $U = U(t)$ that will *minimize* the final time t_1 subject to the constraints (6.2.1), (6.2.2), and (6.2.3). However, we shall assume that the total thrust of the rocket engine over the entire thrust time is limited by the condition

$$\int_0^{t_1} U(t)^2 \, dt = \kappa^2, \tag{6.2.4}$$

where κ is a given positive constant which measures the total amount of fuel available. For example, if the thrust rate is held *constant* with

$$U(t) = U_0 \qquad \text{for all } t > 0,$$

then this condition (6.2.4) implies that $U_0^2 t_1 = \kappa^2$ holds, so that in this case the total thrust time t_1 is small if U_0 is large, and vice versa. Of course, a suitable nonconstant thrust rate may give a smaller time t_1 while satisfying all the given constraints.

The present extremum problem can be formulated in the manner of the preceding section if we take $t_0 = 0$ in (6.1.1) and put the integrand function F in (6.1.1) identically equal to 1, so that

$$J = \int_0^{t_1} F \, dt = \int_0^{t_1} 1 \, dt = t_1.$$

The present second-order differential equation (6.2.1) can be written in the form of (6.1.2) if we set $X(t) = X_1(t)$, with

$$\begin{aligned} X_1'(t) &= X_2(t) \\ X_2'(t) &= -g + U(t). \end{aligned} \tag{6.2.5}$$

[The reader should verify that this system of equations is equivalent to the original equation (6.2.1).] Hence we have $n = 2$ and $m = 1$ in (6.1.2), with

$$\begin{aligned} G_1(t, x_1, x_2, u) &= x_2 \\ G_2(t, x_1, x_2, u) &= -g + u \end{aligned}$$

for any numbers t, x_1, x_2, and u. The extremum problem is to find the control function $U = U(t)$ that will minimize the time functional $J = t_1$ subject to the constraints

$$X_1(0) = 0, \qquad X_2(0) = 0, \qquad X_1(t_1) = \text{h}, \qquad \text{and} \qquad \int_0^{t_1} U(t)^2\, dt = \kappa^2,$$

where the state functions $X_1(t)$ and $X_2(t)$ must satisfy the differential equations (6.2.5). In this case the state function X_1 gives the height of the rocket above the surface of the earth, while the state function $X_2 = X_1'$ gives the speed of the rocket. Hence the state vector $\mathbf{X} = (X_1(t), X_2(t))$ does indeed describe the *state of the "system"* as a function of time.

The second equation of (6.2.5) can be integrated to give

$$X_2(t) = -gt + \int_0^t U(s)\, ds,$$

where we have imposed the initial condition $X_2(0) = 0$. This equation can now be inserted into the right-hand side of the first equation of (6.2.5), and the resulting differential equation can be integrated to give

$$X_1(t) = -\frac{gt^2}{2} + \int_0^t \left\{\int_0^\sigma U(s)\, ds\right\} d\sigma,$$

where we have imposed the initial condition $X_1(0) = 0$. It is convenient to interchange the orders of the iterated integrals in the last equation to find [see (2.5.17)]

$$X_1(t) = -\frac{gt^2}{2} + \int_0^t (t - s)U(s)\, ds,$$

so that the terminal constraint $X_1(t_1) = h$ becomes

$$-\frac{gt_1^2}{2} + \int_0^{t_1} (t_1 - t)U(t)\, dt = h. \tag{6.2.6}$$

We seek to choose the control function $U = U(t)$ and the final time t_1 so as to minimize $J = t_1$ while satisfying both (6.2.6) and the remaining constraint

$$\int_0^{t_1} U(t)^2\, dt = \kappa^2. \tag{6.2.7}$$

Since these constraints involve both the final time value t_1 and the thrust function $U = U(t)$, we shall take as our basic vector space $\mathfrak{X}$ the set of all pairs $(t_1, U) = (t_1, U(t))$, *where t_1 may be any arbitrary number in the vector space $\mathcal{R}$ of all numbers* and *where $U = U(t)$ may be any arbitrary continuous function on $\mathcal{R}$ which vanishes for all large $|t|$*. [It is convenient to require $U(t)$ to vanish for large $|t|$ so that the right-hand side of (6.2.8) below will always be finite.] If (t_1, U) and (t_1^*, U^*) are any two vectors in $\mathfrak{X}$, then we define their sum by

$$(t_1, U) + (t_1^*, U^*) = (t_1 + t_1^*, U + U^*),$$

which again gives a vector in $\mathfrak{X}$. Similarly, we define the product $a(t_1, U)$ by

$$a(t_1, U) = (at_1, aU),$$

which gives a vector in $\mathfrak{X}$ for any number a in $\mathfrak{R}$ and any vector (t_1, U) in $\mathfrak{X}$. One easily checks that all the rules listed in Section 1.2 for vector spaces are satisfied by $\mathfrak{X}$ with these definitions of addition and multiplication by numbers. Finally, we equip $\mathfrak{X}$ with the norm $\|\cdot\|$ defined by

$$\|(t_1, U)\| = |t_1| + \max_{\text{all } t \text{ in } \mathfrak{R}} |U(t)| \tag{6.2.8}$$

for any vector $(t_1, U) = (t_1, U(t))$ in $\mathfrak{X}$; one easily checks that this function $\|\cdot\|$ satisfies all the properties of a norm on $\mathfrak{X}$ as given in Section 1.4. We emphasize that the normed vector space $\mathfrak{X}$ consists of *all* pairs (t_1, U) for all numbers t_1 and for all continuous functions $U = U(t)$. There need *not* be any special relation between the number t_1 and the function U occurring in the pair (t_1, U). In particular, every such pair is in $\mathfrak{X}$ whether or not the pair satisfies the conditions (6.2.6) and (6.2.7).

We now let D be the open subset of $\mathfrak{X}$ consisting of all pairs (t_1, U) for which $t_1 > 0$, and we define functionals J, K_1, and K_2 on D by

$$J(t_1, U) = t_1$$

$$K_1(t_1, U) = -\frac{gt_1^2}{2} + \int_0^{t_1} (t_1 - t)U(t)\, dt$$

$$K_2(t_1, U) = \int_0^{t_1} U(t)^2\, dt$$

for any vector (t_1, U) in D. The extremum problem we are considering is then to find a minimum vector in $D[K_1 = h, K_2 = \kappa^2]$ for the functional J, and we shall use the Euler-Lagrange multiplier theorem of Section 3.6 to solve this problem. For this purpose we need the variations of the functionals J, K_1, and K_2.

Since $J(t_1 + \epsilon\, \Delta t_1, U + \epsilon\, \Delta U) = t_1 + \epsilon\, \Delta t_1$ for any vector $(\Delta t_1, \Delta U)$ in $\mathfrak{X}$ and for any small number ϵ, we easily find from (2.3.4) that

$$\delta J(t_1, U; \Delta t_1, \Delta U) = \Delta t_1$$

for any vector (t_1, U) in D and any vector $(\Delta t_1, \Delta U)$ in $\mathfrak{X}$. For K_1 we find that

$$K_1(t_1 + \epsilon\, \Delta t_1, U + \epsilon\, \Delta U) = -\frac{(t_1 + \epsilon\, \Delta t_1)^2 g}{2}$$
$$+ \int_0^{t_1 + \epsilon \Delta t_1} (t_1 + \epsilon\, \Delta t_1 - t)[U(t) + \epsilon\, \Delta U(t)]\, dt$$

from which we find with (2.3.4) and (4.4.14) that

$$\delta K_1(t_1, U; \Delta t_1, \Delta U) = \Delta t_1 \left[-\mathrm{g}t_1 + \int_0^{t_1} U(t)\,dt \right] + \int_0^{t_1} (t_1 - t)\,\Delta U(t)\,dt$$

for any vector (t_1, U) in D and any vector $(\Delta t_1, \Delta U)$ in $\mathfrak{X}$. (The reader should verify this result.) Similarly, we find for K_2 that

$$\delta K_2(t_1, U; \Delta t_1, \Delta U) = \Delta t_1 U(t_1)^2 + 2 \int_0^{t_1} U(t)\,\Delta U(t)\,dt$$

for any (t_1, U) in D and any $(\Delta t_1, \Delta U)$ in $\mathfrak{X}$.

It is now easy to check that all the hypotheses of the Euler-Lagrange multiplier theorem are satisfied (e.g., the variations of J, K_1, and K_2 are weakly continuous on D) and, moreover, that the *first* possibility (3.6.2) of that theorem fails to hold. (We leave the proofs of these assertions to the reader.) Hence if (t_1, U) is any local minimum vector in $D[K_1 = h, K_2 = \kappa^2]$ for J, it follows from the Euler-Lagrange multiplier theorem that there are constants λ_1 and λ_2 such that [see (3.6.3)]

$$\delta J(t_1, U; \Delta t_1, \Delta U) = \lambda_1\, \delta K_1(t_1, U; \Delta t_1, \Delta U) + \lambda_2\, \delta K_2(t_1, U; \Delta t_1, \Delta U)$$

for *all* vectors $(\Delta t_1, \Delta U)$ in $\mathfrak{X}$. If we insert the previous expressions for the variations of J, K_1, and K_2 into this equation, we can rewrite it as

$$\begin{aligned} \Delta t_1 = \Delta t_1 &\left\{ \lambda_1 \left[-gt_1 + \int_0^{t_1} U(t)\,dt \right] + \lambda_2 U(t_1)^2 \right\} \\ &+ \int_0^{t_1} [\lambda_1(t_1 - t) + 2\lambda_2 U(t)]\,\Delta U(t)\,dt, \end{aligned} \tag{6.2.9}$$

which must hold for all numbers Δt_1 and all continuous functions $\Delta U = \Delta U(t)$ if (t_1, U) is a minimum vector in $D[K_1 = h, K_2 = \kappa^2]$ for J.

If we choose first $\Delta t_1 = 0$ in (6.2.9) and take $\Delta U(t) = \lambda_1(t_1 - t) + 2\lambda_2 U(t)$, we easily find that

$$\lambda_1(t_1 - t) + 2\lambda_2 U(t) = 0$$

or that

$$U(t) = -\frac{\lambda_1(t_1 - t)}{2\lambda_2} \qquad \text{for } 0 \le t \le t_1, \tag{6.2.10}$$

which must be satisfied by any possible extremum vector $(t_1, U) = (t_1, U(t))$. If we use the last result in (6.2.9) and take $\Delta t_1 = 1$, we also find that

$$1 = \lambda_1 \left[-gt_1 + \int_0^{t_1} U(t)\,dt \right],$$

which must also hold for any extremum vector (t_1, U). If we insert $U(t)$ from (6.2.10) into the last equation, we find that

$$\lambda_1 t_1(\lambda_1 t_1 + 4\lambda_2 g) = -4\lambda_2. \tag{6.2.11}$$

On the other hand, the constraint $K_1 = h$ gives with (6.2.10)

$$t_1^2(\lambda_1 t_1 + 3\lambda_2 g) = -6\lambda_2 h, \tag{6.2.12}$$

while the constraint $K_2 = \kappa^2$ gives similarly with (6.2.10)

$$\lambda_1^2 t_1^3 = 12\lambda_2^2 \kappa^2. \tag{6.2.13}$$

The three equations (6.2.11), (6.2.12), and (6.2.13) serve to determine the values of the three unknown quantities λ_1, λ_2, and t_1 in terms of the given known values of the constants g, h, and κ. For example, (6.2.13) implies that

$$\frac{\lambda_1}{\lambda_2} = -\frac{2\sqrt{3}\,\kappa}{t_1^{3/2}}, \tag{6.2.14}$$

where we have taken the negative square root here, since (6.2.10) shows that λ_1/λ_2 must be negative in order to obtain a positive-valued thrust function. If we now use this last result in (6.2.12), we find

$$3gt_1^2 + 6h = 2\sqrt{3}\,\kappa t_1^{3/2}, \tag{6.2.15}$$

and this equation will have a (smallest) positive solution for t_1 *provided that* κ *is large enough compared to* g *and* h.† In practice one would use a suitable numerical algorithm to solve this equation for t_1. For example, Newton's method or the method of successive bisection would work.‡ Once t_1 has been found from (6.2.15), the multipliers λ_1 and λ_2 can be obtained from (6.2.11) and (6.2.14) if they are wanted, while the optimum thrust function U is given by (6.2.10) and (6.2.14) as

$$U(t) = \sqrt{\frac{3}{t_1}}\,\kappa\left(1 - \frac{t}{t_1}\right) \qquad \text{for } 0 \le t \le t_1.$$

The rocket can satisfy all the constraints and achieve the given height h in the *least time* by using just this particular thrust function, which decreases linearly from the value $\kappa\sqrt{3/t_1}$ down to zero as the time increases from 0 to t_1. The

†If both sides of equation (6.2.15) are squared, we obtain a fourth-degree polynomial equation that will have two positive roots for large enough κ. If κ is small, however, the stated extremum problem has no solution. (In this case there is inadequate fuel available for the rocket engine.)

‡See, for example, Peter Henrici, *Elements of Numerical Analysis* (New York: John Wiley & Sons, Inc., 1964), where such methods are discussed.

number t_1 must be chosen to be the smallest positive solution of equation (6.2.15).

Exercises

1. A boatman wishes to steer his boat so as to minimize the transit time required to cross a river of width l. The path of the boat is represented by a curve γ in the (x_1, x_2)-plane, where γ is given parametrically in terms of time as

$$\gamma: \begin{cases} x_1 = X_1(t) \\ x_2 = X_2(t) \qquad \text{for } 0 \leq t \leq t_1 \end{cases}$$

for suitable functions X_1 and X_2. The river has parallel banks, and the left bank coincides with the line $x_1 = 0$, while the right bank coincides with the line $x_1 = l$. The river has no cross currents, so the current velocity is everywhere directed downstream along the x_2-direction. If v_0 is the constant boat speed relative to the surrounding water and if $w = w(t, x_1, x_2)$ denotes the downstream river current speed at the point (x_1, x_2) and at the time t, then the path of the boat will be determined by the differential equations [compare with (1.3.12)] $(d/dt)X_1(t) = v_0 \cos \alpha(t)$ and $(d/dt)X_2(t) = v_0 \sin \alpha(t) + w(t, X_1(t), X_2(t))$, where $\alpha = \alpha(t)$ is the steering angle of the boat at time t. Find the steering control function $\alpha = \alpha(t)$ and the final time t_1 that will transfer the boat from the initial state $(X_1(0), X_2(0)) = (0, 0)$ to the final state $X_1(t_1) = l$ in such a way as to minimize the time functional $J(t_1, \alpha) = t_1$. *Hints:* Show that it is sufficient to minimize J among all pairs (t_1, α) that satisfy the constraint $v_0 \int_0^{t_1} \cos \alpha(t)\, dt = l$, and then use the Euler-Lagrange multiplier theorem to solve this extremum problem.

2. Show directly that the optimal terminal time and control steering function found in Exercise 1 actually *minimize* the transit time among all possible nonnegative transit times subject to the given constraint. *Hints:* If t_1^* and α^* are the optimal terminal time and steering function found in Exercise 1, show that any other admissible pair (t_1, α) must satisfy the condition $\int_0^{t_1} \cos \alpha(t)\, dt = t_1^*$. Use this result to obtain the inequality $t_1^* \leq \int_0^{t_1} 1\, dt = t_1$ for any possible nonnegative transit time t_1.

6.3. A Rocket Control Problem: Minimum Fuel

We consider a sounding rocket that is propelled vertically upward from the earth's surface as in Section 6.2, but we now assume that the distance $X(t)$ of the rocket above the earth's surface satisfies [compare with equation (6.2.1)]

$$X''(t) = -g + U(t) - \alpha X'(t) \qquad \text{for } t > 0, \tag{6.3.1}$$

where α is a given positive constant. The added term $-\alpha X'(t)$ on the right-hand side of this equation represents an atmospheric drag force that opposes the rocket motion and that is proportional to the rocket speed at each time. As in Section 6.2, we impose the initial conditions

$$X(0) = 0, \qquad X'(0) = 0$$

and the final condition

$$X(t_1) = h.$$

In the present case, however, we so *not* seek to minimize the total thrust time t_1, but rather we ask to minimize the functional

$$J = \int_0^{t_1} U(t)^2 \, dt, \tag{6.3.2}$$

which gives a measure of the total thrust force. The thrust time t_1 is itself unimportant in this case provided only that this last integral is as small as possible.

The differential equation (6.3.1) can be written in the form of (6.1.2) if we set $X(t) = X_1(t)$, with

$$\begin{aligned} X_1'(t) &= X_2(t) \\ X_2'(t) &= -g + U(t) - \alpha X_2(t). \end{aligned} \tag{6.3.3}$$

Hence we have $n = 2$ and $m = 1$ in (6.1.2), with

$$\begin{aligned} G_1(t, x_1, x_2, u) &= x_2 \\ G_2(t, x_1, x_2, u) &= -g + u - \alpha x_2 \end{aligned}$$

for any numbers t, x_1, x_2, and u, while the integrand function $F = F(t, x_1, x_2, u)$ appearing in (6.1.1) will be given in this case as

$$F(t, x_1, x_2, u) = u^2.$$

The problem is to find the control function $U = U(t)$ that will minimize the functional J of (6.3.2) subject to the constraints $X_1(0) = 0$, $X_2(0) = 0$, and $X_1(t_1) = h$, where the state functions X_1 and X_2 must satisfy the differential equations (6.3.3).

The second equation of (6.3.3) can be integrated to give†

$$X_2(t) = e^{-\alpha t} \int_0^t e^{\alpha s}[-g + U(s)] \, ds,$$

†See, for example, Chapter 20 of George B. Thomas, Jr., *Calculus and Analytic Geometry*, (Reading, Mass.: Addison-Wesley Publishing Company, Inc., 1968), or any elementary text on differential equations.

where we have imposed the initial condition $X_2(0) = 0$. This result can be inserted into the right-hand side of the first equation of (6.3.3), and the resulting differential equation can be integrated to give

$$X_1(t) = \int_0^t e^{-\alpha\sigma} \int_0^\sigma e^{\alpha s}[-g + U(s)]\, ds\, d\sigma,$$

where we have imposed the initial condition $X_1(0) = 0$. We can interchange the orders of the iterated integrals in the last equation to find [see (2.5.17)]

$$\begin{aligned} X_1(t) &= \int_0^t e^{\alpha s}[-g + U(s)] \int_s^t e^{-\alpha\sigma} d\sigma\, ds \\ &= \int_0^t \left[\frac{1 - e^{-\alpha(t-s)}}{\alpha}\right][-g + U(s)]\, ds, \end{aligned}$$

so that the terminal constraint $X_1(t_1) = h$ becomes

$$\int_0^{t_1} \left[\frac{1 - e^{-\alpha(t_1-t)}}{\alpha}\right][-g + U(t)]\, dt = h. \tag{6.3.4}$$

We must choose the control function $U = U(t)$ and the final time t_1 so as to minimize the functional J of (6.3.2) while satisfying the constraint (6.3.4).

We take the vector space $\mathfrak{X}$ to be the set of all pairs (t_1, U) as in Section 6.2, with norm

$$\|(t_1, U)\| = |t_1| + \max_{\text{all } t \text{ in } \mathcal{R}} |U(t)|,$$

and we let D be the set of all such vectors for which $t_1 > 0$. We then define the functionals J and K on D by

$$J(t_1, U) = \int_0^{t_1} U(t)^2\, dt$$

and

$$K(t_1, U) = \int_0^{t_1} \left[\frac{1 - e^{-\alpha(t_1-t)}}{\alpha}\right][-g + U(t)]\, dt$$

for any positive number t_1 and any continuous function $U = U(t)$. We shall use the Euler-Lagrange multiplier theorem to find a minimum vector in $D[K = h]$ for J.

If we use the same methods as were used in Section 6.2, we find in this case the following expressions for the variations of J and K,

$$\delta J(t_1, U; \Delta t_1, \Delta U) = \Delta t_1 U(t_1)^2 + 2\int_0^{t_1} U(t)\, \Delta U(t)\, dt$$

and

$$\delta K(t_1, U; \Delta t_1, \Delta U) = \Delta t_1 \int_0^{t_1} e^{-\alpha(t_1-t)}[-g + U(t)]\, dt + \int_0^{t_1} \left[\frac{1 - e^{-\alpha(t_1-t)}}{\alpha}\right] \Delta U(t)\, dt$$

for any vector (t_1, U) in D and any vector $(\Delta t_1, \Delta U)$ in $\mathfrak{X}$. (The reader may want to verify these results.) It is now easy to check that all the hypotheses of the Euler-Lagrange multiplier theorem are satisfied, so that if (t_1, U) is any local minimum vector in $D[K = h]$ for J, there must be a constant λ such that

$$\delta J(t_1, U; \Delta t_1, \Delta U) = \lambda\, \delta K(t_1, U; \Delta t_1, \Delta U)$$

holds for all vectors $(\Delta t_1, \Delta U)$ in $\mathfrak{X}$. If we use the previous expressions for the variations of J and K, we can rewrite the last equation as

$$\Delta t_1 \left\{U(t_1)^2 - \lambda \int_0^{t_1} e^{-\alpha(t_1-t)}[-g + U(t)]\, dt\right\} + \int_0^{t_1} \left\{2U(t) - \lambda \left[\frac{1 - e^{-\alpha(t_1-t)}}{\alpha}\right]\right\} \Delta U(t)\, dt = 0, \tag{6.3.5}$$

which must hold for all numbers Δt_1 and all continuous functions $\Delta U = \Delta U(t)$ if (t_1, U) is a minimum vector in $D[K = h]$ for J.

If we choose first $\Delta t_1 = 0$ in (6.3.5) with $\Delta U(t) = 2U(t) - \lambda\{1 - \exp[-\alpha(t_1 - t)]\}/\alpha$, we are led easily to

$$U(t) = \lambda \frac{1 - e^{-\alpha(t_1-t)}}{2\alpha} \qquad \text{for } 0 \leq t \leq t_1, \tag{6.3.6}$$

which must hold for any possible extremum vector (t_1, U). If we use (6.3.6) in (6.3.5) and take $\Delta t_1 = 1$, we also find that

$$\int_0^{t_1} e^{-\alpha(t_1-t)}[-g + U(t)]\, dt = 0, \tag{6.3.7}$$

which must also hold for any extremum vector (t_1, U). If we now insert $U(t)$ from (6.3.6) into the two equations (6.3.4) and (6.3.7), we obtain two equations which can be used to determine (numerical approximations to) the two unknown quantities t_1 and λ. Once t_1 and λ are known, the desired thrust function is given by (6.3.6).

We omit any further details in the general case $\alpha > 0$. However, if we neglect atmospheric drag effects and consider only the special case $\alpha = 0$, it is easy to see that the optimum thrust function is given as

$$U^*(t) = 2g\left(1 - \frac{t}{t_1^*}\right) \qquad \text{for } 0 \leq t \leq t_1^*,$$

with $t_1^* = \sqrt{6h/g}$. This particular pair (t_1^*, U^*) will minimize the functional

$$\int_0^{t_1} U(t)^2\, dt$$

among all pairs (t_1, U) that satisfy the given constraint (in the case $\alpha = 0$). In this case we see that the initial rocket thrust acceleration $U^*(0)$ must be twice the earth's (downward) gravitational acceleration.

Exercises

1. In the case $\alpha = 0$, use the Euler-Lagrange multiplier theorem to verify that the thrust pair (t_1^*, U^*) described above is the only possible pair for which the functional J of (6.3.2) can have a minimum value subject to the given constraint on the final height.

2. Let $X(t)$ be the distance of a space vehicle from its initial position on a straight path along which it moves in free space, and let $U(t)$ be the acceleration furnished by the thrust of the vehicle's engine, with $X''(t) = U(t)$. The vehicle's speed is initially $+1$ at time $t = 0$, $X'(0) = +1$, and $X(0) = 0$ also holds. Find the particular terminal time t_1 and thrust function U that will bring the moving vehicle to rest, with $X'(t_1) = 0$, in such a way as to *minimize* $J(t_1, U) = t_1 + \beta^2 \int_0^{t_1} U(t)^2\, dt$, where β is a given positive constant. What is the resulting total flight distance required to stop the vehicle in this case so as to minimize J, and what is the resulting minimum value of J?

3. Show directly that the optimal terminal time and thrust function found in Exercise 2 actually *minimize* the expression $J(t_1, U)$ among all possible nonnegative times t_1 and among all thrust functions U, subject to the given constraint. *Hints:* If t_1^* and U^* are the optimal terminal time and thrust function found in Exercise 2, show that any other admissible pair (t_1, U) must satisfy the condition $t_1^* - t_1 + \beta \int_0^{t_1} [U(t) - U^*(t)]\, dt = 0$. Use this result to obtain $J(t_1, U) = J(t_1{}^*, U^*) + \beta^2 \int_0^{t_1} [U(t) - U^*(t)]^2\, dt$, from which the desired result follows provided that $t_1 \geq 0$.

4. Let $X(t)$ be the distance of a space vehicle from its initial position on a straight path along which it moves as in Exercise 2, with $X''(t) = U(t)$ for $t > 0$ and $X(0) = 0$, $X'(0) = +1$. Find the particular thrust pair (t_1, U) that will transfer the state of the system from $(X(0), X'(0)) = (0, +1)$ to $(X(t_1), X'(t_1)) = (0, 0)$ in such a way as to minimize $J(t_1, U) = t_1 + \beta^2 \int_0^{t_1} U(t)^2\, dt$. (In this case one is required to bring the moving vehicle to rest while simultaneously bringing it back to its original position, and this is to be done while minimizing J.) If the given functional J is a measure of the cost required to transfer the system from its state at $t = 0$ to its state at $t = t_1$, show that the minimum cost in this case is four times the minimum cost in Exercise 2. Hence the additional constraint $X(t_1) = 0$ increases the minimum cost by a factor of 4.

5. The state $X = X(t)$ of a certain manufacturing system is governed by the differential equation $X'(t) = -X(t) + U(t)$, where $U = U(t)$ is an appropriate control function. If the initial state of the system is given as $X(0) = 0$, find the control function U and terminal time t_1 that will transfer the system to the state $X(t_1) = 1$ while minimizing the cost functional $J(t_1, U) = \int_0^{t_1} [\alpha^2 + U(t)^2]\, dt$, where α is a given positive constant. Show that the optimum value of t_1 must satisfy $e^{t_1} = [(1 + \sqrt{1 + \alpha^2})/2]\,(e^{t_1} - e^{-t_1})$. (This equation always determines a unique positive value of t_1 for each fixed positive value of α. However, the equation must in general be solved numerically.)

6.4. A More General Control Problem

In this section we shall consider the general problem of minimizing (or maximizing) the functional

$$J = \int_{t_0}^{t_1} F(t, X(t), U(t))\, dt \tag{6.4.1}$$

over all functions $X(t)$ which satisfy the differential equation

$$\frac{dX(t)}{dt} = G(t, X(t), U(t)) \qquad \text{for } t_0 < t < t_1 \tag{6.4.2}$$

and the initial condition

$$X(t_0) = x_0. \tag{6.4.3}$$

The numbers t_0, t_1, and x_0 are prescribed in advance and held fixed, while the function U is a control function which is to be determined so as to minimize J. The given real-valued functions $F = F(t, x, u)$ and $G = G(t, x, u)$ are defined for all suitable values of the three real variables t, x, and u and are assumed to be continuous with respect to all these variables and have continuous first-order partial derivatives with respect to x and u. For example, if F and G are defined by

$$\begin{aligned} F(t, x, u) &= x^2 + u^2 \\ G(t, x, u) &= \alpha x + \beta u \end{aligned} \tag{6.4.4}$$

for given constants α and β, then we have

$$J = \int_{t_0}^{t_1} [X(t)^2 + U(t)^2]\, dt$$

and

$$\frac{dX(t)}{dt} = \alpha X(t) + \beta U(t).$$

We have seen other such examples in Section 6.1. (In the formulation of Section 6.1 we are now considering the special case $n = m = 1$. However, analogous results to those of this section can be obtained in the case of any positive integral values for n and m.)

Since the given function $G = G(t, x, u)$ is assumed to be continuous with respect to all its variables and have a continuous first-order partial derivative with respect to x, it is known† that the differential equation (6.4.2) will always have a unique solution $X = X(t)$ satisfying the initial condition (6.4.3), and *this is true for any fixed continuous control function* $U = U(t)$ *that is inserted into the right-hand side of equation* (6.4.2).‡ For example, if G is defined as in (6.4.4), then the solution of (6.4.2) and (6.4.3) is given as

$$X(t) = e^{\alpha(t-t_0)}x_0 + \beta \int_{t_0}^{t} e^{\alpha(t-\tau)}U(\tau)\,d\tau,$$

as is easy to check.

We shall want to study the dependence of the solution X on the control function U, and so we shall write $X = X(t; U)$ for the unique solution of the differential equation (6.4.2) subject to the initial condition (6.4.3). For example, if G is defined as in (6.4.4), we will have

$$X(t; U) = e^{\alpha(t-t_0)}x_0 + \beta \int_{t_0}^{t} e^{\alpha(t-\tau)}U(\tau)\,d\tau, \tag{6.4.5}$$

where U can be any suitable control function. Of course, in many cases it will *not* be possible to obtain such a simple explicit formula for $X(t; U)$. Nevertheless, the function X is always guaranteed to exist, and there are numerical procedures available which can be used to give numerical values for $X(t; U)$ (for any given control function) using the conditions

$$\begin{aligned} \frac{d}{dt}X(t; U) &= G(t, X(t; U), U(t)) \qquad \text{for } t > t_0 \\ X(t_0; U) &= x_0. \end{aligned} \tag{6.4.6}$$

It is now clear that if we impose the given differential equation and initial condition on the state function X, then the functional J of (6.4.1) will depend only on the control function U as

$$J(U) = \int_{t_0}^{t_1} F(t, X(t; U), U(t))\,dt, \tag{6.4.7}$$

†See, for example, Chapter 9 of Tom M. Apostol, *Calculus*, Vol. 2 (Waltham, Mass.: Ginn/Blaisdell, 1962), or almost any text on differential equations.

‡In fact, (6.4.2) and (6.4.3) will always have a unique piecewise continuously differentiable solution X for any piecewise continuous control function U. We say that $U = U(t)$ is *piecewise continuous* on the interval $t_0 \leq t \leq t_1$ if U is continuous except possibly at a finite number of points where U has well-defined limiting values both from the left and from the right.

where $X(t; U)$ is determined by (6.4.6). The extremum problem is to minimize $J(U)$ over all suitable control functions U. We shall consider only continuous control functions, although we could just as easily allow piecewise continuous controls. We could also allow the terminal time t_1 to vary so that J would depend on both t_1 and U, but we shall hold t_1 fixed for the moment.

If U is a local extremum vector for J, then the variation of J at U must vanish with

$$\delta J(U; \Delta U) = 0$$

for all continuous functions $\Delta U = \Delta U(t)$. To calculate the variation of J, we use (6.4.7) to find [compare with (2.4.14)]

$$\begin{aligned}\frac{d}{d\epsilon} J(U + \epsilon\, \Delta U) &= \frac{d}{d\epsilon} \int_{t_0}^{t_1} F(t, X(t; U + \epsilon\, \Delta U), \quad U(t) + \epsilon\, \Delta U(t))\, dt \\ &= \int_{t_0}^{t_1} \frac{\partial}{\partial \epsilon} F(t, X(t; U + \epsilon\, \Delta U), \quad U(t) + \epsilon\, \Delta U(t))\, dt,\end{aligned}$$

where the last equality here can be justified by our stated assumptions on the given functions F and G. The derivative $\partial F/\partial\epsilon$ appearing in the last integral can be calculated by the chain rule of differential calculus; *for each fixed t* we find that

$$\begin{aligned}&\frac{\partial}{\partial \epsilon} F(t, X(t; U + \epsilon\, \Delta U), U(t) + \epsilon\, \Delta U(t)) \\ &= F_x(t, X(t; U + \epsilon\, \Delta U), U(t) + \epsilon\, \Delta U(t)) \frac{\partial}{\partial \epsilon} X(t; U + \epsilon\, \Delta U) \\ &+ F_u(t, X(t; U + \epsilon\, \Delta U), U(t) + \epsilon\, \Delta U(t))\, \Delta U(t),\end{aligned}$$

where $F = F(t, x, u)$, $F_x(t, x, u) = \partial F(t, x, u)/\partial x$, and $F_u(t, x, u) = \partial F(t, x, u)/\partial u$ and we have taken $x = X(t; U + \epsilon\, \Delta U)$ and $u = U(t) + \epsilon\, \Delta U(t)$. If we now evaluate the last equation at $\epsilon = 0$, we get

$$\begin{aligned}&\frac{\partial}{\partial \epsilon} F(t, X(t; U + \epsilon\, \Delta U), U(t) + \epsilon\, \Delta U(t))\Big|_{\epsilon=0} \\ &= F_X(t, X(t; U), U(t)) \frac{\partial}{\partial \epsilon} X(t; U + \epsilon\, \Delta U)\Big|_{\epsilon=0} \\ &+ F_U(t, X(t; U), U(t))\, \Delta U(t),\end{aligned}$$

where

$$\begin{aligned}F_X(t, X(t; U), U(t)) &= \frac{\partial}{\partial x} F(t, x, u)\Big|_{\substack{x=X(t;\,U) \\ u=U(t)}} \\ F_U(t, X(t; U), U(t)) &= \frac{\partial}{\partial u} F(t, x, u)\Big|_{\substack{x=X(t;\,U) \\ u=U(t)}}\end{aligned} \tag{6.4.8}$$

and

$$\frac{\partial}{\partial \epsilon} X(t; U + \epsilon\, \Delta U)\bigg|_{\epsilon=0} = \underset{\epsilon \to 0}{\text{limit}} \frac{X(t; U + \epsilon\, \Delta U) - X(t; U)}{\epsilon}. \tag{6.4.9}$$

For example, if G is defined as in (6.4.4), then we find from (6.4.5) and (6.4.9) that

$$\frac{\partial}{\partial \epsilon} X(t; U + \epsilon\, \Delta U)\bigg|_{\epsilon=0} = \beta \int_{t_0}^{t} e^{\alpha(t-\tau)}\, \Delta U(\tau)\, d\tau \tag{6.4.10}$$

for any fixed t and any fixed functions U and ΔU. In the general case these results along with (2.3.4) now lead to the following equation for the variation of J:

$$\delta J(U; \Delta U) = \int_{t_0}^{t_1} [F_X(t, X(t; U), U(t)) \frac{\partial}{\partial \epsilon} X(t; U + \epsilon\, \Delta U)\bigg|_{\epsilon=0} + F_U(t, X(t; U), U(t))\, \Delta U(t)]\, dt \tag{6.4.11}$$

for any continuous function $\Delta U = \Delta U(t)$. The function $X = X(t; U)$ is determined by (6.4.6), while the other expressions appearing in (6.4.11) are defined by (6.4.8) and (6.4.9).

To use (6.4.11), we shall need an explicit formula for

$$\frac{\partial}{\partial \epsilon} X(t; U + \epsilon\, \Delta U)\bigg|_{\epsilon=0}.$$

We would like a general result analogous to the special result (6.4.10), which holds in the special case in which $G(t, x, u) = \alpha x + \beta u$. And, indeed, in the general case we can use (6.4.9) along with the differential equation and initial condition given by (6.4.6) to find that

$$\frac{\partial}{\partial \epsilon} X(t; U + \epsilon\, \Delta U)\bigg|_{\epsilon=0} = \int_{t_0}^{t} e^{\int_\tau^t A(s)\, ds}\, B(\tau)\, \Delta U(\tau)\, d\tau, \tag{6.4.12}$$

where

$$\begin{aligned} A(t) &= G_X(t, X(t; U), U(t)) = \frac{\partial}{\partial x} G(t, x, u)\bigg|_{\substack{x = X(t;\, U) \\ u = U(t)}} \\ B(t) &= G_U(t, X(t; U), U(t)) = \frac{\partial}{\partial u} G(t, x, u)\bigg|_{\substack{x = X(t;\, U) \\ u = U(t)}}. \end{aligned} \tag{6.4.13}$$

We include a proof of this result in Section A10 of the Appendix. We might observe here, however, that (6.4.12) does reduce to the special case (6.4.10) if $G = \alpha x + \beta u$. Indeed, in this special case we find from (6.4.13) that

$$A(t) = \alpha$$
$$B(t) = \beta \qquad \text{for all } t,$$

and then it is easy to check that (6.4.12) agrees with (6.4.10). We refer the reader to Section A10 of the Appendix for the derivation of (6.4.12) in the general case.

If we now use (6.4.12) in (6.4.11), we find that

$$\delta J(U; \Delta U) = \int_{t_0}^{t_1} \Big[F_X(t, X(t; U), U(t)) \int_{t_0}^{t} e^{\int_\tau^t A(s)ds} B(\tau)\, \Delta U(\tau)\, d\tau + F_U(t, X(t; U), U(t))\, \Delta U(t) \Big] dt,$$

or if we interchange the orders of the iterated integrals here [see (2.5.17)], we obtain

$$\delta J(U; \Delta U) = \int_{t_0}^{t_1} \Big\{ F_U(t, X(t; U), U(t)) + \int_t^{t_1} F_X(\tau, X(\tau; U), U(\tau))\, e^{-\int_\tau^t A(s)ds} B(t)\, d\tau \Big\} \Delta U(t)\, dt,$$

where the functions A and B are defined by (6.4.13).

Since the variation of J must vanish at any local extremum vector U, we find that

$$0 = \int_{t_0}^{t_1} \Big[F_U(t, X(t; U), U(t)) + \int_t^{t_1} F_X(\tau, X(\tau; U), U(\tau))\, e^{-\int_\tau^t A(s)ds} B(t)\, d\tau \Big] \Delta U(t)\, dt,$$

which must hold for all continuous functions ΔU if U is a local extremum vector for J. In particular, we can take ΔU to be the function in square brackets appearing in the integrand here, in which case we find that

$$0 = \int_{t_0}^{t_1} \Big[F_U(t, X(t; U), U(t)) + \int_t^{t_1} F_X(\tau, X(\tau; U), U(\tau))\, e^{-\int_\tau^t A(s)ds} B(t)\, d\tau \Big]^2 dt,$$

from which we conclude that

$$F_U(t, X(t; U), U(t)) + B(t) \int_t^{t_1} F_X(\tau, X(\tau; U), U(\tau))\, e^{-\int_\tau^t A(s)ds}\, d\tau = 0 \qquad \text{for } t_0 \leq t \leq t_1, \tag{6.4.14}$$

which must be satisfied by an local extremum function U for J. The functions A and B are defined by (6.4.13), while the function $X = X(t; U)$ is determined by (6.4.6). The last necessary condition (6.4.14) along with the original specified differential equation and initial condition of (6.4.6) together serve to determine the extremum control function U and the corresponding state function X.

It is sometimes useful to replace the integral equation (6.4.14) with a corresponding differential equation. Indeed, if we differentiate both sides of (6.4.14) with respect to t and use (6.4.13) and (6.4.14) to simplify the result, we find the following first-order differential equation,

$$\frac{d}{dt}\left[\frac{F_U(t, X(t; U), U(t))}{G_U(t, X(t; U), U(t))}\right] + \left[\frac{F_U(t, X(t; U), U(t))}{G_U(t, X(t; U), U(t))}\right] G_X(t, X(t; U), U(t))$$
$$= F_X(t, X(t; U), U(t)), \tag{6.4.15}$$

which must be satisfied by any extremum function, at least if $G_U \neq 0$. Along with this differential equation we have the natural boundary condition

$$F_U(t_1, X(t_1; U), U(t_1)) = 0, \tag{6.4.16}$$

obtained by putting $t = t_1$ in the integral equation (6.4.14). The two coupled differential equations appearing in (6.4.6) and (6.4.15) along with the two boundary conditions appearing in (6.4.6) and (6.4.16) serve to determine the two functions $U = U(t)$ and $X = X(t; U)$.

Example. To illustrate the above results, we shall consider the problem of maximizing the functional

$$J = \int_0^\pi X(t)^2 \cos^2 U(t)\, dt \tag{6.4.17}$$

among all functions $X(t)$ which satisfy the differential equation

$$\frac{d}{dt} X(t) = \frac{\sin U(t)}{2} \qquad \text{for } 0 < t < \pi \tag{6.4.18}$$

and the initial condition

$$X = \frac{\pi}{2} \qquad \text{when } t = 0.$$

In keeping with our previous notation, we should write $X = X(t; U)$, where in this case we find from the last two conditions that

$$X(t; U) = \frac{\pi}{2} + \frac{1}{2}\int_0^t \sin U(\tau)\, d\tau$$

for any fixed control function U. The present functional $J = J(U)$ is clearly bounded from above; for example, the reader should be able to derive the crude estimate

$$J(U) \leq \frac{7\pi^3}{12},$$

which holds for all (piecewise) continuous control functions U. We shall see below that the maximum value of J is actually $J_{max} = \pi^3/3$, and we shall determine an extremum control function that leads to this maximum value.

In the present case the functions F and G are defined by

$$F(t, x, u) = x^2 \cos^2 u$$

$$G(t, x, u) = \frac{\sin u}{2}$$

for any numbers t, x, and u, from which we find that

$$F_X(t, X(t; U), U(t)) = \frac{\partial}{\partial x} F(t, x, u) \bigg|_{\substack{x = X(t;\, U) \\ u = U(t)}} = 2X(t; U) \cos^2 U(t)$$

$$F_U(t, X(t; U), U(t)) = \frac{\partial}{\partial u} F(t, x, u) \bigg|_{\substack{x = X(t;\, U) \\ u = U(t)}} = -2X(t; U)^2 \cos U(t) \sin U(t)$$

$$G_X(t, X(t; U), U(t)) = \frac{\partial}{\partial x} G(t, x, u) \bigg|_{\substack{x = X(t;\, U) \\ u = U(t)}} = 0$$

$$G_U(t, X(t; U), U(t)) = \frac{\partial}{\partial u} G(t, x, u) \bigg|_{\substack{x = X(t;\, U) \\ u = U(t)}} = \frac{\cos U(t)}{2}.$$

Hence in this case the extremum equation (6.4.15) gives the differential equation

$$2\frac{d}{dt}(X^2 \sin U) + X \cos^2 U = 0, \tag{6.4.19}$$

while the natural boundary condition (6.4.16) becomes (with $t_1 = \pi$)

$$X^2 \cos U \cdot \sin U = 0 \qquad \text{at } t = \pi. \tag{6.4.20}$$

We can now solve the extremum differential equation (6.4.19) along with the original differential equation (6.4.18) to find the extremum functions U and X. Indeed, we can use (6.4.18) and (6.4.19) to obtain the alternative equation

$$X\left[2\frac{d}{dt}(X \sin U) + 1\right] = 0,$$

which leads in turn to

$$2\frac{d}{dt}(X \sin U) + 1 = 0.$$

(We have excluded the possibility $X = 0$ at least over some initial time period since $X = \pi/2$ at $t = 0$.) The last equation can be integrated to give

$$X(t) \sin U(t) = -\frac{t}{2} + c \tag{6.4.21}$$

for some constant of integration c. [For brevity we are writing $X(t)$ for $X(t; U)$ here.] But now this result and the original differential equation (6.4.18) imply that

$$\frac{dX}{dt} = \frac{-(t/2) + c}{2X},$$

which can be integrated (by separation of variables) to give

$$X(t) = \sqrt{-\frac{t^2}{4} + ct + \frac{\pi^2}{4}}, \tag{6.4.22}$$

where we have imposed the initial condition $X = \pi/2$ at $t = 0$. If we use (6.4.22) in (6.4.21), we find for the extremum control function U that

$$\sin U(t) = \frac{-t/2 + c}{\sqrt{(\pi^2/4) + ct - (t^2/4)}}. \tag{6.4.23}$$

The value of the remaining constant c can be determined by the natural boundary condition (6.4.20). Alternatively, we can use (6.4.22) and (6.4.23) to find that

$$X^2 \cos^2 U = \frac{\pi^2}{4} - c^2 + 2ct - \frac{t^2}{2}$$

for $0 \leq t \leq \pi$, so that (6.4.17) gives for any possible extremum control

$$\begin{aligned} J &= \int_0^\pi \left(\frac{\pi^2}{4} - c^2 + 2ct - \frac{t^2}{2}\right) dt \\ &= \left(\frac{\pi^2}{4} - c^2\right)\pi + c\pi^2 - \frac{\pi^3}{6}. \end{aligned}$$

It is easy to check that this expression takes on its maximum value when $c = \pi/2$, and, indeed, this value of c agrees with the value obtained from the natural boundary condition (6.4.20).

Hence the extremum control function U is determined by equation (6.4.23) with $c = \pi/2$, while the extremum state function is given similarly by

(6.4.22) with $c = \pi/2$. The resulting maximum value of J is then found to be $J_{\max} = \pi^3/3$.

We should mention that in the general case it is often *not possible* in practice to obtain simple explicit formulas for the solutions of the two given differential equations [see (6.4.2) and (6.4.15)]

$$\frac{d}{dt}X = G(t, X, U)$$

$$\frac{d}{dt}\left[\frac{F_U(t, X, U)}{G_U(t, X, U)}\right] + G_X(t, X, U)\left[\frac{F_U(t, X, U)}{G_U(t, X, U)}\right] = F_X(t, X, U)$$

subject to the two boundary conditions [see (6.4.3) and (6.4.16)]

$$X(t_0) = x_0, \qquad F_U(t_1, X(t_1), U(t_1)) = 0.$$

In such cases recourse must be had to numerical procedures which allow one to obtain numerical approximations to the solutions. However, the numerical solution of such *two-point boundary value problems* is by no means easy. We refer the interested reader to H. Keller,† where such procedures are discussed.

Exercises

1. The state $X = X(t)$ of a certain manufacturing system is governed by the differential equation $dX/dt = -X + U$ for $0 < t < 1$, where $U = U(t)$ is an appropriate control function. If the initial state of the system is given as $X = e^{\sqrt{2}}(1 + \sqrt{2}) + e^{-\sqrt{2}}(-1 + \sqrt{2})$ at $t = 0$, find the control function U that will minimize the cost functional $J = \frac{1}{2}\int_0^1 [X(t)^2 + U(t)^2]\, dt$.

2. Use the Euler-Lagrange multiplier theorem to derive the necessary condition [compare with (6.4.14)] $F_U(t, X(t; U), U(t)) - \lambda H_U(t, X(t; U), U(t)) + B(t) \int_t^{t_1} [F_X(\tau, X(\tau; U), U(\tau)) - \lambda H_X(\tau, X(\tau; U), U(\tau))]e^{-\int_\tau^t A(s)ds}\, d\tau = 0$ for $t_0 \leq t \leq t_1$, which must hold for any control function U that minimizes or maximizes the functional $J(U) = \int_{t_0}^{t_1} F(t, X(t; U), U(t))\, dt$ subject to the conditions

$$\begin{aligned} &\frac{d}{dt}X(t; U) = G(t, X(t; U), U(t)) \qquad \text{for } t_0 < t < t_1 \\ &X(t_0; U) = x_0 \end{aligned} \tag{6.4.24}$$

and $\int_{t_0}^{t_1} H(t, X(t; U), U(t))\, dt = h_0$. Here λ is an Euler-Lagrange multiplier, while t_0, t_1, x_0, and h_0 are given fixed constants. The given real-valued functions $F = F(t, x, u)$, $G = G(t, x, u)$, and $H = H(t, x, u)$ are defined for all suitable

†Herbert B. Keller, *Numerical Methods for Two-Point Boundary-Value Problems* (Waltham, Mass.: Ginn/Blaisdell, 1968).

values of the three real variables t, x, and u and are assumed to have continuous first-order partial derivatives with respect to these variables. The functions A and B are defined by (6.4.13), while the function $X = X(t; U)$ is determined by the conditions (6.4.24) for any fixed control function U.

3. Let U be any control function that minimizes or maximizes the functional $J(U) = \int_{t_0}^{t_1} F(t, X(t; U), U(t))\, dt$ subject to the conditions $dX/dt = G(t, X, U)$ for $t_0 < t < t_1$, $X = x_0$ for $t = t_0$, and $\int_{t_0}^{t_1} H(t, X(t; U), U(t))\, dt = h_0$, as in Exercise 2. Show that U must satisfy the differential equation [compare with (6.4.15)] $(d/dt)\,[(F_U - \lambda H_U)/G_U] + [(F_U - \lambda H_U)/G_U]\, G_X = F_X - \lambda H_X$ for $t_0 < t < t_1$ if $G_U \neq 0$, and derive a suitable natural boundary condition which must hold at $t = t_1$.

4. The sales rate $S = S(t)$ of a certain company is (assumed to be) governed by the differential equation $dS/dt = -\alpha S + \beta A$, where $A = A(t)$ is the rate of advertising and where α and β are given positive constants. (Hence the sales rate decreases at a rate proportional to sales but increases at a rate proportional to advertising.) Find the particular advertising policy which maximizes the sales functional $J = \int_0^{t_1} S(t)\, dt$ if the initial sales rate is given as $S = s_0$ at $t = 0$, and if the total advertising is limited by the condition $\int_0^{t_1} A(t)^2\, dt = a^2$. Here s_0, t_1, and a are given fixed positive constants.† *Hint:* Use Exercise 3.

6.5. A Simple Bang-Bang Problem

We shall now consider certain special control problems in which the actual extremum control functions are *discontinuous.* More precisely, the extremum functions will be continuous everywhere with the exception of a finite number of points where they will have well-defined limiting values both from the left and from the right; that is, the control functions will be *piecewise continuous.*

The problems of interest here will involve control functions U that are required to satisfy pointwise inequality constraints of the form

$$\varphi(t) \leq U(t) \leq \psi(t) \qquad \text{for } t_0 \leq t \leq t_1, \tag{6.5.1}$$

where φ and ψ are given known piecewise continuous functions with $\psi(t) \geq \varphi(t)$ for $t_0 \leq t \leq t_1$. For example, we may consider a space vehicle moving in free space with

$$X''(t) = U(t),$$

†Exercise 4 is patterned after Problem 14-P on p. 365 of Michael D. Intriligator, *Mathematical Optimization and Economic Theory* (Englewood Cliffs, N.J.: Prentice-Hall, Inc., 1971).

where $X(t)$ is the distance of the vehicle from some fixed point on a straight path along which it moves, and where $U(t)$ is the acceleration furnished by the thrust of the vehicle's engine. In practice the engine system will be designed to operate within certain prescribed bounds so that the acceleration U will be constrained to satisfy a condition of the type (6.5.1). For example, a particular engine system might be limited by the condition

$$-M \leq U(t) \leq M \qquad \text{for} \quad t \geq t_0, \tag{6.5.2}$$

where M is a given positive constant which gives the magnitude of the largest possible thrust acceleration which the system is capable of achieving. The condition (6.5.2) allows both positive and *negative* thrust accelerations, as might be permitted during a docking operation by a vehicle with rocket engines located both fore and aft.

We can automatically account for the inequality constraint (6.5.2) by writing

$$U(t) = M \sin V(t)$$

for some suitable function V. It is clear that the constraint (6.5.2) will be satisfied by any function U of this last type for *any* piecewise continuous function V. In the general case of the inequality constraint (6.5.1) we can write

$$U(t) = \frac{\psi(t) + \varphi(t)}{2} + \frac{\psi(t) - \varphi(t)}{2} \sin V(t) \tag{6.5.3}$$

for some suitable function V, and again it is clear that (6.5.1) will automatically hold for any function U of this last type for any piecewise continuous function V.†

A Vehicular Docking Problem. By way of illustration we shall consider a docking problem for a space vehicle moving along a straight path in free space subject to the inequality constraint (6.5.2) on the rocket acceleration function U. Hence we have

$$X''(t) = M \sin V(t),$$

where the control function V is related to the rocket acceleration U by the equation $U = M \sin V$. We let $X(t)$ be the distance of the vehicle from its docking target, and we let a_1 and a_2 be the given initial position and speed of the vehicle at time $t_0 = 0$,

$$X(0) = a_1, \qquad X'(0) = a_2.$$

†If we were to consider a control problem with several different control functions and if several of these control functions were required to satisfy pointwise inequality constraints, we would then use several different relations of the type (6.5.3).

We seek to dock the vehicle in minimum time; i.e., we seek a thrust function that will minimize the final time t_1 while bringing the vehicle to rest at the origin, with

$$X(t_1) = 0, \qquad X'(t_1) = 0.$$

To put it yet another way, we seek to transfer the state of the system from $(X(0), X'(0)) = (a_1, a_2)$ to $(X(t_1), X'(t_1)) = (0, 0)$ in such a way as to minimize the time t_1.

We shall use X_1 and X_2 for the state functions, with $X_1(t) = X(t)$ and

$$\begin{aligned} X_1'(t) &= X_2(t) \\ X_2'(t) &= M \sin V(t) \qquad \text{for} \qquad t > 0. \end{aligned} \tag{6.5.4}$$

The initial conditions become $X_1(0) = a_1$ and $X_2(0) = a_2$, while the final conditions are $X_1(t_1) = 0$ and $X_2(t_1) = 0$.

The second equation of (6.5.4) can be integrated to give

$$X_2(t) = a_2 + M\int_0^t \sin V(\tau)\, d\tau,$$

where we have imposed the initial condition $X_2(0) = a_2$. This last equation can be inserted into the right-hand side of the first equation of (6.5.4), and the resulting equation can be integrated to give

$$X_1(t) = a_1 + a_2 t + M \int_0^t (t - \tau) \sin V(\tau)\, d\tau,$$

where we have imposed the remaining initial condition $X_1(0) = a_1$. It follows that the constraint $X_1(t_1) = 0$ becomes

$$a_1 + a_2 t_1 + M\int_0^{t_1} (t_1 - t) \sin V(t)\, dt = 0,$$

while the constraint $X_2(t_1) = 0$ becomes

$$a_2 + M \int_0^{t_1} \sin V(t)\, dt = 0.$$

The last two conditions must be satisfied by all admissible control functions V and terminal times t_1.

We take as our basic vector space $\mathfrak{X}$ the set of all pairs $(t_1, V) = (t_1, V(t))$, where t_1 may be any arbitrary number and where $V = V(t)$ may be any arbitrary piecewise continuous function defined on $\mathcal{R}$ which vanishes for all large $|t|$. We define addition of vectors in $\mathfrak{X}$ and multiplication of vectors by numbers as usual, and we use the norm given by

$$\|(t_1, V)\| = |t_1| + \max_{\text{all } t \text{ in } \mathcal{R}} |V(t)|.$$

We now let D be the open subset of $\mathfrak{X}$ consisting of all pairs (t_1, V) for which $t_1 > 0$, and we define functionals J, K_1, and K_2 on D by

$$J(t_1, V) = t_1$$

$$K_1(t_1, V) = a_1 + a_2 t_1 + M \int_0^{t_1} (t_1 - t) \sin V(t)\, dt$$

$$K_2(t_1, V) = a_2 + M \int_0^{t_1} \sin V(t)\, dt$$

for any vector (t_1, V) in D. The present extremum problem is to find a minimum vector in $D[K_1 = 0, K_2 = 0]$ for the functional J. We shall use the Euler-Lagrange multiplier theorem to study this problem.

If we use the same methods as were used in Section 6.2, we find in this case the following expressions for the variations of J, K_1, and K_2:

$$\delta J(t_1, V; \Delta t_1, \Delta V) = \Delta t_1$$

$$\delta K_1(t_1, V; \Delta t_1, \Delta V) = \Delta t_1 \left[a_2 + M \int_0^{t_1} \sin V(t)\, dt \right] + M \int_0^{t_1} (t_1 - t) \cos V(t)\, \Delta V(t)\, dt$$

and

$$\delta K_2(t_1, V; \Delta t_1, \Delta V) = \Delta t_1\, [M \sin V(t_1)] + M \int_0^{t_1} \cos V(t)\, \Delta V(t)\, dt$$

for any vector (t_1, V) in D and any vector $(\Delta t_1, \Delta V)$ in $\mathfrak{X}$.† It is easy to check that all the hypotheses of the Euler-Lagrange multiplier theorem are satisfied, and so if (t_1, V) is any local minimum vector in $D[K_1 = 0, K_2 = 0]$ for J, there must be constants λ_1 and λ_2 such that

$$\delta J(t_1, V; \Delta t_1, \Delta V) = \lambda_1\, \delta K_1(t_1, V; \Delta t_1, \Delta V) + \lambda_2\, \delta K_2(t_1, V; \Delta t_1, \Delta V)$$

holds for all vectors $(\Delta t_1, \Delta V)$ in $\mathfrak{X}$. If we use the previous expressions for the variations of J, K_1, and K_2, we can rewrite the last equation as

$$\Delta t_1 = \Delta t_1 \left\{ \lambda_1 \left[a_2 + M \int_0^{t_1} \sin V(t)\, dt \right] + \lambda_2 M \sin V(t_1) \right\} + M \int_0^{t_1} \left[\lambda_1 (t_1 - t) + \lambda_2 \right] \cos V(t)\, \Delta V(t)\, dt, \tag{6.5.5}$$

which must hold for all numbers Δt_1 and all piecewise continuous functions $\Delta V = \Delta V(t)$ if (t_1, V) is a minimum vector in $D[K_1 = 0, K_2 = 0]$ for J.

†Strictly speaking, the result given here for the variation of K_2 holds only if the function V is continuous at the number t_1. We shall see that the extremum pair (t_1, V) can always be taken to satisfy this condition.

If we choose first $\Delta t_1 = 0$ in (6.5.5) and take $\Delta V(t) = [\lambda_1(t_1 - t) + \lambda_2]$ $\cos V(t)$, we are led easily to

$$[\lambda_1(t_1 - t) + \lambda_2] \cos V(t) = 0 \qquad \text{for } 0 \leq t \leq t_1, \tag{6.5.6}$$

which must hold for any possible extremum vector (t_1, V). If we use (6.5.6) in (6.5.5) and take $\Delta t_1 = 1$, we also find the condition

$$1 = \lambda_1\left[a_2 + M \int_0^{t_1} \sin V(t)\,dt\right] + \lambda_2 M \sin V(t_1),$$

which must also hold for any extremum vector (t_1, V). We can use the constraint $X_2(t_1) = 0$ to rewrite the last condition as

$$\lambda_2 M \sin V(t_1) = 1. \tag{6.5.7}$$

It follows now from the two conditions (6.5.6) and (6.5.7) that any extremum function V must satisfy

$$\cos V(t) = 0 \tag{6.5.8}$$

for all t for which V is continuous. Indeed, the function $\lambda_1(t_1 - t) + \lambda_2$ can vanish for at most one value of t unless both λ_1 and λ_2 are zero, and the latter possibility is excluded by (6.5.7). Hence (6.5.6) implies (6.5.8).

We can now conclude from (6.5.8) that any piecewise continuous extremum control function V must satisfy (for each t) either the condition

$$\sin V(t) = +1$$

or the condition

$$\sin V(t) = -1$$

except possibly for a finite number of values of t at which $\sin V(t)$ may switch from the value $+1$ to the value -1 or vice versa. Therefore in the present case any possible (piecewise continuous) extremum rocket acceleration function $U = M \sin V$ must satisfy either $U(t) = +M$ or $U(t) = -M$ except possibly at a finite number of points where U switches from $+M$ to $-M$ or from $-M$ to $+M$. Hence if the rocket is to be transferred from its initial state to the specified final state in minimum time with a limited source of thrust ($|U| \leq M$), then *the engine should operate at full power at all times* except possibly for a finite number of switching times. Indeed, if some of the power were not being used, it is reasonable to expect that the transfer could be speeded up by using the additional power suitably.

Such a piecewise continuous control function U that switches from one extreme value to another is called a *bang-bang* control function. It can be shown that the solution of the present control problem with the values of U

restricted to the interval $-M \leq U \leq M$ is actually the same as the bang-bang solution to the analogous problem with the values of the control function restricted only to the two extreme values $+M$ and $-M$.† This is a special case of a more general result known as the *bang-bang principle*.‡ As a consequence of this result, in the present case the rocket designer need only provide an engine docking system that is capable of switching between the extreme thrust values, and this may be cheaper than providing a system that can produce accelerations within the full range $-M \leq U \leq M$.

We can gain further insight into the bang-bang solution of the present minimum time problem by sketching the path of the extremum state vector $\mathbf{X} = (X_1(t), X_2(t))$ in the (x_1, x_2)-plane. For this purpose it will be useful first to consider separately the two cases $U = +M$ and $U = -M$.

If $U = +M$ with $\sin V = +1$, we can integrate the resulting differential equations of (6.5.4) to find

$$X_1(t) = c_1 + c_2 t + M\frac{t^2}{2}$$

$$X_2(t) = c_2 + Mt$$

for suitable integration constants c_1 and c_2, from which we find that

$$x_1 = \frac{x_2^2}{2M} + c_1 - \frac{c_2^2}{2M} \tag{6.5.9}$$

for $x_1 = X_1(t)$ and $x_2 = X_2(t)$. Several of these parabolas (6.5.9) are shown in Figure 1 for several different choices of the constants c_1 and c_2. There will be a unique such parabola passing through each point in the (x_1, x_2)-plane. The arrows on the parabolas in Figure 1 indicate the motion of the system with increasing time. Indeed, since $X_2'(t) = +M$ in this case, the motion of the system will be in the positive x_2-direction, as shown. Moreover, $x_1 = X_1(t)$ will tend toward $+\infty$ along each path as $t \longrightarrow +\infty$, so that the rocket will eventually recede from the target as expected in the case of constant positive thrust $U = +M$.

Since only one of the parabolas (6.5.9) passes through the origin, it is clear that in general it is not possible to transfer the state of the system from its initial state directly to the origin along such an extremum path for which $U = +M$. This will only be possible if the initial state $(X_1(0), X_2(0)) =$

†See, for example, L. C. Young, *Lectures on the Calculus of Variations and Optimal Control Theory* (Philadelphia: W. B. Saunders Company, 1969), pp. 235–236. We indicate the proof of this result in a special case in Exercise 2 at the end of this section.

‡J. P. LaSalle, "The Time Optimal Control Problem," in *Contributions to the Theory of Nonlinear Oscillations* (L. Cesari, J. P. LaSalle, and S. Lefschetz, eds.), (Princeton, N.J.: Princeton University Press, 1960) Vol. 5. See also Hubert Halkin, "A Generalization of LaSalle's 'Bang-Bang' Principle," *Journal SIAM Control*, **2** (1965), 199–202.

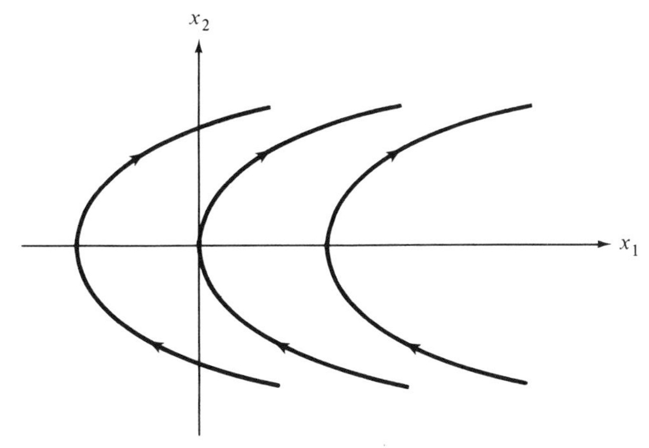

State paths for $U = +M$

Figure 1

(a_1, a_2) satisfies the condition $a_1 = a_2^2/2M$, with $a_2 < 0$, as the reader should be able to verify. (See Exercise 1.)

We turn now to the case $U = -M$ with $\sin V = -1$. Similarly, we find in this case that

$$x_1 = -\frac{x_2^2}{2M} + d_1 + \frac{d_2^2}{2M},$$

which must hold for $x_1 = X_1(t)$ and $x_2 = X_2(t)$ for suitable integration constants d_1 and d_2. We have illustrated several of these parabolas in Figure 2 for several different choices of the constants d_1 and d_2. In this case $X_2'(t) = -M$, and so the motion of the system is in the negative x_2-direction. Also,

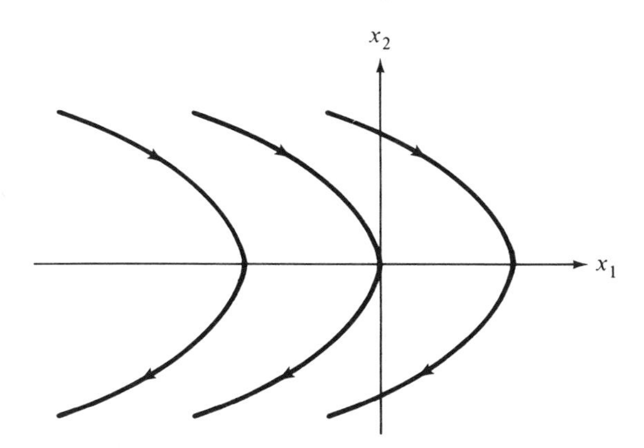

State paths for $U = -M$

Figure 2

$x_1 = X_1(t)$ tends toward $-\infty$ along each extremum path as $t \longrightarrow +\infty$, as expected in the case of constant negative thrust $U = -M$. In this case it will be possible to transfer the state of the system from its initial state $(X_1(0), X_2(0)) = (a_1, a_2)$ directly to the origin along an extremum path with $U = -M$ if and only if the initial state satisfies the condition $a_1 = -a_2^2/2M$ with $a_2 > 0$.

In general the extremum path which transfers the state of the system to the origin in minimum time will consist of an initial portion of one of the parabolas of type (6.5.9) or type (6.5.10), followed by a portion of the parabola of the other type passing through the origin. For example, if the initial point $(X_1(0), X_2(0)) = (a_1, a_2)$ lies *above* the *switching curve AOB* shown in Figure 3, then the extremum control U will be given initially as $U = -M$ so as to drive the system down along the parabola

$$x_1 = -\frac{x_2^2}{2M} + a_1 + \frac{a_2^2}{2M}$$

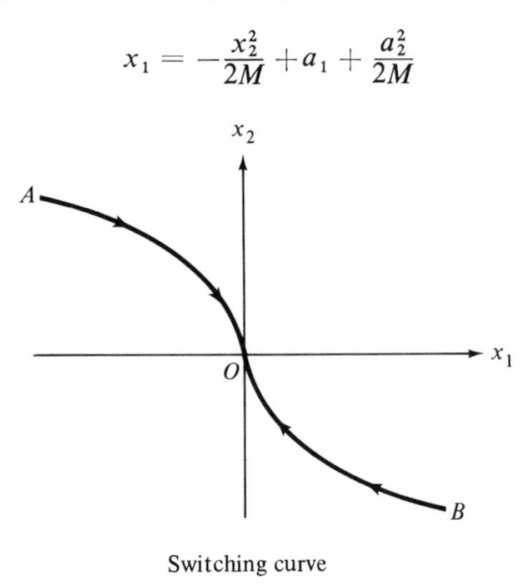

Switching curve

Figure 3

until the switching curve OB is reached. The extremum control function then switches to $U = +M$ so as to drive the system to the origin along the parabola $x_1 = x_2^2/2M$. The situation is illustrated in Figure 4. In the other case in which the initial point lies *below* the switching curve AOB, initially $U = +M$ will hold; the resulting situation is illustrated in Figure 5. Several different extremum paths starting at several different initial points are shown in Figure 6.

We refer the interested reader to Lee and Markus† and Young‡ for further details on this control problem and other similar bang-bang problems.

†E. B. Lee and L. Markus, *Foundations of Optimal Control Theory* (New York: John Wiley & Sons, Inc., 1967), pp. 4–14.

‡Young, *op. cit*., pp. 233–242.

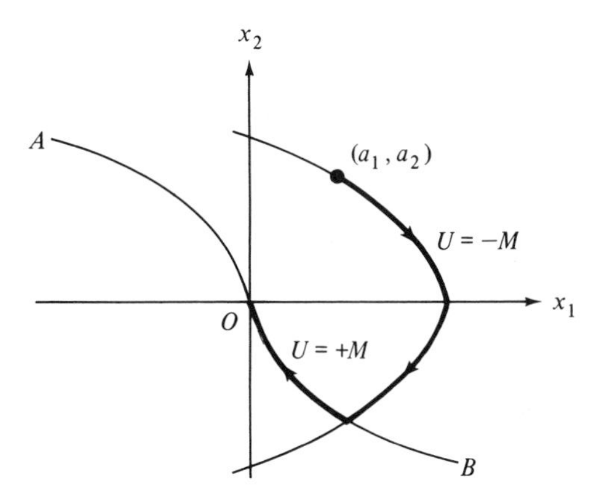

Figure 4

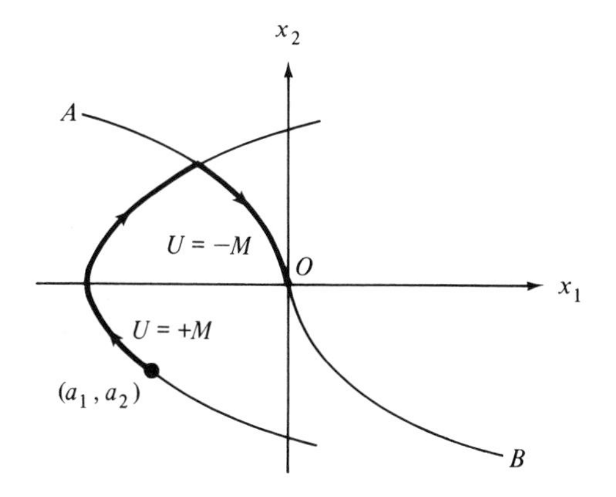

Figure 5

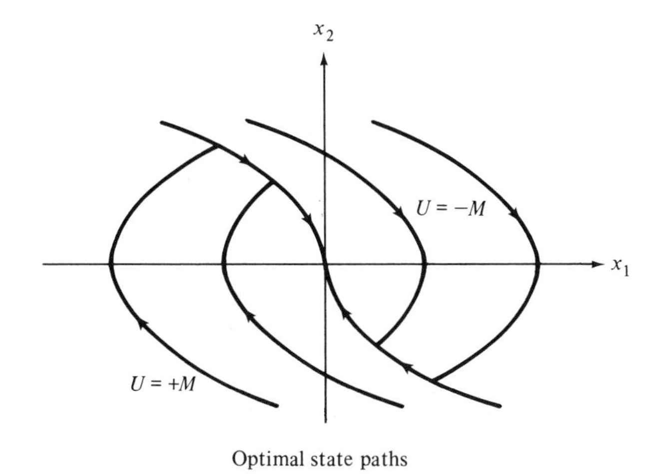

Optimal state paths

Figure 6

Exercises

1. Show that the state of the vehicle discussed above can be transferred directly to the origin using the control $U = +M$ if and only if the initial state $(X_1(0), X_2(0)) = (a_1, a_2)$ satisfies $a_1 = a_2^2/2M$ with $a_2 < 0$. Find the resulting optimal docking time in this case.

2. Show directly that the optimal docking time and thrust function found in Exercise 1 actually minimize the docking time in that case among all possible nonnegative docking times subject to the given constraints. *Hints:* Show that any admissible pair (t_1, U) must satisfy the condition $-a_2 = \int_0^{t_1} U(t)\,dt$. Use this result to obtain the inequality $|a_2| \leq Mt_1$ for any possible nonnegative time t_1 and any admissible U. You should be able to use this last inequality to prove the result $t_1 \geq t_1^*$, where t_1^* is the optimal docking time found in Exercise 1.

3. If the initial state of the vehicle discussed above satisfies $a_1 = a_2^2/2M$, as in Exercise 1, but with $a_2 = X_2(0) > 0$, show that the resulting minimum docking time is three times as great as that found in Exercise 1.

4. Find the shortest time in which the state of a space vehicle initially at rest can be transferred from $(X_1(0), X_2(0)) = (a_1, 0)$ to $(X_1(t_1), X_2(t_1)) = (0, 0)$ subject to the conditions $X_1'(t) = X_2(t)$, $X_2'(t) = U(t)$, and $|U(t)| \leq M$ for all t. Compute the resulting minimum transfer time in the special case in which the maximum permitted acceleration is taken to be $M = 32$ feet per square second (which is approximately the acceleration due to gravity at the earth's surface) and the initial distance to the target is taken to be $a_1 = 240{,}000$ miles (which is approximately the distance from the earth to the moon). What is the maximum speed of the vehicle relative to the target during this transfer from the initial to the final state?

5. Find a suitable control function U that will transfer the state of a vehicle from $(X_1(0), X_2(0)) = (\frac{7}{2}, 1)$ to $(X_1(t_1), X_2(t_1)) = (0, 0)$ in such a way as to minimize the transfer time t_1 subject to the conditions $X_1'(t) = X_2(t)$, $X_2'(t) = U(t)$, and $|U(t)| \leq 1$ for all t. Find the resulting minimum transfer time.

6. The state $X = X(t)$ of a certain system is governed by the differential equation $dX/dt = U$ for $0 < t < 1$. Find a control function U_1 that will transfer the state of the system from $X(0) = x_0$ to $X(1) = x_1$ in such a way as to *maximize* the functional $J(U) = \int_0^1 U(t)^2\,dt$ subject to the inequality constraint $-1 \leq U(t) \leq +1$ for all t. Similarly, find a control function U_2 that will transfer the state of the system from $X(0) = x_0$ to $X(1) = x_1$ so as to *minimize* the same functional J subject to the same inequality constraint. Verify directly the result $J(U_2) \leq J(U) \leq J(U_1)$ for all admissible piecewise continuous control functions U. The given fixed numbers x_0 and x_1 are assumed to satisfy $|x_1 - x_0| \leq 1$. *Hints:* Use (6.5.3) to account for the inequality constraint, and then apply the result of Exercise 2 of Section 6.4 to obtain candidates for U_1 and U_2. (U_1 will be a bang-bang control.) You should be able to use Schwarz's inequality (1.4.7) to prove that $J(U_2) \leq J(U)$ for all admissible U, while $J(U) \leq J(U_1)$ should be evident for all admissible U once you calculate $J(U_1)$.

6.6 Some Remarks on the Maximum Principle and Dynamic Programming

In the previous sections we have seen how to solve various kinds of optimal control problems using the Euler-Lagrange multiplier theorem. And, indeed, there are many other classes of control problems not discussed here that can also be solved with this same multiplier method. (Strictly speaking, of course, the multiplier theorem only provides *candidates* for the solutions of the various problems.) These same problems and other more difficult extremum problems are often solved today using *Pontryagin's maximum principle*.† This principle is in a certain sense a generalization of the Euler-Lagrange multiplier theorem in which the multipliers are allowed to vary as functions of time. The maximum principle yields a certain coupled system of ordinary differential equations which involves both the extremum control functions and the time-varying multipliers and which can often be used to determine the optimum control functions. In certain bang-bang control problems the maximum principle provides direct information about the optimum number of switching times. The maximum principle is well suited to handle problems involving a wide range of different types of constraints, and it often provides certain limited information about the nature of the optimum control functions even when the given problem is too difficult to be solved completely in any simple form. In such cases in which the extremum problem cannot be completely solved in any simple form, recourse must be had to numerical methods which can be used to obtain approximations to the solutions of the given differential equations. Unfortunately the resulting numerical problems are themselves quite formidable.

Another method that can be used to solve extremum problems is *Bellman's method of dynamic programming*.‡ This approach leads to a certain *partial* differential equation which must be solved to effect the solution of the control problem, and again this partial differential equation must be solved numerically in many cases.

We shall not pursue these more powerful methods in this book, although we do encourage the interested reader to refer to the references cited. It is certainly true that these more powerful methods are needed in certain cases to

†L. S. Pontryagin, V. G. Boltyanskii, R. V. Gamkrelidze, and E. F. Mishchenko, *The Mathematical Theory of Optimal Processes*, translated and edited by K. N. Trirogoff and L. W. Neustadt [New York: John Wiley & Sons., Inc. (Interscience Division) 1962]. See also Hubert Halkin, "Mathematical Foundations of Systems Optimization," in *Topics in Optimization*, ed. George Leitman (New York: Academic Press, Inc., 1967), and Henry Hermes and Joseph P. LaSalle, *Functional Analysis and Time Optimal Control* (New York: Academic Press, Inc., 1969).

‡R. Bellman, *Dynamic Programming* (Princeton, N.J.: Princeton University Press, 1959).

solve more difficult extremum problems. However, it is also true, as we have seen, that *the use of the Euler-Lagrange multiplier theorem alone is capable of leading to the solutions of a strikingly wide range of extremum problems.* Moreover, the underlying theory behind the Euler-Lagrange multiplier theorem is substantially simpler than the corresponding theories behind the maximum principle and dynamic programming. On this basis alone the Euler-Lagrange multiplier method may be preferred for those problems to which it applies.

7. *The Variational Description of Sturm-Liouville Eigenvalues*

In this chapter we shall show how to use variational procedures to study the eigenvalues of certain Sturm-Liouville problems which arise in practice. We shall show that each eigenvalue is an extreme value of the Rayleigh quotient subject to certain constraints, and we shall discuss how to use the Rayleigh-Ritz method to find approximate values for these eigenvalues. We shall also discuss the Courant minimax principle and several of its important implications. Finally, we shall give a brief general discussion of the use of the Ritz method as a direct method for the approximate minimization or maximization of functionals.

7.1. Introduction to Sturm-Liouville Problems

Many physical problems, including problems in heat conduction, fluid dynamics, elasticity, electromagnetism, and quantum mechanics, lead to various partial differential equations which must be solved subject to certain auxiliary conditions such as boundary conditions or initial conditions. The resulting partial differential equations are often solved in practice using D'Alembert's method of *separation of variables*, which nearly always leads to an ordinary differential equation of the type

$$\frac{d}{dx}\left[\tau(x)\frac{dW}{dx}\right] + q(x)W = -\lambda\rho(x)W, \tag{7.1.1}$$

which must then be solved for a function $W = W(x)$ for $x_0 < x < x_1$ subject to certain specified end-point conditions at $x = x_0$ and $x = x_1$. The functions $\tau = \tau(x)$, $q = q(x)$, and $\rho = \rho(x)$ are given known functions in each particular case, and the number λ is a parameter which will be discussed below. Many of the important ordinary differential equations of mathematical physics such as Legendre's equation, Hermite's equation, and Bessel's equation are of the type (7.1.1) and arise in just this manner. Certain special values of the parameter λ, called *eigenvalues*, are of great importance in such problems, and we shall see that these eigenvalues can be obtained as the solutions of certain extremum problems. This result is useful since it leads to practical methods of calculating the eigenvalues, as we shall see later in this chapter.

As an illustration of the way in which equations such as (7.1.1) arise in practice, we shall consider briefly the problem of solving the vibrating string equation (see Section 4.11)

$$\rho(x)\frac{\partial^2 Z}{\partial t^2} = \frac{\partial}{\partial x}\left[\tau(x)\frac{\partial Z}{\partial x}\right] \qquad \text{for } 0 \leq x \leq l,\ t \geq 0, \tag{7.1.2}$$

subject to the fixed end-point conditions

$$Z(0, t) = 0, \qquad Z(l, t) = 0 \qquad \text{for } t \geq 0 \tag{7.1.3}$$

and the initial conditions

$$Z(x, 0) = \phi(x), \qquad \frac{\partial}{\partial t}Z(x, 0) = \psi(x) \qquad \text{for } 0 \leq x \leq l. \tag{7.1.4}$$

The given functions $\phi = \phi(x)$ and $\psi = \psi(x)$ represent the prescribed initial shape and initial speed of the string, which are specified in advance. The material properties of the string are allowed to vary with position and are specified through the given nonnegative functions $\rho = \rho(x)$ and $\tau = \tau(x)$ which represent the linear mass density and tension of the string. The problem is to find the particular displacement function $Z = Z(x, t)$ which satisfies the partial differential equation (7.1.2) along with the auxiliary conditions (7.1.3) and (7.1.4).

We shall indicate briefly how this problem can be solved using D'Alembert's separation method.† The idea is first to seek special product solutions of the form $Z(x, t) = X(x)T(t)$ for suitable functions $X = X(x)$ and $T = T(t)$, and then later these special solutions will be combined in a certain way so as to obtain the particular solution actually sought.

If we insert $Z = XT$ into the partial differential equation (7.1.2), we find after some manipulation that

$$\frac{1}{T(t)}\frac{d^2T(t)}{dt^2} = \frac{1}{\rho(x)X(x)}\frac{d}{dx}\left[\tau(x)\frac{dX(x)}{dx}\right],$$

which must hold for $0 < x < l$ and for $t > 0$. Since the left-hand side of this equation is a function of t alone, while the right-hand side is a function of x alone, it follows that this equation can hold only if both sides are *constant*. Hence

$$\frac{1}{T(t)}\frac{d^2T(t)}{dt^2} = \frac{1}{\rho(x)X(x)}\frac{d}{dx}\left[\tau(x)\frac{dX(x)}{dx}\right] = -\lambda$$

must hold for some constant λ, so that we find the two ordinary differential equations

$$\frac{d^2T(t)}{dt^2} = -\lambda T(t) \qquad \text{for } t > 0 \tag{7.1.5}$$

and

$$\frac{d}{dx}\left[\tau(x)\frac{dX(x)}{dx}\right] = -\lambda\rho(x)X(x) \qquad \text{for } 0 < x < l. \tag{7.1.6}$$

Hence D'Alembert's method reduces the study of the original partial differential equation to the study of certain ordinary differential equations.

The product function $Z = XT$ is now required to satisfy the fixed end-point conditions (7.1.3), and this leads to the following boundary conditions for X:

$$X(0) = 0, \qquad X(l) = 0. \tag{7.1.7}$$

Hence we must solve the differential equation (7.1.6) for X subject to these boundary conditions at $x = 0$ and $x = l$. In the present case we also impose the additional condition

$$\int_0^l \rho(x)X(x)^2\,dx > 0, \tag{7.1.8}$$

†D'Alembert introduced his method in 1750. See p. 241 of C. Truesdell, "The Rational Mechanics of Flexible or Elastic Bodies, 1638–1788," L. Euleri *Opera Omnia*, (2) **11**$_2$ (Zurich: Fussli, 1960).

which serves to eliminate the uninteresting trivial solution given by the zero function $X = 0$.

The differential equation (7.1.6) is a special case of equation (7.1.1) with $q = 0$, and indeed the present *boundary value problem* for (7.1.6) is an example of just the sort of problem that often arises in this way for various different equations of the type (7.1.1). Such boundary value problems are often called *Sturm-Liouville problems* after the two men who began the systematic study of these problems in 1836–1837.

It can be shown under quite general assumptions that *these Sturm-Liouville problems have nontrivial solutions which satisfy* (7.1.8) *only for certain special values of the numerical parameter* λ occurring in (7.1.1). For example, if we consider the above vibrating string problem for a string of *constant* tension and density, then equation (7.1.6) becomes

$$X'' = -\frac{\lambda}{c^2}X \qquad \left(c^2 = \frac{\tau}{\rho}\right)$$

with general solution given as†

$$X(x) = A \sin \sqrt{\lambda}\,\frac{x}{c} + B \cos \sqrt{\lambda}\,\frac{x}{c}$$

for arbitrary constants of integration A and B. The boundary condition $X(0) = 0$ implies that $B = 0$, and then the remaining boundary condition $X(l) = 0$ implies that

$$A \sin \sqrt{\lambda}\,\frac{l}{c} = 0,$$

which will hold if either $A = 0$ or $\sqrt{\lambda}\, l/c = n\pi$ for some integer n. The condition $A = 0$ yields only the zero solution for X, and so to get nontrivial solutions we must require that $A \neq 0$. This leads to a *sequence* of admissible values for λ, namely $\lambda_1, \lambda_2, \lambda_3, \ldots$, with

$$\lambda_n = \left(\frac{n\pi c}{l}\right)^2 = \left(\frac{n\pi}{l}\right)^2 \frac{\tau}{\rho} \qquad \text{for } n = 1, 2, 3, \ldots. \tag{7.1.9}$$

Corresponding to the value $\lambda = \lambda_n$, we find the nontrivial solution $X = X_n$ given as

$$X_n(x) = \sin \frac{n\pi x}{l} \tag{7.1.10}$$

or any (nonzero) constant multiple of this solution. These special values of

†See, for example, Chapter 20 of George B. Thomas, Jr., *Calculus and Analytic Geometry* (Reading, Mass.: Addison-Wesley Publishing Company, Inc., 1968), or any elementary text on differential equations. We leave as an exercise for the reader the verification that λ must be nonnegative.

λ for which the given Sturm-Liouville problem has nontrivial solutions are called *eigenvalues*, and the corresponding nontrivial solutions are called *eigenfunctions*. It is clear from this example that *the boundary conditions play an important role in determining the eigenvalues.*

If we insert the nth eigenvalue $\lambda = \lambda_n$ back into the remaining equation (7.1.5), we find the most general solution $T = T_n$ of the resulting equation to be given as

$$\begin{aligned} T_n(t) &= a_n \cos \sqrt{\lambda_n} t + b_n \sin \sqrt{\lambda_n} t \\ &= a_n \cos \frac{n\pi ct}{l} + b_n \sin \frac{n\pi ct}{l} \end{aligned}$$

for arbitrary constants of integration a_n and b_n. The resulting product solution $Z = X_n T_n$ becomes

$$X_n(x)T_n(t) = \sin \frac{n\pi x}{l}\left[a_n \cos \frac{n\pi ct}{l} + b_n \sin \frac{n\pi ct}{l}\right].$$

Of course, this special solution will in general *not* satisfy the prescribed initial conditions of (7.1.4). However, it is possible to form *sums* of these special solutions, say

$$\sum_{n=1}^{N} \sin \frac{n\pi x}{l}\left[a_n \cos \frac{n\pi ct}{l} + b_n \sin \frac{n\pi ct}{l}\right],$$

and such sums still satisfy the original partial differential equation (7.1.2) along with the fixed end-point conditions (7.1.3) *for any positive integer N and for any choice of the constants $a_1, b_1, a_2, b_2, a_3, \ldots, a_{N-1}, b_{N-1}, a_N, b_N$.* It can now be shown under quite general conditions that *it is possible to satisfy the specified initial conditions by taking infinitely many terms in the sum while at the same time making suitable special choices for the constants a_n and b_n.*† The resulting displacement function for the vibrating string has the form

$$Z(x, t) = \sum_{n=1}^{\infty} \sin \frac{n\pi x}{l}\left[a_n \cos \frac{n\pi ct}{l} + b_n \sin \frac{n\pi ct}{l}\right],$$

with the constants a_n and b_n defined as

$$\begin{aligned} a_n &= \frac{2}{l}\int_0^l \phi(x) \sin \frac{n\pi x}{l}\, dx \\ b_n &= \frac{2}{nc\pi}\int_0^l \psi(x) \sin \frac{n\pi x}{l}\, dx \end{aligned}$$

†See Hans Sagan, *Boundary and Eigenvalue Problems in Mathematical Physics* (New York: John Wiley & Sons, Inc., 1961), pp. 37–42 and 132–143.

and with $c = \sqrt{\tau/\rho}$. We refer the interested reader to the book by Sagan for a more complete discussion of this procedure.

The nth special product solution $Z_n = X_n(x)T_n(t)$ found above represents the nth *fundamental vibration* for the given string. The eigenfunction $X_n(x) = \sin(n\pi x/l)$ gives the shape or form of this fundamental vibration at each instant t, while the function $T_n(t) = a_n \cos\sqrt{\lambda_n}t + b_n \sin\sqrt{\lambda_n}t$ appearing in the product $Z_n = X_nT_n$ causes each point of the string to vary periodically with time with a period equal to $2\pi/\sqrt{\lambda_n}$,

$$\frac{2\pi}{\sqrt{\lambda_n}} = \frac{2l}{n}\sqrt{\frac{\rho}{\tau}} = n\text{th fundamental period.}$$

This *period* is the time required for one complete vibration to occur. The reciprocal of the period is called the *frequency* of the vibration and gives the number of complete vibrations occurring per unit time. In this case we find that

$$\frac{\sqrt{\lambda_n}}{2\pi} = \frac{n}{2l}\sqrt{\frac{\tau}{\rho}} = n\text{th fundamental frequency.}$$

The frequency of the vibration is often taken to be an indication of the pitch or tone of the resulting sound. For example, the lowest fundamental tone corresponds to the frequency $\sqrt{\lambda_1}/2\pi = (2l)^{-1}\sqrt{\tau/\rho}$, while higher fundamental frequencies correspond to higher fundamental tones.

7.2. The Relation Between the Lowest Eigenvalue and the Rayleigh Quotient

We shall now consider the Sturm-Liouville problem for the following general equation [see equation (7.1.1)]:

$$\frac{d}{dx}\left[\tau(x)\frac{dW}{dx}\right] + q(x)W = -\lambda\rho(x)W \qquad \text{for } x_0 < x < x_1.$$

We assume that the given functions τ and ρ are *positive-valued*, as is generally the case in actual applications, and we assume that both ρ and q are continuous while τ is continuously differentiable. Moreover, for definiteness and simplicity we shall only consider the boundary conditions [compare with (7.1.7)]

$$W(x_0) = 0 \qquad \text{and} \qquad W(x_1) = 0, \tag{7.2.1}$$

although other boundary conditions also arise and can be handled similarly.†

It can be shown‡ that this Sturm-Liouville problem always has infinitely

†See Sagan, *op. cit.*, pp. 253–260.

‡*Ibid.*, pp. 168–175.

many eigenvalues which can be arranged in an increasing sequence as [compare with (7.1.9)]

$$\lambda_1 < \lambda_2 < \lambda_3 < \cdots < \lambda_{n-1} < \lambda_n < \cdots, \tag{7.2.2}$$

with

$$\operatorname*{limit}_{k \to \infty} \lambda_k = +\infty.$$

Corresponding to each eigenvalue $\lambda = \lambda_n$ there is an eigenfunction $W = W_n(x)$ which is a nontrivial solution to the given boundary value problem. Of course, any constant (nonzero) multiple of W_n is also an eigenfunction corresponding to the same eigenvalue λ_n. The converse is also true: Any two eigenfunctions which correspond to the same eigenvalue must be constant multiples of each other. In various physical applications the eigenfunctions often represent certain distinguished or special states of the given system, while the eigenvalues are related to the energy levels of these special states.

Hence the results in the present general case are quite similar to those in the previous special case involving the vibrating string in which the eigenvalues and eigenfunctions were given by (7.1.9) and (7.1.10). There is, however, one important difference. In the general case *it is usually not possible to find any simple expressions for the eigenvalues and/or eigenfunctions.* In such cases it is often important in practice that one be able to find *approximate* expressions at least for the eigenvalues, and for this purpose it is fortunate that each eigenvalue satisfies a certain extremum property, which we shall now begin to describe.

For this purpose we introduce two functionals $D = D(W)$ and $H = H(W)$ defined by

$$D(W) = \int_{x_0}^{x_1} [\tau(x)W'(x)^2 - q(x)W(x)^2]\, dx$$

and

$$H(W) = \int_{x_0}^{x_1} \rho(x)W(x)^2\, dx$$

for any vector W in the vector space $\mathfrak{X}$, where $\mathfrak{X}$ consists of all continuously differentiable functions on the interval $[x_0, x_1]$ *which vanish at the end points* [see (7.2.1)]. We shall also need to consider the *quotient* of D and H, denoted as R,

$$R(W) = \frac{D(W)}{H(W)}$$

for any nonzero vector W in $\mathfrak{X}$. The denominator $H(W)$ in this quotient is always positive for nonzero functions W since we are assuming that the given function ρ is positive-valued.

The quotient $R = D/H$ is called the *Rayleigh quotient* after the English mathematical physicist John William Strutt (1842–1919), who became the third Lord Rayleigh in 1873. Beginning in 1870, Strutt showed how to use a certain variational principle to calculate approximate values for the lowest (smallest) eigenvalues of certain special Sturm-Liouville problems arising in the fields of acoustics and elasticity. The method of Strutt is based on the fact that *the lowest eigenvalue* $\lambda = \lambda_1$ *is actually equal to the minimum value of the quotient* $R = D/H$.

Indeed, if W_1 is a minimum vector in $\mathfrak{X}$ for the functional $R = R(W)$, then the basic theorem on the vanishing of the variation at an extremum vector implies that

$$\delta R(W_1; \Delta W) = 0$$

for all vectors ΔW in $\mathfrak{X}$. We can use the formula $R = D/H$ along with Exercise 12 of Section 2.3 to find the variation of R as

$$\delta R(W; \Delta W) = \frac{1}{H(W)}\{\delta D(W; \Delta W) - R(W)\,\delta H(W; \Delta W)\}, \tag{7.2.3}$$

so that any such minimum vector W_1 must satisfy

$$\delta D(W_1; \Delta W) = R(W_1)\,\delta H(W_1; \Delta W)$$

for all vectors ΔW in $\mathfrak{X}$. On the other hand, we easily find from (2.3.4) and the definitions of D and H that

$$\begin{aligned}\delta D(W; \Delta W) &= 2\int_{x_0}^{x_1} [\tau(x)W'(x)\,\Delta W'(x) - q(x)W(x)\,\Delta W(x)]\,dx \\ &= -2\int_{x_0}^{x_1} \left\{\frac{d}{dx}\left[\tau(x)\frac{dW(x)}{dx}\right] + q(x)W(x)\right\}\Delta W(x)\,dx\end{aligned}$$

and

$$\delta H(W; \Delta W) = 2\int_{x_0}^{x_1} p(x)W(x)\,\Delta W(x)\,dx$$

for any function ΔW in $\mathfrak{X}$. (We integrated by parts to obtain the final expression for δD.) Hence we finally obtain for the minimum vector W_1

$$\int_{x_0}^{x_1} \left\{\frac{d}{dx}\left[\tau(x)\frac{dW_1(x)}{dx}\right] + q(x)W_1(x) + \lambda^* p(x)W_1(x)\right\}\Delta W(x)\,dx = 0$$

for all continuously differentiable functions ΔW which vanish at x_0 and x_1, where we have set

$$\lambda^* = R(W_1) = \frac{D(W_1)}{H(W_1)}.$$

It follows now from this last result and the lemma of Section A4 of the

Appendix that W_1 must satisfy the following differential equation:

$$\frac{d}{dx}\left[\tau(x)\frac{dW_1(x)}{dx}\right] + q(x)W_1(x) = -\lambda^* \rho(x)W_1(x)$$

for $x_0 < x < x_1$. (This is just the Euler-Lagrange equation for the given extremum problem.) Hence $\lambda^* = R(W_1)$ is an eigenvalue for the original Sturm-Liouville problem, and W_1 is the corresponding eigenfunction.† Moreover, since W_1 is a *minimum* vector in $\mathfrak{X}$ for $R = R(W)$, we can conclude that $\lambda^* = R(W_1)$ must be the *lowest* (smallest) eigenvalue appearing in (7.2.2), that is,

$$\lambda^* = \lambda_1.$$

Indeed, if λ is *any* eigenvalue with corresponding eigenfunction W, then we can multiply both sides of the differential equation

$$\frac{d}{dx}\left[\tau(x)\frac{dW}{dx}\right] + q(x)W = -\lambda\rho(x)W$$

by W and integrate the resulting equation over $x_0 < x < x_1$ to find after a brief calculation that

$$\lambda = R(W).$$

Since we have assumed all along that W_1 is a minimum vector for R, with $R(W_1) \leq R(W)$, it now follows from these results that $\lambda^* \leq \lambda$ for any eigenvalue λ. Hence λ^* is the lowest eigenvalue, and since λ^* is also the minimum value of the Rayleigh quotient, we conclude that *the lowest eigenvalue λ_1 is equal to the minimum value of the Rayleigh quotient.*‡ We call this result *Rayleigh's principle.*

We can use Rayleigh's principle to obtain certain qualitative results concerning the lowest fundamental tone or frequency of a vibrating string of variable density $\rho = \rho(x)$ and variable tension $\tau = \tau(x)$. In particular, it follows from Rayleigh's principle that the lowest fundamental frequency $\sqrt{\lambda_1}/2\pi$ will increase if the Rayleigh quotient $R(W)$ is made to increase for every function W, while this frequency will decrease if the Rayleigh quotient decreases. Thus if the tension $\tau(x)$ increases or if the density $\rho(x)$ decreases, then the lowest fundamental tone must increase, while if the tension decreases or the density increases, then the lowest tone must decrease. In Section 7.6 we shall see that the same results are actually true for *all* the fundamental tones, including the higher tones.

†The function W_1 is only determined up to an arbitrary multiplicative constant since any nonzero constant multiple of any such function W_1 will give the same minimum value λ^* for the Rayleigh quotient and will also be an eigenfunction corresponding to λ^*. It is unimportant here how this multiplicative constant is chosen.

‡We have used all along here without proof the known fact that the Rayleigh quotient is bounded below and actually achieves a minimum value on $\mathfrak{X}$.

Exercises

1. Use Rayleigh's principle to obtain $(1/2l)\sqrt{\tau_{\min}/\rho_{\max}} \leq \sqrt{\lambda_1}/2\pi \leq (1/2l)\sqrt{\tau_{\max}/\rho_{\min}}$ for the lowest fundamental frequency $\sqrt{\lambda_1}/2\pi$ of a vibrating string of length l, tension $\tau = \tau(x)$, and density $\rho = \rho(x)$ (for $0 \leq x \leq l$), where we have set $\tau_{\max} = \max_{0 \leq x \leq l} \tau(x)$, $\tau_{\min} = \min_{0 \leq x \leq l} \tau(x)$ and similarly for $\rho_{\max}$ and $\rho_{\min}$. We are assuming that both $\tau_{\min}$ and $\rho_{\min}$ are positive.

2. Show that the lowest eigenvalue λ_1 for the Sturm-Liouville problem $W''(x) - xW(x) = -\lambda W(x)$ for $0 < x < 1$ and $W(0) = 0$, $W(1) = 0$ satisfies the inequalities $9.87 \leq \lambda_1 \leq 10.87$. *Hints:* Derive

$$\frac{\int_0^1 W'(x)^2\,dx}{\int_0^1 W(x)^2\,dx} \leq \frac{\int_0^1 [W'(x)^2 + xW(x)^2]\,dx}{\int_0^1 W(x)^2\,dx} \leq \frac{\int_0^1 W'(x)^2\,dx}{\int_0^1 W(x)^2\,dx} + 1$$

for any nonzero smooth function W, and then use Rayleigh's principle to conclude the result $\mu_1 \leq \lambda_1 \leq \mu_1 + 1$, were μ_1 is the lowest eigenvalue for the problem $W''(x) = -\mu W(x)$ for $0 < x < 1$ and $W(0) = 0$, $W(1) = 0$. The stated result should follow if you calculate μ_1.

7.3. The Rayleigh-Ritz Method for the Lowest Eigenvalue

The exercises at the end of Section 7.2 indicate one way in which Rayleigh's principle can be used to calculate approximate values for the lowest eigenvalue of a given Sturm-Liouville problem. In this section we shall describe another method which is also based on Rayleigh's principle and which yields upper bounds on the lowest eigenvalue.

We shall consider the following Sturm-Liouville problem:

$$\frac{d}{dx}\left[\tau(x)\frac{dW}{dx}\right] + q(x)W = -\lambda\rho(x)W \qquad \text{for } x_0 < x < x_1$$

with $W(x_0) = 0$ and $W(x_1) = 0$. The general procedure we shall describe is due to Walter Ritz (1878–1909), who developed this method in 1908 and 1909. Today the general procedure is called the *Rayleigh-Ritz method.*

The idea of Ritz was to replace the problem of minimizing the quotient D/H over the entire vector space $\mathfrak{X}$ by the simpler problem of minimizing D/H over some small subset of $\mathfrak{X}$. In particular, Ritz considered the problem of minimizing D/H over all functions W in the *subspace* $\mathfrak{M}_n$ *spanned by a given collection of fixed functions* $\psi_1, \psi_2, \ldots, \psi_n$ taken from $\mathfrak{X}$. Here $\mathfrak{M}_n$ is the subspace which consists of all functions W of the form

$$W = c_1\psi_1 + c_2\psi_2 + \cdots + c_n\psi_n = \sum_{i=1}^{n} c_i\psi_i$$

for arbitrary constants $c_1, c_2, \ldots, c_n$, where $\psi_1, \psi_2, \ldots, \psi_n$ are any given *fixed* functions in $\mathfrak{X}$. The latter problem of minimizing the Rayleigh quotient $R = D/H$ over $\mathfrak{M}_n$ involves only a suitable choice for the n constants c_1, $c_2, \ldots, c_n$ and is a much simpler problem than that of minimizing R over the entire vector space $\mathfrak{X}$.

To solve this simpler problem, it is convenient to combine the n numbers $c_1, c_2, \ldots, c_n$ into a single n-tuple $\mathbf{c} = (c_1, c_2, \ldots, c_n)$, and it is also useful to define two real-valued functions $d = d(\mathbf{c})$ and $h = h(\mathbf{c})$ on $\mathfrak{R}_n$ by

$$d(\mathbf{c}) = D\left(\sum_{i=1}^{n} c_i \psi_i\right)$$
$$h(\mathbf{c}) = H\left(\sum_{i=1}^{n} c_i \psi_i\right)$$

for any n-tuple $\mathbf{c} = (c_1, c_2, \ldots, c_n)$ in $\mathfrak{R}_n$, where D and H are the same functionals which occur in the Rayleigh quotient. Then we have the obvious identity

$$R(W) = \frac{d(\mathbf{c})}{h(\mathbf{c})}$$

for any nonzero function W of the form $W = \sum_{i=1}^{n} c_i \psi_i$, and so the problem of minimizing the Rayleigh quotient over the subspace $\mathfrak{M}_n$ is equivalent to the problem of minimizing the ratio $d(\mathbf{c})/h(\mathbf{c})$ over $\mathfrak{R}_n$. It is the latter problem which we now consider. (It is not difficult to prove that d/h actually achieves a minimum value on $\mathfrak{R}_n$, but we shall simply use this result without proof in the following discussion.)

If $\mathbf{c}^* = (c_1^*, c_2^*, \ldots, c_n^*)$ is a minimum vector in $\mathfrak{R}_n$ for the ratio $d(\mathbf{c})/h(\mathbf{c})$, then the first-order partial derivatives of the function d/h must vanish at $\mathbf{c}^*$, with

$$\frac{\partial}{\partial c_k}\left[\frac{d(\mathbf{c})}{h(\mathbf{c})}\right] = 0 \qquad \text{for } k = 1, 2, \ldots, n \qquad \text{at } \mathbf{c} = \mathbf{c}^*,$$

from which we find that

$$\frac{\partial}{\partial c_k} d(\mathbf{c}^*) = r \frac{\partial}{\partial c_k} h(\mathbf{c}^*) \qquad \text{for } k = 1, 2, \ldots, n, \tag{7.3.1}$$

where we have set $r = d(\mathbf{c}^*)/h(c^*)$. We shall see that the n equations of (7.3.1) can be used to determine the desired minimum vector $\mathbf{c}^*$ in $\mathfrak{R}_n$. (Of course, the vector $\mathbf{c}^*$ is not unique. Any constant multiple of $\mathbf{c}^*$ will also be a minimum vector in $\mathfrak{R}_n$ for d/h.)

We can use the definitions of the functions $d(\mathbf{c})$ and $h(c)$ to find

$$\begin{aligned} d(\mathbf{c}) &= \sum_{i,j=1}^{n} a_{ij} c_i c_j \\ h(\mathbf{c}) &= \sum_{i,j=1}^{n} b_{ij} c_i c_j \end{aligned} \tag{7.3.2}$$

for any n-tuple $\mathbf{c} = (c_1, c_2, \ldots, c_n)$, where the fixed numbers a_{ij} and b_{ij} are defined by

$$a_{ij} = \int_{x_0}^{x_1} [\tau(x)\psi_i'(x)\psi_j'(x) - q(x)\psi_i(x)\psi_j(x)]\,dx$$
$$b_{ij} = \int_{x_0}^{x_1} \rho(x)\psi_i(x)\psi_j(x)\,dx \qquad \text{for } i, j = 1, 2, \ldots, n. \tag{7.3.3}$$

We can now use (7.3.2) along with the obvious symmetry properties $a_{ij} = a_{ji}$ and $b_{ij} = b_{ji}$ to calculate the required partial derivatives as

$$\frac{\partial}{\partial c_k} d(\mathbf{c}) = 2 \sum_{j=1}^{n} a_{kj} c_j$$
$$\frac{\partial}{\partial c_k} h(\mathbf{c}) = 2 \sum_{j=1}^{n} b_{kj} c_j.$$

If we insert these results into (7.3.1), we find that

$$\sum_{j=1}^{n} a_{kj} c_j^* = r \sum_{j=1}^{n} b_{kj} c_j^* \qquad \text{for } k = 1, 2, \ldots, n, \tag{7.3.4}$$

from which we hope to determine the desired minimum vector $\mathbf{c}^*$. The latter system of equations can be rewritten in matrix form as

$$A\mathbf{c}^* = rB\mathbf{c}^*, \tag{7.3.5}$$

where A and B are the $n \times n$ matrices given as $A = (a_{ij})$ and $B = (b_{ij})$. The expression $A\mathbf{c}^*$ denotes the usual *matrix product* of the matrix A and the vector $\mathbf{c}^*$, and similarly for the product $B\mathbf{c}^*$. For example, the ith component of the resulting vector $A\mathbf{c}^*$ in $\mathcal{R}_n$ is given as $\sum_{j=1}^{n} a_{ij} c_j^*$.

If $\mathbf{c}^*$ is any nonzero solution vector in $\mathcal{R}_n$ for (7.3.5) for any constant r, then necessarily

$$r = \frac{d(\mathbf{c}^*)}{h(\mathbf{c}^*)}$$

will hold, as follows by multiplying (7.3.4) by c_k^* and summing over k. We also note that the zero vector $\mathbf{c}^* = 0$ clearly satisfies (7.3.5), but, of course, we are seeking *nonzero* solutions in $\mathcal{R}_n$. Any such nonzero solution vector $\mathbf{c}^*$ for (7.3.5) is called an *eigenvector* in $\mathcal{R}_n$ for the matrix A relative to the matrix B, and the corresponding number r is said to be an *eigenvalue*. Hence the Rayleigh-Ritz method for the approximation of the lowest eigenvalue λ_1 of the original Sturm-Liouville problem leads to the consideration of the matrix eigenvalue problem (7.3.5).

It is well known that the matrix equation (7.3.5) can have a nonzero solution $\mathbf{c}^*$ if and only if the matrix $A - rB$ is *singular*†—that is, if and only if

†See Ben Noble, *Applied Linear Algebra* (Englewood Cliffs, N. J.: Prentice-Hall, Inc., 1969), or almost any text on linear algebra or matrix theory.

the determinant of $A - rB$ is zero,

$$\det(A - rB) = 0. \tag{7.3.6}$$

Equation (7.3.6) will be a polynomial equation of degree n in the variable r with at most n distinct solutions. It is easy to prove in the present case that all solutions r of this equation (7.3.6) will be *real* numbers since the given matrices $A = (a_{ij})$ and $B = (b_{ij})$ are *symmetric* with $a_{ij} = a_{ji}$ and $b_{ij} = b_{ji}$. For each solution r of (7.3.6) there will be a nonzero solution vector $\mathbf{c}^*$ for (7.3.5) with $r = d(\mathbf{c}^*)/h(\mathbf{c}^*)$. Hence each solution r of (7.3.6) is an eigenvalue for the matrix equation (7.3.5).

The *smallest* such eigenvalue $r = r_1$ will be the desired minimum value of d/h over $\mathcal{R}_n$, and consequently this same eigenvalue r_1 will also be the minimum value of the Rayleigh quotient D/H over the subspace of $\mathfrak{X}$ spanned by the given fixed functions $\psi_1, \psi_2, \ldots, \psi_n$. It is this smallest eigenvalue r_1 of the matrix problem (7.3.5) that the Rayleigh-Ritz method provides as an approximation to the lowest eigenvalue λ_1 of the original Sturm-Liouville problem. Here r_1 is taken to be the smallest solution of the equation (7.3.6). (We are for the moment suppressing the obvious dependency of r_1 on n. See Exercise 4.)

If r_1 is the resulting minimum value of the Rayleigh quotient R over $\mathfrak{M}_n$, then Rayleigh's principle ensures that r_1 will give an *upper* bound for the lowest eigenvalue λ_1 of the Sturm-Liouville problem,

$$\lambda_1 \leq r_1.$$

Indeed, it is clear that the minimum value r_1 of R over the subspace $\mathfrak{M}_n$ cannot be less than the minimum value λ_1 of R over the entire vector space $\mathfrak{X}$. It can actually be proved that the number r_1 will provide an accurate approximation to λ_1 for large *n provided* that the given functions $\psi_1, \psi_2, \ldots, \psi_n, \ldots$ satisfy certain conditions which we shall not go into here.† In practice the Rayleigh-Ritz method often gives remarkably good approximations to λ_1 even for small values of n. Moreover, for each fixed n the higher (larger) eigenvalues of the matrix equation (7.3.5) also provide upper bounds for the corresponding higher eigenvalues of the Sturm-Liouville problem, as we shall see in Section 7.6.

Example. We close this section with a numerical example illustrating the Rayleigh-Ritz method. We shall consider the vibrating string problem discussed in Section 7.1 for a string of length l and constant tension and density. Hence we have

$$\tau W'' = -\lambda \rho W \qquad \text{for } 0 < x < l,$$

†See Sagan, *op. cit.*, pp. 281–283. In practice the success of the method depends on being able to make an intelligent guess for the functions $\psi_1, \psi_2, \ldots, \psi_n, \ldots$.

with $W(0) = 0$ and $W(l) = 0$. Here τ and ρ are given positive constants. In this case the smallest eigenvalue λ_1 is given by equation (7.1.9) as

$$\lambda_1 = \frac{\pi^2}{l^2}\frac{\tau}{\rho} \approx \frac{9.87\tau}{l^2\rho}, \tag{7.3.7}$$

where we have used the approximation $\pi^2 \approx 9.87$, which is correct to three digits.

We shall calculate the Rayleigh-Ritz value r_1 for the case $n = 2$ and for the functions ψ_1 and ψ_2 defined as

$$\psi_1(x) = x(l - x), \qquad \psi_2(x) = x(l^2 - x^2) \qquad \text{for } 0 \le x \le l.$$

Note that these functions ψ_1 and ψ_2 have been chosen so as to vanish at the end points $x = 0$ and $x = l$. If we use these functions ψ_1 and ψ_2 in (7.3.3) with $q(x) = 0$, $x_0 = 0$, $x_1 = l$, and τ and ρ constant, we find that

$$A = \begin{pmatrix} a_{11} & a_{12} \\ a_{21} & a_{22} \end{pmatrix} = \tau l^3 \begin{pmatrix} \dfrac{1}{3} & \dfrac{l}{2} \\ \dfrac{l}{2} & \dfrac{4l^2}{5} \end{pmatrix}$$

$$B = \begin{pmatrix} b_{11} & b_{12} \\ b_{21} & b_{22} \end{pmatrix} = \frac{\rho l^5}{5} \begin{pmatrix} \dfrac{1}{6} & \dfrac{l}{4} \\ \dfrac{l}{4} & \dfrac{8l^2}{21} \end{pmatrix}.$$

Hence we obtain for $A - rB$

$$A - rB = \begin{pmatrix} \dfrac{\tau l^3}{3} - \dfrac{r\rho l^5}{30} & \dfrac{\tau l^4}{2} - \dfrac{r\rho l^6}{20} \\ \dfrac{\tau l^4}{2} - \dfrac{r\rho l^6}{20} & \dfrac{4\tau l^5}{5} - \dfrac{r\rho l^7 8}{105} \end{pmatrix},$$

with determinant

$$\det(A - rB) = \frac{\rho^2 l^{12} r^2 - 52\rho l^{10}\tau r + 420 l^8 \tau^2}{25{,}200}.$$

It follows that equation (7.3.6) leads in this case to the following quadratic equation for r:

$$\rho^2 l^4 r^2 - 52\rho l^2 \tau r + 420\tau^2 = 0.$$

The two solutions of this equation are easily found to be $r = 10\tau/l^2\rho$ and $r = 42\tau/l^2\rho$. These are the eigenvalues of the matrix equation (7.3.5) in the present case, and we take r_1 to be the smallest of these eigenvalues,

$$r_1 = \frac{10\tau}{l^2\rho}.$$

We see that r_1 does indeed give an upper bound for the lowest Sturm-Liouville eigenvalue λ_1 of (7.3.7), and indeed the accuracy of the approximation is remarkable for such a low choice of n. (The accuracy would be even better with larger n—at the expense of a more difficult calculation.)

Exercises

1. Show that the lowest eigenvalue λ_1 for the Sturm-Liouville problem $W''(x) - xW(x) = -\lambda W(x)$ for $0 < x < 1$ and $W(0) = 0$, $W(1) = 0$ satisfies the inequality $\lambda_1 \leq 10.5$. Use the Rayleigh-Ritz method with $n = 1$ and with $\psi_1(x) = x(1 - x)$. (It is remarkable that even with $n = 1$ the Rayleigh-Ritz method gives a better upper bound for λ_1 than was obtained in Exercise 2 of Section 7.2. It can be shown that the exact value of λ_1 is about 10.4 in this case, and so the present approximation is really quite good.)

2. Show that the lowest eigenvalue of the Sturm-Liouville problem $W''(x) = -\lambda W(x)$ for $-1 < x < 1$ and $W(-1) = 0$, $W(1) = 0$ satisfies the inequality $\lambda_1 \leq 2.5$. Use the Rayleigh-Ritz method with $n = 2$ and with $\psi_1(x) = (1 - x)(1 + x)$ and $\psi_2(x) = x(1 - x)(1 + x)$. (In this case you should also be able to calculate the eigenvalues and eigenfunctions explicitly so as to find $\lambda_n = n^2\pi^2/4$. Hence $\lambda_1 = \pi^2/4 \approx 2.47$.)

3. Show that the lowest eigenvalue of the Sturm-Liouville problem $(d/dx)[x^2 (dW/dx)] = -\lambda x^2 W$ for $\frac{1}{2} < x < 1$ and $W(\frac{1}{2}) = 0$, $W(1) = 0$ satisfies the inequality $\lambda_1 \leq 40$. Use the Rayleigh-Ritz method with $n = 1$ and with $\psi_1(x) = [(2x - 1)(1 - x)/x]$. [The exact eigenvalues in this case are given as $\lambda_n = 4\pi^2 n^2$ with corresponding eigenfunctions $W_n(x) = (1/x)\sin 2\pi n x$. Hence $\lambda_1 = 4\pi^2 \approx 39.5$.]

4. Let $r_1 = r_1(n)$ be the Rayleigh-Ritz approximation to the lowest eigenvalue of the Sturm-Liouville problem $(d/dx)[\tau(x)(dW/dx)] + q(x)W = -\lambda\rho(x)W$ for $x_0 < x < x_1$ and $W(x_0) = 0$, $W(x_1) = 0$ obtained using the first n functions $\psi_1, \psi_2, \ldots, \psi_n$ of a given infinite sequence of functions $\psi_1, \psi_2, \ldots, \psi_n, \psi_{n+1}, \ldots$ in the vector space $\mathfrak{X}$. Show that the resulting sequence of Rayleigh-Ritz numbers $r_1(1), r_1(2), \ldots, r_1(n), r_1(n+1), \ldots$ *decreases monotonically;* i.e., $r_1(m) \leq r_1(n)$ if $m \geq n$. (This Rayleigh-Ritz sequence will actually *converge* monotonically down to the lowest eigenvalue λ_1 as n tends toward infinity *provided* that the original sequence of functions $\psi_1, \psi_2, \ldots, \psi_n, \psi_{n+1}, \ldots$ satisfies certain conditions. See Sagan, *op. cit.*, pp. 281–283.)

7.4. Higher Eigenvalues and the Rayleigh Quotient

We have seen that the lowest eigenvalue λ_1 for the Sturm-Liouville problem

$$\frac{d}{dx}\left[\tau(x)\frac{dW}{dx}\right] + q(x)W = -\lambda\rho(x)W \qquad \text{for } x_0 < x < x_1$$

$$W(x_0) = 0, \qquad W(x_1) = 0$$

can be characterized as the minimum value of the Rayleigh quotient $R = D/H$, where

$$D(W) = \int_{x_0}^{x_1} [\tau(x)W'(x)^2 - q(x)W(x)^2]\,dx$$

and

$$H(W) = \int_{x_0}^{x_1} \rho(x)W(x)^2\,dx$$

for any function W in the vector space $\mathfrak{X}$. We now wish to show that the higher eigenvalues $\lambda_2, \lambda_3, \ldots, \lambda_n, \ldots$ can also be characterized as extreme values of the Rayleigh quotient subject to certain constraints.

We begin by showing that the second eigenvalue λ_2 occurring in (7.2.2) is equal to the minimum value of the Rayleigh quotient $R(W)$ over $\mathfrak{X}$ *subject to the added constraint*

$$\int_{x_0}^{x_1} \rho(x)W_1(x)W(x)\,dx = 0. \tag{7.4.1}$$

Here W_1 is the eigenfunction corresponding to the lowest eigenvalue λ_1, and this constraint serves to *eliminate* W_1 (and all constant multiples of W_1) from any further consideration since $W = W_1$ violates (7.4.1),

$$\int_{x_0}^{x_1} \rho(x)W_1(x)W_1(x)\,dx = \int_{x_0}^{x_1} \rho(x)W_1(x)^2\,dx > 0.$$

[The minimum value of $R(W)$ *without* this added constraint has already been seen to be equal to λ_1, with $R(W_1) = \lambda_1$.]

If W_2 is a solution to the last constrained extremum problem, then according to the Euler-Lagrange multiplier theorem, [see (7.2.3)]

$$\delta D(W_2; \Delta W) - R(W_2)\,\delta H(W_2; \Delta W) = \text{constant} \int_{x_0}^{x_1} \rho(x)W_1(x)\,\Delta W(x)\,dx$$

must hold for all vectors ΔW in $\mathfrak{X}$, where the last integral appearing here on the right-hand side is the variation of the constraint functional of (7.4.1), and where the multiplicative constant is (essentially) an Euler-Lagrange multiplier. If we now use the previous formulas for the variations of D and H found in Section 7.2, we find for W_2 the condition

$$\int_{x_0}^{x_1} \left\{\frac{d}{dx}\left[\tau(x)\frac{dW_2(x)}{dx}\right] + q(x)W_2(x) + \lambda^*\rho(x)W_2(x) + \mu\rho(x)W_1(x)\right\} \Delta W(x)\,dx = 0,$$

which must hold for some suitable constant μ and for all continuously differentiable functions ΔW which vanish at x_0 and x_1, where in this case we have set

$$\lambda^* = R(W_2) = \frac{D(W_2)}{H(W_2)}.$$

It follows now from this result and the lemma of Section A4 of the Appendix that W_2 must satisfy the following differential equation:

$$\frac{d}{dx}\left[\tau(x)\frac{dW_2(x)}{dx}\right] + q(x)W_2(x) = -\lambda^* \rho(x)W_2(x) - \mu\rho(x)W_1(x)$$

for $x_0 < x < x_1$. Moreover, if we now multiply this equation on both sides by the function W_1 and integrate both sides over $x_0 < x < x_1$, we find with the constraint (7.4.1) (for $W = W_2$) that

$$-\mu\int_{x_0}^{x_1} \rho(x)W_1(x)^2\,dx = \int_{x_0}^{x_1}\left\{\frac{d}{dx}\left[\tau(x)\frac{dW_2(x)}{dx}\right]W_1(x) + q(x)W_2(x)W_1(x)\right\}dx = \int_{x_0}^{x_1}\left\{\frac{d}{dx}\left[\tau(x)\frac{dW_1(x)}{dx}\right] + q(x)W_1(x)\right\}W_2(x)\,dx,$$

where two integrations by parts are required to demonstrate the validity of the last equality here. Since W_1 is an eigenfunction for the Sturm-Liouville problem corresponding to the eigenvalue λ_1, we find now that

$$-\mu\int_{x_0}^{x_1} \rho(x)W_1(x)^2\,dx = -\lambda_1\int_{x_0}^{x_1} \rho(x)W_1(x)W_2(x)\,dx,$$

and since $W = W_2$ satisfies the constraint (7.4.1), we finally have

$$-\mu\int_{x_0}^{x_1} \rho(x)W_1(x)^2\,dx = 0.$$

It follows that $\mu = 0$ (why?), and thus W_2 satisfies

$$\frac{d}{dx}\left[\tau(x)\frac{dW_2}{dx}\right] + q(x)W_2 = -\lambda^*\rho(x)W_2$$

for $x_0 < x < x_1$.

Hence $\lambda^* = R(W_2)$ is an eigenvalue for the Sturm-Liouville problem, and W_2 (or any constant multiple of W_2) is the corresponding eigenfunction. Moreover, since W_2 is a *minimum* vector in $\mathfrak{X}$ for $R(W)$ subject to the constraint (7.4.1), we can conclude that $\lambda^* = R(W_2)$ must be the *second* eigenvalue appearing in (7.2.2),

$$\lambda^* = \lambda_2.$$

Indeed, if λ is any eigenvalue other than λ_1, with corresponding eigenfunction W, then (see Exercise 1)

$$\int_{x_0}^{x_1} \rho(x)W_1(x)W(x)\,dx = 0$$

will hold, so that *all* eigenfunctions except W_1 satisfy the constraint imposed in the present extremum problem. Since we have assumed that W_2 is a minimum vector in $\mathfrak{X}$ for R subject to this constraint, it follows that $R(W_2) \leq R(W)$ for every eigenfunction W other than W_1. But we have already seen in Section 7.2 that $\lambda = R(W)$ holds for every eigenvalue λ and corresponding eigenfunction W. Hence we conclude that $\lambda^* = R(W_2) \leq R(W) = \lambda$, or $\lambda^* \leq \lambda$, for every eigenvalue λ other than λ_1. Thus λ^* must be the second eigenvalue occurring in (7.2.2), $\lambda^* = \lambda_2$.

Since λ^* is also the minimum value of the Rayleigh quotient subject to the given constraint, it follows now that *the second eigenvalue λ_2 occurring in* (7.2.2) *is equal to the minimum value of the Rayleigh quotient $R(W)$ subject to the constraint*

$$\int_{x_0}^{x_1} \rho(x)W_1(x)W(x)\,dx = 0,$$

where W_1 is the eigenfunction corresponding to the lowest eigenvalue λ_1.

We can now continue this argument inductively. If we already have the first $n - 1$ eigenvalues $\lambda_1 < \lambda_2 < \cdots < \lambda_{n-1}$ appearing in (7.2.2) along with their corresponding eigenfunctions $W_1, W_2, \ldots, W_{n-1}$, then we can show that *the* nth *eigenvalue λ_n occurring* in (7.2.2) *is equal to the minimum value of the Rayleigh quotient $R(W)$ subject to the $n - 1$ simultaneous constraints*

$$\int_{x_0}^{x_1} \rho(x)W_i(x)W(x)\,dx = 0 \qquad \text{for } i = 1, 2, \ldots, n-1.$$

Indeed, we have already proved this result in the case $n = 2$. If we now assume that the result is true for $n = 2, 3, \ldots, N$, then it is easy to prove that it must also be true for $n = N + 1$. (The latter proof can be patterned after the above proof already given for the case $n = 2$. We leave the details in this case as an exercise for the reader.) Hence we conclude by induction that the result must be true for all $n = 2, 3, \ldots$.

Thus *each* eigenvalue of the given Sturm-Liouville problem can be characterized as an extreme value of the Rayleigh quotient subject to the constraints given above, which involve the earlier eigenfunctions.

Exercises

1. Let W^* and W^{**} be any two eigenfunctions for the Sturm-Liouville problem $(d/dx)[\tau(x)(dW/dx)] + q(x)W = -\lambda\rho(x)W$ for $x_0 < x < x_1$ and $W(x_0) = 0$, $W(x_1) = 0$, corresponding to *distinct* eigenvalues λ^* and λ^{**}. Prove that W^* and W^{**} must satisfy the *orthogonality condition* $\int_{x_0}^{x_1} \rho(x)W^*(x)W^{**}(x)\,dx = 0$. *Hints:* Multiply $(d/dx)[\tau(x)(dW^*/dx)] + q(x)W^* = -\lambda^*\rho(x)W^*$ by W^{**} and integrate the resulting equation over $x_0 < x < x_1$ to find $\int_{x_0}^{x_1} \{(d/dx)[\tau(x)(dW^*(x)/dx)] + q(x)W^*(x)\}W^{**}(x)\,dx = -\lambda^* \int_{x_0}^{x_1} \rho(x)W^*(x)W^{**}(x)\,dx$. Similarly, multiply $(d/dx)[\tau(x)\,(dW^{**}/dx)] + q(x)W^{**} = -\lambda^{**}\rho(x)W^{**}$ by W^* and find $\int_{x_0}^{x_1} \{(d/dx)[\tau(x)(dW^{**}(x)/dx)] + q(x)W^{**}(x)\}W^*(x)\,dx = -\lambda^{**} \int_{x_0}^{x_1} \rho(x) W^*(x)W^{**}(x)\,dx$. Now use two integrations by parts to obtain $\int_{x_0}^{x_1} \{(d/dx)[\tau(x)(dW^*/dx)] + q(x)W^*\}W^{**}\,dx = \int_{x_0}^{x_1} \{(d/dx)[\tau(x)(dW^{**}/dx)] + q(x)W^{**}\}W^*\,dx$. From all these results you should be able to conclude that $(\lambda^* - \lambda^{**}) \int_{x_0}^{x_1} \rho(x)W^*(x)W^{**}(x)\,dx = 0$, and the desired result should follow for any two distinct eigenvalues $\lambda^* \neq \lambda^{**}$.

2. Let $\mathfrak{M}_n$ be the subspace of $\mathfrak{X}$ spanned by a given collection of fixed functions $\psi_1, \psi_2, \ldots, \psi_n$ as in Section 7.3, and let $r_1, r_2, \ldots, r_n$ be the eigenvalues of the matrix equation (7.3.5) obtained from the Rayleigh-Ritz procedure. The numbers $r_1, r_2, \ldots, r_n$ are the solutions of equation (7.3.6), and we index them in increasing size so that $r_1 \leq r_2 \leq r_3 \leq \cdots \leq r_n$. Let $\mathbf{c}^{(1)}, \mathbf{c}^{(2)}, \ldots, \mathbf{c}^{(n)}$ be the corresponding eigenvectors in $\mathfrak{R}_n$, with $A\mathbf{c}^{(i)} = r_i B\mathbf{c}^{(i)}$ for $i = 1, 2, \ldots, n$, where $\mathbf{c}^{(i)} = (c_1^{(i)}, c_2^{(i)}, \ldots, c_n^{(i)})$, and let $V_i = \sum_{j=1}^{n} c_j^{(i)}\psi_j$ be the corresponding functions in $\mathfrak{M}_n$. Show that the second eigenvalue r_2 is equal to the minimum value of the Rayleigh quotient $R(W)$ over $\mathfrak{M}_n$ subject to the constraint $\int_{x_0}^{x_1} \rho(x)V_1(x)W(x)\,dx = 0$. Similarly, show that the kth eigenvalue r_k is equal to the minimum value of the Rayleigh quotient $R(W)$ over $\mathfrak{M}_n$ subject to the $k - 1$ simultaneous constraints $\int_{x_0}^{x_1} \rho(x)V_i(x)W(x)\,dx = 0$ for $i = 1, 2, \ldots, k - 1$.

7.5. The Courant Minimax Principle

We have seen in the previous section that the nth eigenvalue λ_n appearing in (7.2.2) can be characterized as the minimum value of the Rayleigh quotient R subject to certain constraints which involve the earlier $n - 1$ eigenfunctions. Except in the simplest cases, however, these earlier eigenfunctions themselves are unknown, so the *usefulness* of this characterization is often quite limited (except in the case of the lowest eigenvalue λ_1). For this reason it is fortunate that each eigenvalue can also be characterized directly by certain related methods which do *not* involve the eigenfunctions. Richard Courant (1888–

1972) discovered one such method early in the twentieth century,† while G. Pólya and M. Schiffer discovered a different method in 1953.‡ The two methods tend to complement each other for various different reasons; each method has its own advantages for problems in certain different areas. We shall only discuss Courant's method here.

To describe Courant's method, we let $C(\phi_1, \phi_2, \ldots, \phi_{n-1})$ denote the *minimum value of the Rayleigh quotient* $R(W)$ *subject to the* $n - 1$ *constraints*

$$\int_{x_0}^{x_1} \phi_i(x) W(x)\, dx = 0 \qquad \text{for } i = 1, 2, \ldots, n-1,$$

where $\phi_1, \phi_2, \ldots, \phi_{n-1}$ are any given $n - 1$ functions from the vector space $\mathfrak{X}$. Here, as before, $\mathfrak{X}$ is the vector space consisting of all continuously differentiable functions on the interval $x_0 \leq x \leq x_1$ which vanish at the end points.

It follows from the results of Section 7.4 that if, in particular, we take the functions $\phi_1, \phi_2, \ldots, \phi_{n-1}$ to be given as

$$\phi_i = \rho(x) W_i(x) \qquad \text{for } i = 1, 2, \ldots, n-1,$$

where $W_1, W_2, \ldots, W_{n-1}$ are the first $n - 1$ eigenfunctions, then

$$C(\rho W_1, \rho W_2, \ldots, \rho W_{n-1}) = \lambda_n$$

will hold. Moreover, we shall prove the *inequality*

$$C(\phi_1, \phi_2, \ldots, \phi_{n-1}) \leq \lambda_n, \tag{7.5.1}$$

which is valid for all possible choices of the functions $\phi_1, \phi_2, \ldots, \phi_{n-1}$ in $\mathfrak{X}$. From these two results it will follow that *the* nth *eigenvalue* λ_n *is equal to the maximum value of the expression* $C(\phi_1, \phi_2, \ldots, \phi_{n-1})$ *over all possible functions* $\phi_1, \phi_2, \ldots, \phi_{n-1}$ *in the vector space* $\mathfrak{X}$. If we recall the definition of the quantity $C(\phi_1, \phi_2, \ldots, \phi_{n-1})$, we can restate the last result as follows: *the* nth *eigenvalue* λ_n *is equal to the maximum value, over all possible functions* $\phi_1, \phi_2, \ldots, \phi_{n-1}$, *of the minimum value of the Rayleigh quotient* $R(W)$ *subject to the* $n - 1$ *constraints*

$$\int_{x_0}^{x_1} \phi_i(x) W(x)\, dx = 0 \qquad \text{for } i = 1, 2, \ldots, n-1. \tag{7.5.2}$$

This result is often referred to as *Courant's minimax principle.*

†R. Courant, "Uber die Eigenwerte bei den Differentialgleichungen der mathematischen Physik," *Mathematische Zeitschrift*, **7** (1920), 1–57. These results may also be found in Chapter 6 of R. Courant and D. Hilbert, *Methods of Mathematical Physics*, Vol. 1 [New York: John Wiley & Sons, Inc. (Interscience Division), 1953].

‡G. Pólya and M. Schiffer, "Convexity of Functionals by Transplantation," *Journal D'Analyse Mathématique*, **3** (1953), 245–345. The method of Pólya and Schiffer is based on a theorem published by H. Poincaré in 1890.

From the above discussion it is clear that the validity of Courant's minimax principle will follow from the inequality (7.5.1). On the other hand, the validity of (7.5.1) will be guaranteed for any fixed choice of the functions $\phi_1, \phi_2, \ldots, \phi_{n-1}$ if we can find some particular function W^* in $\mathfrak{X}$ which satisfies the given $n - 1$ constraints of (7.5.2) and for which

$$R(W^*) \leq \lambda_n \tag{7.5.3}$$

holds. This is clear since $C(\phi_1, \phi_2, \ldots, \phi_{n-1})$ is the *minimum* value of $R(W)$ subject to the given constraints. Hence we need only show how to find such a function W^*.

We shall try to find such a function W^* as a suitable linear combination of the first n eigenfunctions $W_1, W_2, \ldots, W_{n-1}, W_n$. (The eigenfunctions are used here in the proof of the minimax principle even though they do *not* appear in the final statement of the principle itself.) Hence we seek W^* in the form

$$W^* = \sum_{j=1}^{n} c_j W_j \tag{7.5.4}$$

for suitable constants $c_1, c_2, \ldots, c_{n-1}, c_n$ which must be determined so that W^* satisfies the conditions

$$\int_{x_0}^{x_1} \phi_i(x) W^*(x)\, dx = 0 \qquad \text{for } i = 1, 2, \ldots, n - 1.$$

These conditions will hold in the present case if the constants $c_1, c_2, \ldots, c_{n-1}, c_n$ satisfy

$$\sum_{j=1}^{n} \gamma_{ij} c_j = 0 \qquad \text{for } i = 1, 2, \ldots, n - 1,$$

where the numbers γ_{ij} are defined as

$$\gamma_{ij} = \int_{x_0}^{x_1} \phi_i(x) W_j(x)\, dx.$$

But it is well known that such a system of $n - 1$ linear homogeneous equations in n unknowns always has a nontrivial solution.† Hence it is always possible to find suitable constants $c_1, c_2, \ldots, c_{n-1}, c_n$, not all zero, such that the resulting function W^* will satisfy the given $n - 1$ constraints.

We shall now show that the resulting function W^* will satisfy the inequality (7.5.3). Indeed, if W^* is any (nonzero) function of the form (7.5.4),

†See, for example, Noble, *op. cit.*, pp. 77 and 91.

then we find that

$$R(W^*) = \frac{D\left(\sum_{j=1}^{n} c_j W_j\right)}{H\left(\sum_{j=1}^{n} c_j W_j\right)} = \frac{\sum_{i,j=1}^{n} c_i c_j \alpha_{ij}}{\sum_{i,j=1}^{n} c_i c_j \beta_{ij}},$$

where

$$\alpha_{ij} = \int_{x_0}^{x_1} [\tau(x) W_i'(x) W_j'(x) - q(x) W_i(x) W_j(x)]\, dx$$

and

$$\beta_{ij} = \int_{x_0}^{x_1} \rho(x) W_i(x) W_j(x)\, dx$$

for $i, j = 1, 2, \ldots, n$. An integration by parts gives for α_{ij}

$$\alpha_{ij} = -\int_{x_0}^{x_1} \left\{\frac{d}{dx}\left[\tau(x)\frac{dW_i(x)}{dx}\right] + q(x) W_i(x)\right\} W_j(x)\, dx,$$

and since W_i is the eigenfunction of the original Sturm-Liouville problem corresponding to the eigenvalue λ_i, we find that

$$\alpha_{ij} = \lambda_i \beta_{ij}.$$

Hence $R(W^*)$ will satisfy

$$R(W^*) = \frac{\sum_{i,j=1}^{n} c_i c_j \lambda_i \beta_{ij}}{\sum_{i,j=1}^{n} c_i c_j \beta_{ij}}.$$

On the other hand, $\beta_{ij} = 0$ for $i \neq j$, as follows from Exercise 1 of Section 7.4, so that

$$R(W^*) = \frac{\sum_{i=1}^{n} c_i^2 \lambda_i \beta_{ii}}{\sum_{i=1}^{n} c_i^2 \beta_{ii}}, \tag{7.5.5}$$

where

$$\beta_{ii} = \int_{x_0}^{x_1} \rho(x) W_i(x)^2\, dx > 0.$$

Since $\lambda_i \leq \lambda_n$ for $i = 1, 2, \ldots, n - 1$, we have $\sum_{i=1}^{n} c_i^2 \lambda_i \beta_{ii} \leq \lambda_n \sum_{i=1}^{n} c_i^2 \beta_{ii}$, and thus (7.5.5) implies the desired result $R(W^*) \leq \lambda_n$. This completes the proof of the Courant minimax principle.

Exercise

1. Let $\mathfrak{M}_n$ be the subspace of $\mathfrak{X}$ spanned by a given collection of fixed functions $\psi_1, \psi_2, \ldots, \psi_n$ as in Section 7.3, and let $r_1, r_2, \ldots, r_n$ be the eigenvalues of the matrix equation (7.3.5) obtained from the Rayleigh-Ritz procedure, indexed in

increasing order as $r_1 \leq r_2 \leq \cdots \leq r_n$. Show that *the* kth *eigenvalue* r_k *is equal to the maximum value, over all possible functions* $\phi_1, \phi_2, \ldots, \phi_{k-1}$ *in* $\mathfrak{X}$, *of the minimum value of the Rayleigh quotient* $R(W)$ *over* $\mathfrak{M}_n$ *subject to the* $k-1$ *constraints* $\int_{x_0}^{x_1} \phi_i(x)W(x)\,dx = 0$ for $i = 1, 2, \ldots, k-1$. *Hint:* Use the methods of this section along with the results of Exercise 2 of Section 7.4.

7.6. Some Implications of the Courant Minimax Principle

The minimax principle is well suited for the purpose of comparing the eigenvalues of *different* Sturm-Liouville problems. Indeed, if λ_n and λ_n^* denote the respective eigenvalues for the problems

$$\frac{d}{dx}\left[\tau(x)\frac{dW}{dx}\right] + q(x)W = -\lambda\rho(x)W \qquad \text{for } x_0 < x < x_1$$

$$W(x_0) = 0, \qquad W(x_1) = 0$$

and

$$\frac{d}{dx}\left[\tau^*(x)\frac{dW}{dx}\right] + q^*(x)W = -\lambda\rho^*(x)W \qquad \text{for } x_0 < x < x_1$$

$$W(x_0) = 0, \qquad W(x_1) = 0,$$

with $\lambda_1 < \lambda_2 < \cdots < \lambda_n < \cdots$ and $\lambda_1^* < \lambda_2^* < \cdots < \lambda_n^* < \cdots$, then λ_n and λ_n^* can be effectively compared by comparing the corresponding Rayleigh quotients

$$R(W) = \frac{\int_{x_0}^{x_1} [\tau(x)W'(x)^2 - q(x)W(x)^2]\,dx}{\int_{x_0}^{x_1} \rho(x)W(x)^2\,dx}$$

and

$$R^*(W) = \frac{\int_{x_0}^{x_1} [\tau^*(x)W'(x)^2 - q^*(x)W(x)^2]\,dx}{\int_{x_0}^{x_1} \rho^*(x)W(x)^2\,dx}.$$

Here $\tau(x)$, $q(x)$, $\rho(x)$, $\tau^*(x)$, $q^*(x)$, and $\rho^*(x)$ are given functions defined for $x_0 \leq x \leq x_1$.

For example, we can conclude that the eigenvalues must satisfy

$$\lambda_n \leq \lambda_n^* \qquad \text{for all } n = 1, 2, 3, \ldots$$

if we already know that the Rayleigh quotients satisfy the condition

$$R(W) \leq R^*(W) \qquad \text{for all functions } W \text{ in } \mathfrak{X}.$$

Indeed, if the last condition holds, then clearly

$$C(\phi_1, \phi_2, \ldots, \phi_{n-1}) \leq C^*(\phi_1, \phi_2, \ldots, \phi_{n-1}) \tag{7.6.1}$$

follows, where $C(\phi_1, \phi_2, \ldots, \phi_{n-1})$ denotes the minimum value of $R(W)$ subject to the $n - 1$ constraints

$$\int_{x_0}^{x_1} \phi_i(x) W(x)\, dx = 0 \qquad \text{for } i = 1, 2, \ldots, n-1$$

and, similarly, $C^*(\phi_1, \phi_2, \ldots, \phi_{n-1})$ denotes the minimum value of $R^*(W)$ subject to the same $n - 1$ constraints. Moreover, (7.6.1) will hold in this case for *any* choice of the functions $\phi_1, \phi_2, \ldots, \phi_{n-1}$ in $\mathfrak{X}$, and the desired result $\lambda_n \leq \lambda_n^*$ then follows by the minimax principle, for all $n = 1, 2, 3, \ldots$.

Hence if the Rayleigh quotient *increases* from $R(W)$ to $R^*(W)$, then *every* eigenvalue will increase (or at least not decrease). In particular, if the given functions τ, q, ρ, τ^*, q^*, and ρ^* satisfy the inequalities

$$\tau(x) \leq \tau^*(x), \qquad q(x) \geq q^*(x) \qquad \text{and} \qquad \rho(x) \geq \rho^*(x) \qquad \text{for } x_0 \leq x \leq x_1, \tag{7.6.2}$$

then clearly $R(W) \leq R^*(W)$ will hold for all W, so that we obtain the result $\lambda_n \leq \lambda_n^*$ for all $n = 1, 2, \ldots$ whenever (7.6.2) holds. If the inequalities of (7.6.2) are reversed, then we find, similarly, that $\lambda_n \geq \lambda_n^*$.

If, in particular, we take τ^*, q^*, and ρ^* to be *constant* functions, with

$$\begin{aligned} \tau^* &= \max_{x_0 \leq x \leq x_1} \tau(x) = \tau_M \\ q^* &= \min_{x_0 \leq x \leq x_1} q(x) = q_m \\ \rho^* &= \min_{x_0 \leq x \leq x_1} \rho(x) = \rho_m, \end{aligned} \tag{7.6.3}$$

then clearly (7.6.2) holds, and we have $\lambda_n \leq \lambda_n^*$. But the Sturm-Liouville problem characterizing the eigenvalues λ_n^* is easy to solve in the latter case since the functions τ^*, q^*, and ρ^* are *constants*. Indeed, the given problem for λ^* becomes

$$W''(x) = -\frac{q_m + \lambda^* \rho_m}{\tau_M} W(x) \qquad \text{for } x_0 < x < x_1$$

$$W(x_0) = 0, \qquad W(x_1) = 0,$$

with eigenvalues λ_n^* and eigenfunctions $W_n^*(x)$ given as [see the calculation leading to (7.1.9)]

$$\lambda_n^* = -\frac{q_m}{\rho_m} + \frac{n^2\pi^2}{(x_1 - x_0)^2} \frac{\tau_M}{\rho_m}$$

$$W_n^* = \sin\sqrt{\frac{q_m + \lambda_n^* \rho_m}{\tau_M}}\,(x - x_0) \qquad \text{for } x_0 \le x \le x_1$$

and for $n = 1, 2, 3, \ldots$. Since $\lambda_n \le \lambda_n^*$, we have

$$\lambda_n \le -\frac{q_m}{\rho_m} + \frac{n^2\pi^2}{(x_1 - x_0)^2}\frac{\tau_M}{\rho_m} \qquad \text{for all } n = 1, 2, 3, \ldots, \tag{7.6.4}$$

where λ_n is the nth eigenvalue of the Sturm-Liouville problem corresponding to the functions $\tau(x)$, $q(x)$, and $\rho(x)$, and where the constants τ_M, q_m, and ρ_m are defined by (7.6.3). Similarly, if we define constants τ_m, q_M, and ρ_M by

$$\min_{x_0 \le x \le x_1} \tau(x) = \tau_m$$

$$\max_{x_0 \le x \le x_1} q(x) = q_M$$

$$\max_{x_0 \le x \le x_1} \rho(x) = \rho_M,$$

we find the inequality

$$-\frac{q_M}{\rho_M} + \frac{n^2\pi^2}{(x_1 - x_0)^2}\frac{\tau_m}{\rho_M} \le \lambda_n \qquad \text{for all } n = 1, 2, \ldots. \tag{7.6.5}$$

The estimates given by (7.6.4) and (7.6.5) are often somewhat crude. It is nevertheless remarkable that such estimates can be so easily obtained for *all* the eigenvalues $\lambda_1, \lambda_2, \lambda_3, \ldots$. These estimates verify the earlier stated result that the eigenvalues λ_n of the given Sturm-Liouville problem must become *unbounded* as $n \to +\infty$. Indeed, we now have λ_n squeezed between two expressions, each of which becomes unbounded.

A slightly more sophisticated analysis can be given, again based on the minimax principle, to obtain the following related result:

$$\lambda_n = \frac{n^2\pi^2}{\left(\int_{x_0}^{x_1} \sqrt{\rho(x)/\tau(x)}\, dx\right)^2} + E_n$$

for certain constants $E_1, E_2, \ldots$ *which remain bounded*, $|E_n| \le$ constant. Moreover, a similar *asymptotic relation* can be obtained for the corresponding eigenfunction $W_n(x)$. We refer the reader to Smirnov† for these important results.

If we consider in particular the vibrating string problem of Section 7.1, with $q(x) = 0$ and with $\tau(x)$ and $\rho(x)$ representing the (variable) tension and

†V. I. Smirnov, *A Course of Higher Mathematics*, Vol. 4: *Integral Equations and Partial Differential Equations*, trans. D. E. Brown (Reading, Mass.: Addison-Wesley Publishing Company, Inc., 1964), pp. 558–565.

density of the string for $0 \leq x \leq l$, we can use the above results now to conclude that *any increase in the tension and any decrease in the density of the string will cause the tone of the string to increase, including all the fundamental tones.*† *Similarly, any added constraints imposed on the string* (as, for example, keeping the string pegged at some point or points) *will cause all the fundamental tones to increase.* Indeed, in the latter case the added constraints will have the effect of decreasing the size of the class of admissible functions $\mathfrak{X}$ so as to *increase* the minimum value of the Rayleigh quotient over this smaller class as compared with the corresponding minimum value of the Rayleigh quotient over the larger (unconstrained) admissible class. For example, any *shortening* of the string can be viewed as arising from the imposition of certain additional constraints, so that the fundamental tones of the string will all increase if the string is shortened. Conversely, if any constraints are removed from the string, or if the tension decreases or the density increases, then all the fundamental tones will decrease (or at least not increase).

Finally, we shall indicate how the Courant minimax principle can be used to prove that the Rayleigh-Ritz procedure described in Section 7.3 actually furnishes upper estimates for each of the first n lowest eigenvalues $\lambda_1, \lambda_2, \ldots, \lambda_n$. Indeed, if $r_1, r_2, \ldots, r_n$ are the Rayleigh-Ritz values as described in Exercise 1 of Section 7.5, then (according to that exercise) r_k is equal to the maximum value of $C^{(n)}(\phi_1, \phi_2, \ldots, \phi_{k-1})$ over all possible functions $\phi_1, \phi_2, \ldots, \phi_{k-1}$ in $\mathfrak{X}$, where $C^{(n)}(\phi_1, \phi_2, \ldots, \phi_{k-1})$ is the minimum value of the Rayleigh quotient $R(W)$ over $\mathfrak{M}_n$ subject to the $k-1$ constraints

$$\int_{x_0}^{x_1} \phi_i(x)W(x)\,dx = 0 \qquad \text{for } i = 1, 2, \ldots, k-1.$$

On the other hand, $C(\phi_1, \phi_2, \ldots, \phi_{k-1})$ is the minimum value of $R(W)$ over the larger set $\mathfrak{X}$ subject to the same $k-1$ constraints. Since $\mathfrak{M}_n$ is a subset of $\mathfrak{X}$, it is clear then that

$$C(\phi_1, \phi_2, \ldots, \phi_{k-1}) \leq C^{(n)}(\phi_1, \phi_2, \ldots, \phi_{k-1})$$

for any choice of the functions $\phi_1, \phi_2, \ldots, \phi_{k-1}$. The last inequality along with the minimax principle implies the desired result $\lambda_k \leq r_k$ for $k = 1, 2, \ldots, n$.

By way of illustration we recall again the example considered at the end of Section 7.3 for the problem

$$\tau W'' = -\lambda \rho W \qquad \text{for } 0 < x < l,$$

with $W(0) = 0$ and $W(l) = 0$. Here τ and ρ are fixed positive constants. The

†This result and similar results were known to Rayleigh himself as early as 1873. The proofs given by Rayleigh are based on ideas similar to those which lie behind the Courant minimax principle.

eigenvalues are given as $\lambda_n = (n^2\pi^2\tau/l^2\rho)$ for $n = 1, 2, 3, \ldots$, while the Rayleigh-Ritz values r_1 and r_2 were found in Section 7.3 to be given as $r_1 = (10\tau/l^2\rho)$ and $r_2 = (42\tau/l^2\rho)$ for the case $n = 2$ and for the functions $\psi_1(x) = x(l - x)$ and $\psi_2(x) = x(l^2 - x^2)$. Since $\lambda_1 = \pi^2\tau/l^2\rho \approx 9.87\tau/l^2\rho$ and $\lambda_2 = 4\lambda_1 \approx 39.5\tau/l^2\rho$, we find that r_1 and r_2 do, indeed, give upper bounds for the lowest two eigenvalues λ_1 and λ_2.

In 1931 the mathematician N. M. Krylov showed how to estimate the *difference* between the kth Rayleigh-Ritz value $r_k(n)$ and the kth eigenvalue λ_k in the special cases in which the Rayleigh-Ritz subspace $\mathfrak{M}_n$ is spanned by the n trigonometric functions $\psi_j(x) = \sin[j\pi(x - x_0)/(x_1 - x_0)]$ for $j = 1, 2, \ldots, n$ or by the n polynomials $\psi_j(x) = (x - x_0)(x - x_1)x^{j-1}$ for $j = 1, 2, \ldots, n$. We refer the reader to Smirnov† for these important results.

Exercises

1. Show that the eigenvalues for the Sturm-Liouville problem $W''(x) - xW(x) = -\lambda W(x)$ for $0 < x < 1$ and $W(0) = 0$, $W(1) = 0$ satisfy the estimates $n^2\pi^2 \leq \lambda_n \leq 1 + n^2\pi^2$ for $n = 1, 2, 3, \ldots$.
2. Use the Rayleigh-Ritz method to obtain estimates for the lowest two eigenvalues λ_1 and λ_2 of the problem $W''(x) = -\lambda W(x)$ for $-1 < x < 1$ and $W(-1) = 0$, $W(1) = 0$. Take $n = 2$ with $\psi_1(x) = (1 - x)(1 + x)$ and $\psi_2(x) = x(1 - x)(1 + x)$. (See Exercise 2 of Section 7.3.)
3. Use the Rayleigh-Ritz method to obtain estimates for the lowest two eigenvalues λ_1 and λ_2 of the problem $(d/dx)[x^2(dW/dx)] = -\lambda x^2 W$ for $\frac{1}{2} < x < 1$ and $W(\frac{1}{2}) = 0$, $W(1) = 0$. (See Exercise 3 of Section 7.3.)

7.7. Further Extensions of the Theory

The previous results of this chapter for the Sturm-Liouville equation

$$\frac{d}{dx}\left[\tau(x)\frac{dW}{dx}\right] + q(x)W = -\lambda\rho(x)W$$

subject to the boundary conditions

$$W(x_0) = 0, \qquad W(x_1) = 0$$

can be extended to other boundary conditions which also arise in practice. We refer the interested reader to Courant‡ and Sagan§ for these extensions.

†Smirnov, *op. cit.*, pp. 565–566.

‡Courant and Hilbert, *op. cit.*, Chapters 5 and 6.

§Sagan, *op, cit.*, pp. 253–260.

Similarly, the results can be extended to Sturm-Liouville equations involving more than one independent variable. By way of illustration we shall consider briefly the following equation in two independent variables,

$$\frac{\partial}{\partial x}\left[\tau(x, y)\frac{\partial W}{\partial x}\right] + \frac{\partial}{\partial y}\left[\tau(x, y)\frac{\partial W}{\partial y}\right] + q(x, y)W = -\lambda\rho(x, y)W \qquad \text{for } (x, y) \text{ in } G, \tag{7.7.1}$$

where G is a given open region in the (x, y)-plane, and where the functions $\tau = \tau(x, y)$, $q = q(x, y)$, and $\rho = \rho(x, y)$ are known functions defined on G. Several different types of boundary conditions actually arise in practice for the unknown function W, but we shall only consider the condition

$$W(x, y) = 0 \qquad \text{for all } (x, y) \text{ on } \partial G,$$

where ∂G denotes the boundary of the open region G.

The prototype of the latter problem arises upon using D'Alembert's method of separation of variables to solve the vibrating membrane equation (see Section 4.11)

$$\rho(x, y)\frac{\partial^2 Z}{\partial t^2} = \frac{\partial}{\partial x}\left[\tau(x, y)\frac{\partial Z}{\partial x}\right] + \frac{\partial}{\partial y}\left[\tau(x, y)\frac{\partial Z}{\partial y}\right]$$

subject to the fixed boundary condition

$$Z(x, y, t) = 0 \qquad \text{for } (x, y) \text{ on } \partial G \text{ and for } t \geq 0.$$

Indeed, if we seek a product solution of this membrane problem in the form

$$Z(x, y, t) = W(x, y)T(t),$$

we are led to

$$\frac{1}{T(t)}\frac{d^2T(t)}{dt^2} = \frac{1}{\rho(x, y)W}\left\{\frac{\partial}{\partial x}\left[\tau(x, y)\frac{\partial W}{\partial x}\right] + \frac{\partial}{\partial y}\left[\tau(x, y)\frac{\partial W}{\partial y}\right]\right\},$$

which must hold for all (x, y) in G and for all $t > 0$. It follows that both sides of this equation must be equal to some fixed constant, say $-\lambda$, from which we find

$$\frac{d^2T(t)}{dt^2} = -\lambda T(t) \qquad \text{for } t > 0 \tag{7.7.2}$$

and

$$\frac{\partial}{\partial x}\left[\tau(x, y)\frac{\partial W}{\partial x}\right] + \frac{\partial}{\partial y}\left[\tau(x, y)\frac{\partial W}{\partial y}\right] = -\lambda\rho(x, y)W \qquad \text{for } (x, y) \text{ in } G. \tag{7.7.3}$$

Equation (7.7.3) is a special case of equation (7.7.1) with $q = 0$.

For this membrane problem we seek nontrivial solutions of (7.7.3) subject to the boundary condition

$$W(x, y) = 0 \qquad \text{for } (x, y) \text{ on } \partial G.$$

Such nontrivial solutions can be shown to exist only for certain special values of the parameter λ, and corresponding to each such eigenvalue λ there are one or more eigenfunctions W for the given boundary value problem.

For example, if τ and ρ are positive *constants* and if G is the square region

$$G = \{(x, y): 0 < x < l, 0 < y < l\},$$

then the given boundary value problem for (7.7.3) can itself be solved by separation of variables, and the eigenvalues are found to be given as

$$\lambda = (n^2 + m^2)\frac{\pi^2\tau}{l^2\rho} \qquad \text{for } n, m = 1, 2, 3, \ldots \tag{7.7.4}$$

with corresponding eigenfunctions

$$W = \sin\frac{n\pi x}{l}\sin\frac{m\pi y}{l}.$$

The lowest eigenvalue λ_1 is obtained by taking $n = 1$ and $m = 1$ in (7.7.4) and is found to be $\lambda_1 = 2\pi^2\tau/l^2\rho$. This lowest eigenvalue has the single eigenfunction $W = \sin(\pi x/l)\sin(\pi y/l)$. However, the second lowest eigenvalue λ_2 is obtained by taking *either* $n = 1, m = 2$ *or* $n = 2, m = 1$ in (7.7.4) and is found to be $\lambda_2 = 5\pi^2\tau/l^2\rho$. There are *two* independent eigenfunctions in this case, given as $\sin(\pi x/l)\sin(2\pi y/l)$ and $\sin(2\pi x/l)\sin(\pi y/l)$.

For each eigenvalue λ_n of the given membrane boundary value problem for (7.7.3) there is a corresponding solution T_n to equation (7.7.2) given as

$$T_n(t) = a_n\cos\sqrt{\lambda_n}t + b_n\sin\sqrt{\lambda_n}t.$$

Hence for each eigenfunction $W_n(x, y)$ which corresponds to the eigenvalue λ_n we obtain a *fundamental state* for the membrane, represented by the product solution $Z = W_nT_n$. The eigenfunction $W_n(x, y)$ gives the shape or form of this fundamental state, while the function $T_n(t)$ causes each point of the membrane to vibrate periodically with the frequency $\sqrt{\lambda_n}/2\pi$.

To study the original Sturm-Liouville problem for the general equation (7.7.1) and for a general region G, we can use the same variational methods described in the earlier sections of this chapter. In this case the Rayleigh quotient is defined by

$$R(W) = \frac{\iint_G \{\tau(x, y)[(\partial W/\partial x)^2 + (\partial W/\partial y)^2] - q(x, y)W(x, y)^2\}\,dx\,dy}{\iint_G \rho(x, y)W(x, y)^2\,dx\,dy}$$

for any suitable function W, and again the various eigenvalues can be characterized as extreme values of this Rayleigh quotient. In particular, the lowest eigenvalue is equal to the minimum value of $R(W)$. The Courant minimax principle is again valid and can be used to discover many useful results concerning the eigenvalues. For example, the eigenvalues will all *decrease* if the region G is enlarged or if the function $\tau(x, y)$ decreases or if the functions $p(x, y)$ and $q(x, y)$ increase. We refer the reader to Courant† for these important developments.

7.8. Some General Remarks on the Ritz Method of Approximate Minimization

We have seen that the method of Ritz can be used to find approximate values for the minimum value of the Rayleigh quotient. In fact, the Ritz method is well suited for the approximate minimization (or maximization) of many different functionals.

We shall consider briefly the general problem of minimizing a given functional H over a given normed vector space $\mathfrak{X}$ subject to the given constraints $K_i(x) = 0$ for $i = 1, 2, \ldots, m$, where $H, K_1, \ldots, K_m$ are given functionals defined for all vectors x in $\mathfrak{X}$ (or for all vectors x in some given open subset of $\mathfrak{X}$). We assume for simplicity that there is a unique vector x^* which actually minimizes H subject to the given constraints, and we seek a *direct* method that can be used to find approximate values for the minimum value $H(x^*)$ subject to the given constraints.

We let $\mathfrak{M}_n$ be the n-dimensional subspace of $\mathfrak{X}$ spanned by the first n vectors of a given infinite sequence of (independent) vectors $x_1, x_2, \ldots, x_n, x_{n+1}, \ldots$ taken from $\mathfrak{X}$. Hence $\mathfrak{M}_n$ consists of all vectors x in $\mathfrak{X}$ of the form

$$x = \sum_{j=1}^{n} c_j x_j$$

for arbitrary constants $c_1, c_2, \ldots, c_n$.

We now consider the related problem of minimizing H over the subspace $\mathfrak{M}_n$ subject to the same given constraints $K_i = 0$ for $i = 1, 2, \ldots, m$. This problem involves only a suitable choice for the n constants $c_1, c_2, \ldots, c_n$ and is usually a simpler problem than that of minimizing H over the entire vector space $\mathfrak{X}$ subject to the given constraints. (We are tacitly assuming that $\mathfrak{X}$ is an infinite dimensional vector space, whereas $\mathfrak{M}_n$ is always finite dimensional.)

We let x_n^* denote a solution vector for the latter problem in $\mathfrak{M}_n$; i.e., x_n^* is a minimum vector in $\mathfrak{M}_n$ for H subject to the given constraints involving $K_1, \ldots, K_m$. (We assume that such a vector x_n^* exists, at least for all large n.) The value $H(x_n^*)$ can be viewed as an approximation to the exact minimum

†Courant and Hilbert, *op. cit.*, Chapters 5 and 6.

value $H(x^*)$. In fact, it can be proved that this minimum value of H over $\mathfrak{M}_n$ will converge to the desired minimum value of H over $\mathfrak{X}$ as n tends toward infinity *provided* that the given functionals $H, K_1, \ldots, K_m$ and the given sequence $x_1, x_2, \ldots, x_n, x_{n+1}, \ldots$ satisfy certain conditions which we shall not go into here.†

We wish to indicate briefly how the vector x_n^* can be found in general, and we shall then consider a specific example illustrating the method.

To minimize H over all vectors x of the stated form

$$x = \sum_{j=1}^{n} c_j x_j$$

subject to the constraints $K_i = 0$ for $i = 1, 2, \ldots, m$, it is convenient to combine the n numbers $c_1, c_2, \ldots, c_n$ into a single n-tuple $\mathbf{c} = (c_1, c_2, \ldots, c_n)$. It is also useful to define real-valued functions $h = h(\mathbf{c})$ and $k_i = k_i(\mathbf{c})$ on $\mathfrak{R}_n$ by

$$\begin{aligned} h(\mathbf{c}) &= H\Big(\sum_{j=1}^{n} c_j x_j\Big) \\ k_i(\mathbf{c}) &= K_i\Big(\sum_{j=1}^{n} c_j x_j\Big) \qquad \text{for } i = 1, 2, \ldots, m \end{aligned} \tag{7.8.1}$$

for any n-tuple $\mathbf{c} = (c_1, c_2, \ldots, c_n)$ in $\mathfrak{R}_n$. Then the problem of minimizing H over the subspace $\mathfrak{M}_n$ subject to the given constraints is equivalent to the problem of minimizing the function h over $\mathfrak{R}_n$ subject to the constraints

$$k_i(\mathbf{c}) = 0 \qquad \text{for } i = 1, 2, \ldots, m. \tag{7.8.2}$$

It is the latter problem which we now consider.

If $\mathbf{c}^* = (c_1^*, c_2^*, \ldots, c_n^*)$ is a minimum vector in $\mathfrak{R}_n$ for the function h subject to the constraints (7.8.2), then the Euler-Lagrange multiplier theorem implies that

$$\delta h(\mathbf{c}^*; \Delta\mathbf{c}) = \sum_{i=1}^{m} \lambda_i \, \delta k_i(\mathbf{c}^*; \Delta\mathbf{c}) \tag{7.8.3}$$

for all vectors $\Delta\mathbf{c}$ in $\mathfrak{R}_n$ and for some suitable multipliers $\lambda_1, \ldots, \lambda_m$, where the variations of h and k_i are given as

$$\delta h(\mathbf{c}^*; \Delta\mathbf{c}) = \sum_{j=1}^{n} \frac{\partial h(\mathbf{c}^*)}{\partial c_j} \Delta c_j$$

$$\delta k_i(\mathbf{c}^*; \Delta\mathbf{c}) = \sum_{j=1}^{n} \frac{\partial k_i(\mathbf{c}^*)}{\partial c_j} \Delta c_j.$$

The partial derivatives $\partial h/\partial c_j$ and $\partial k_i/\partial c_j$ appearing here are calculated from (7.8.1). If we use these expressions for the variations of h and k_i, we can

†For an indication of these conditions, see James W. Daniel, *The Approximate Minimization of Functionals* (Englewood Cliffs, N.J.: Prentice-Hall, Inc., 1971), pp. 65–69.

rewrite the condition (7.8.3) in the form

$$\sum_{j=1}^{n} \left\{ \frac{\partial h(\mathbf{c}^*)}{\partial c_j} - \sum_{i=1}^{m} \lambda_i \frac{\partial k_i(\mathbf{c}^*)}{\partial c_j} \right\} \Delta c_j = 0,$$

and this condition must hold for all vectors $\Delta \mathbf{c} = (\Delta c_1, \Delta c_2, \ldots, \Delta c_n)$ in $\mathcal{R}_n$. If, in particular, we take Δc_j to be given as

$$\Delta c_j = \frac{\partial h(\mathbf{c}^*)}{\partial c_j} - \sum_{i=1}^{m} \lambda_i \frac{\partial k_i(\mathbf{c}^*)}{\partial c_j} \qquad \text{for } j = 1, 2, \ldots, n,$$

we are easily led to the conditions

$$\frac{\partial h(\mathbf{c}^*)}{\partial c_j} = \sum_{i=1}^{m} \lambda_i \frac{\partial k_i(\mathbf{c}^*)}{\partial c_j} \qquad \text{for } j = 1, 2, \ldots, n, \tag{7.8.4}$$

which must then be satisfied by any extremum vector $\mathbf{c}^*$ in $\mathcal{R}_n$ for h subject to the constraints (7.8.2).

The n equations of (7.8.4) along with the m constraint equations

$$k_i(\mathbf{c}^*) = 0 \qquad \text{for } i = 1, 2, \ldots, m \tag{7.8.5}$$

furnish $n + m$ equations from which the $n + m$ unknown numbers c_1^*, $c_2^*, \ldots, c_n^*, \lambda_1, \lambda_2, \ldots, \lambda_m$ may be determined. Once $\mathbf{c}^* = (c_1^*, c_2^*, \ldots, c_n^*)$ has been found, the desired vector x_n^* in $\mathfrak{M}_n$ is given as

$$x_n^* = \sum_{j=1}^{n} c_j^* x_j.$$

In practice one would use some appropriate numerical procedure to solve the simultaneous (algebraic) equations (7.8.4) and (7.8.5). Of course, these algebraic equations may be highly nonlinear in many cases, and they may be quite difficult to solve in practice—particularly if n is large. Even so, these algebraic equations are usually considerably simpler than the corresponding functional equation

$$\delta H(x^*; \Delta x) = \sum_{i=1}^{m} \mu_i \, \delta K_i(x^*; \Delta x),$$

which characterizes the minimizing vector x^* in the (infinite dimensional) vector space $\mathfrak{X}$.

Example. By way of illustration we shall consider the problem of minimizing the functional

$$H(Y) = \int_1^2 t^2 Y'(t)^2 \, dt$$

over the vector space $\mathcal{C}^1[1, 2]$ subject to the boundary constraints $Y(1) = 1$ and $Y(2) = 2$. There is a unique vector Y^* which minimizes H in $\mathcal{C}^1[1, 2]$ subject to the given constraints. Indeed, we have seen in Exercise 6 of Section 4.1 that this minimizing function Y^* is defined in this case by

$$Y^*(t) = -\frac{2}{t} + 3 \qquad \text{for } 1 \leq t \leq 2.$$

The resulting minimum value of the functional H subject to the given boundary constraints is given as $H(Y^*) = 2$.

We shall use the Ritz method to obtain an approximate value for this minimum value of H subject to the given constraints. For this purpose we shall use the fixed vectors $Y_1, Y_2, \ldots, Y_n, Y_{n+1}, \ldots$ in $\mathcal{C}^1[1, 2]$, which are defined by

$$Y_i(t) = t^{i-1} \qquad \text{for } 1 \leq t \leq 2 \text{ and for } i = 1, 2, \ldots,$$

and we then let $\mathfrak{M}_n$ be the subspace of $\mathcal{C}^1[1, 2]$ consisting of all functions $Y(t)$ of the form

$$Y(t) = \sum_{j=1}^{n} c_j Y_j(t) = \sum_{j=1}^{n} c_j t^{j-1}$$

for arbitrary constants $c_1, c_2, \ldots, c_n$.

We now consider the problem of minimizing H over this subspace $\mathfrak{M}_n$ subject to the constraints

$$K_1(Y) = 0 \qquad \text{and} \qquad K_2(Y) = 0,$$

where the functionals K_1 and K_2 are defined by

$$K_1(Y) = Y(1) - 1$$
$$K_2(Y) = Y(2) - 2$$

for any function $Y = Y(t)$ in $\mathcal{C}^1[1, 2]$.

In the present case the functions $h(\mathbf{c})$, $k_1(\mathbf{c})$, and $k_2(\mathbf{c})$ are found with (7.8.1) to be given as

$$h(\mathbf{c}) = \sum_{i,j=1}^{n} \left\{\frac{2^{i+j-1} - 1}{i + j - 1}\right\} (i - 1)(j - 1)c_i c_j$$

$$k_1(\mathbf{c}) = -1 + \sum_{j=1}^{n} c_j$$

$$k_2(\mathbf{c}) = -2 + \sum_{j=1}^{n} 2^{j-1} c_j$$

for any n-tuple $\mathbf{c} = (c_1, c_2, \ldots, c_n)$ in $\mathcal{R}_n$. (The reader should verify these results.) The required partial derivatives of these functions are easily found to be given as

$$\frac{\partial h(\mathbf{c})}{\partial c_j} = 2\sum_{i=1}^{n}\left\{\frac{2^{i+j-1}-1}{i+j-1}\right\}(i-1)(j-1)c_i$$

$$\frac{\partial k_1(\mathbf{c})}{\partial c_j} = 1$$

$$\frac{\partial k_2(\mathbf{c})}{\partial c_j} = 2^{j-1} \qquad \text{for } j = 1, 2, \ldots, n.$$

Hence in this case the equations of (7.8.4) become

$$2\sum_{i=1}^{n}\left\{\frac{2^{i+j-1}-1}{i+j-1}\right\}(i-1)(j-1)c_i^* = \lambda_1 + \lambda_2 2^{j-1} \qquad \text{for } j = 1, 2, \ldots, n,$$

while the constraint equations of (7.8.5) become

$$\sum_{i=1}^{n} c_i^* = 1 \qquad \text{and} \qquad \sum_{i=1}^{n} 2^{i-1}c_i^* = 2.$$

Hence we have $n + 2$ equations with which to determine the $n + 2$ numbers $c_1^*, c_2^*, \ldots, c_n^*, \lambda_1$, and λ_2. For example, if we consider the case $n = 3$, we can solve the resulting five equations to find

$$c_1^* = -\frac{5}{4}, \qquad c_2^* = \frac{23}{8}, \qquad c_3^* = -\frac{5}{8}, \qquad \lambda_1 = -\frac{239}{48}, \qquad \text{and}$$

$$\lambda_2 = \frac{239}{48} \qquad \text{in the case } n = 3.$$

(The reader should check these results.) Hence in this case the minimizing vector Y_3^* in $\mathfrak{M}_3$ becomes $Y_3^* = c_1^*Y_1 + c_2^*Y_2 + c_3^*Y_3$ or

$$Y_3^*(t) = -\frac{5}{4} + \frac{23t}{8} - \frac{5t^2}{8} \qquad \text{for } 1 \le t \le 2.$$

The corresponding minimum value of the functional H over $\mathfrak{M}_3$ subject to the given boundary constraints is then found to be

$$H(Y_3^*) = 2\tfrac{1}{48},$$

which gives a remarkably good approximation to the corresponding minimum value over $\mathcal{C}^1[1, 2]$, $H(Y^*) = 2$.

8. Some Remarks on the Use of the Second Variation in Extremum Problems

We want to generalize the second derivative test for a maximum or minimum. This will force us to define the analogue of the second derivative, namely, the *second variation*, of a functional. We shall discuss briefly some of the uses of the second variation in the study of extremum problems, and we shall also indicate some of the difficulties that can arise in this regard in the study of functionals on *infinite* dimensional vector spaces.

8.1. Higher-Order Variations

If J is a functional defined on an open subset D of a normed vector space $\mathfrak{X}$, then the variation of J at a vector x in D has been defined by

$$\delta J(x; \Delta x) = \frac{d}{d\epsilon} J(x + \epsilon\, \Delta x)\bigg|_{\epsilon=0}$$

provided that the expression $J(x + \epsilon\, \Delta x)$ is differentiable with respect to the numerical parameter ϵ at $\epsilon = 0$ for each vector Δx in $\mathfrak{X}$. For reasons which will appear shortly, it will be convenient now to refer to this variation as the *first* variation of J at x. We have seen in the previous chapters that this first variation of J can be used to solve various types of extremum problems which involve the given functional J.

We now wish to introduce the notion of *higher-order variations* of a functional. If J is defined on an open subset D of $\mathfrak{X}$ as before, then the ***n*th variation** of J at a vector x in D is defined by

$$\delta^n J(x; \Delta x) = \frac{d^n}{d\epsilon^n} J(x + \epsilon\, \Delta x)\bigg|_{\epsilon=0} \tag{8.1.1}$$

provided that the expression $J(x + \epsilon\, \Delta x)$ is n times differentiable with respect to the numerical parameter ϵ at $\epsilon = 0$ for every vector Δx in $\mathfrak{X}$. It follows directly from this definition that the earlier $n - 1$ variations δJ, $\delta^2 J$, $\delta^3 J$, $\ldots, \delta^{n-1} J$ must all exist at x if the nth variation exists at x. Moreover, it is easy to check that the nth variation must satisfy the *homogeneity relation*

$$\delta^n J(x; a\, \Delta x) = a^n\, \delta^n J(x; \Delta x) \tag{8.1.2}$$

for any number a since

$$\begin{aligned} a^n\, \delta^n J(x; \Delta x) &= a^n \frac{d^n}{d\sigma^n} J(x + \sigma\, \Delta x)\bigg|_{\sigma=0} \\ &= \frac{d^n}{d\epsilon^n} J(x + \epsilon a\, \Delta x)\bigg|_{\epsilon=0} = \delta^n J(x; a\, \Delta x), \end{aligned}$$

where we have put $\sigma = \epsilon a$ with $dJ/d\epsilon = a\, dJ/d\sigma$, etc. In particular, the *second variation of J at x* is defined as

$$\delta^2 J(x; \Delta x) = \frac{d^2}{d\epsilon^2} J(x + \epsilon\, \Delta x)\bigg|_{\epsilon=0}$$

for any vector Δx in $\mathfrak{X}$ and satisfies the relation

$$\delta^2 J(x; a\, \Delta x) = a^2\, \delta^2 J(x; \Delta x)$$

for any vector Δx and for any number a.

We shall see later in this chapter that the second variation has certain uses in the study of various extremum problems. First, however, we shall see how to calculate the second variations of several different functionals.

The simplest case occurs if J is an ordinary real-valued function defined for all *numbers* x in some given open interval $a < x < b$. In this case we find by the chain rule of differential calculus that

$$\frac{d}{d\epsilon} J(x + \epsilon \, \Delta x) = J'(x + \epsilon \, \Delta x) \, \Delta x$$

and

$$\frac{d^2}{d\epsilon^2} J(x + \epsilon \, \Delta x) = J''(x + \epsilon \, \Delta x)(\Delta x)^2$$

for any suitable numbers x, Δx, and ϵ, from which

$$\delta^2 J(x; \Delta x) = J''(x) \, (\Delta x)^2$$

for any real number Δx. Here $J''(x)$ denotes the ordinary second derivative of the function J at x.

Another simple case occurs if $J = J(x)$ is a real-valued function defined for all n-tuples $x = (x_1, x_2, \ldots, x_n)$ in some given open region of $\mathcal{R}_n$. In this case the chain rule gives

$$\frac{d}{d\epsilon} J(x + \epsilon \, \Delta x) = \sum_{i=1}^{n} \left[\frac{\partial}{\partial x_i} J(x + \epsilon \, \Delta x) \right] \Delta x_i$$

and

$$\frac{d^2}{d\epsilon^2} J(x + \epsilon \, \Delta x) = \sum_{i,j=1}^{n} \frac{\partial^2 J(x + \epsilon \, \Delta x)}{\partial x_i \, \partial x_j} \Delta x_i \, \Delta x_j,$$

from which we find that

$$\delta^2 J(x; \Delta x) = \sum_{i,j=1}^{n} \frac{\partial^2 J(x)}{\partial x_i \, \delta x_j} \Delta x_i \, \Delta x_j$$

for any vector $\Delta x = (\Delta x_1, \Delta x_2, \ldots, \Delta x_n)$ in $\mathcal{R}_n$.

Finally, if the functional J has the form

$$J(Y) = \int_{t_0}^{t_1} F(t, Y(t), Y'(t)) \, dt$$

for any vector $Y = Y(t)$ in the vector space $\mathcal{C}^1[t_0, t_1]$ (or any Y in some given open subset of $\mathcal{C}^1[t_0, t_1]$ relative to some given norm), then

$$J(Y + \epsilon \, \Delta Y) = \int_{t_0}^{t_1} F(t, Y(t) + \epsilon \, \Delta Y(t), Y'(t) + \epsilon \, \Delta Y'(t)) \, dt,$$

and we find that

$$\begin{aligned} \frac{d}{d\epsilon} J(Y + \epsilon \, \Delta Y) = \int_{t_0}^{t_1} & [F_Y(t, Y(t) + \epsilon \, \Delta Y(t), Y'(t) + \epsilon \, \Delta Y'(t)) \, \Delta Y(t) \\ & + F_{Y'}(t, Y(t) + \epsilon \, \Delta Y(t), Y'(t) + \epsilon \, \Delta Y'(t)) \, \Delta Y'(t)] \, dt \end{aligned}$$

and

$$\frac{d^2}{d\epsilon^2}J(Y + \epsilon\,\Delta Y) = \int_{t_0}^{t_1} \{F_{YY}(t, Y(t) + \epsilon\,\Delta Y(t), Y'(t) + \epsilon\,\Delta Y'(t))[\Delta Y(t)]^2$$
$$+ 2F_{YY'}(t, Y(t) + \epsilon\,\Delta Y(t), Y'(t) + \epsilon\,\Delta Y'(t))\,\Delta Y(t)\,\Delta Y'(t)$$
$$+ F_{Y'Y'}(t, Y(t) + \epsilon\,\Delta Y(t), Y'(t) + \epsilon\,\Delta Y'(t))[\Delta Y'(t)]^2\}\,dt.$$

Hence in this case the second variation at Y is given as

$$\delta^2 J(Y; \Delta Y) = \int_{t_0}^{t_1} \{F_{YY}(t, Y(t), Y'(t))[\Delta Y(t)]^2$$
$$+ 2F_{YY'}(t, Y(t), Y'(t))\,\Delta Y(t)\,\Delta Y'(t)$$
$$+ F_{Y'Y'}(t, Y(t), Y'(t))[\Delta Y'(t)]^2\}\,dt$$

for any vector $\Delta Y = \Delta Y(t)$ in $\mathcal{C}^1[t_0, t_1]$. Here $F = F(t, u, w)$ is assumed to be a given smooth function of three real variables, and the expressions F_{YY}, $F_{YY'}$, and $F_{Y'Y'}$ appearing in this formula for the second variation are defined as

$$F_{YY}(t, Y(t), Y'(t)) = \frac{\partial^2}{\partial u^2}F(t, u, w)\Big|_{\substack{u=Y(t)\\ w=Y'(t)}}$$

$$F_{YY'}(t, Y(t), Y'(t)) = \frac{\partial^2}{\partial u\,\partial w}F(t, u, w)\Big|_{\substack{u=Y(t)\\ w=Y'(t)}}$$

$$F_{Y'Y'}(t, Y(t), Y'(t)) = \frac{\partial^2}{\partial w^2}F(t, u, w)\Big|_{\substack{u=Y(t)\\ w=Y'(t)}}.$$

Exercises

1. Calculate the second variation of the functional J defined on the vector space $\mathcal{C}^1[t_0, t_1]$ by $J(Y) = \int_{t_0}^{t_1} [Y(t)^2 + Y'(t)^2 - 2Y(t)\sin t]\,dt$.

2. Calculate the second variation of the functional J defined on the vector space $\mathcal{C}^1[t_0, t_1]$ by $J(Y) = \int_{t_0}^{t_1} t^2\,Y'(t)^3\,dt$.

3. Let J be a given functional defined on an open subset D of a normed vector space $\mathfrak{X}$ and assume that the expression $J(x + \epsilon\,\Delta x)$ is n times continuously differentiable with respect to the numerical parameter ϵ for all numbers ϵ near $\epsilon = 0$, for a fixed vector x in D and for any vector Δx in $\mathfrak{X}$. Show that $J(x + \epsilon\,\Delta x)$ can be given as

$$J(x + \epsilon\,\Delta x) = \sum_{k=0}^{n} \frac{\epsilon^k}{k!}\,\delta^k J(x; \Delta x) + R_n(x; \Delta x; \epsilon),$$

with

$$R_n(x; \Delta x; \epsilon) = \int_0^{\epsilon} \frac{(\epsilon - \sigma)^{n-1}}{(n-1)!}\left[\frac{d^n}{d\sigma^n}J(x + \sigma\,\Delta x) - \delta^n J(x; \Delta x)\right]d\sigma.$$

Hint: The integral defining $R_n(x; \Delta x; \epsilon)$ can be evaluated through n integrations by parts.

4. Show that the expression $R_n(x; \Delta x; \epsilon)$ of Exercise 3 satisfies the estimate

$$|R_n(x; \Delta x; \epsilon)| \leq \frac{|\epsilon|^n}{n!} \max_{|\sigma| \leq |\epsilon|} \left| \frac{d^n}{d\sigma^n} J(x + \sigma \Delta x) - \delta^n J(x; \Delta x) \right|$$

for all small numbers ϵ. (Hence $\operatorname{limit}_{\epsilon \to 0} [R_n(x; \Delta x; \epsilon)/\epsilon^n] = 0$ will hold provided that $J(x + \epsilon \Delta x)$ is n times continuously differentiable near $\epsilon = 0$.)

5. Let J be any functional defined on an open subset D of a normed vector space $\mathfrak{X}$ and assume that the expression $J(x + \epsilon \Delta x)$ is twice continuously differentiable with respect to the numerical parameter ϵ for all ϵ near $\epsilon = 0$. Show that the second variation of J at x can be given as $\delta^2 J(x; \Delta x) = \operatorname{limit}_{\epsilon \to 0} \{[J(x + \epsilon \Delta x) + J(x - \epsilon \Delta x) - 2J(x)]/\epsilon^2\}$. *Hint:* Use the results of Exercises 3 and 4.

8.2. A Necessary Condition Involving the Second Variation at an Extremum

We let J be a functional defined on an open subset D of a normed vector space $\mathfrak{X}$ as before. If x^* is a local extremum vector in D for J with

$$\delta J(x^*; \Delta x) = 0 \qquad \text{for every vector } \Delta x \text{ in } \mathfrak{X}$$

and if the expression $J(x^* + \epsilon \Delta x)$ is twice continuously differentiable near $\epsilon = 0$, then the results of Exercises 3 and 4 of Section 8.1 imply that

$$J(x^* + \epsilon \Delta x) - J(x^*) = \frac{\epsilon^2}{2} \delta^2 J(x^*; \Delta x) + R_2(x^*; \Delta x; \epsilon), \qquad (8.2.1)$$

where $R_2(x^*; \Delta x; \epsilon)$ satisfies

$$\operatorname*{limit}_{\epsilon \to 0} \frac{R_2(x^*; \Delta x; \epsilon)}{\epsilon^2} = 0. \qquad (8.2.2)$$

If x^* is a local *minimum* vector in D for J, then the inequality

$$J(x^* + \epsilon \Delta x) - J(x^*) \geq 0 \qquad \text{for all small numbers } \epsilon$$

will hold, so that in this case (8.2.1) implies that

$$\delta^2 J(x^*; \Delta x) + \frac{2}{\epsilon^2} R_2(x^*; \Delta x; \epsilon) \geq 0$$

for all small nonzero numbers ϵ. If we let ϵ tend toward zero in the last result, we find with (8.2.2) the necessary condition

$$\delta^2 J(x^*; \Delta x) \geq 0 \qquad \text{for every vector } \Delta x \text{ in } \mathfrak{X}, \qquad (8.2.3)$$

which must hold if x^* is a local *minimum* vector in D for J. Similarly, if x^* is a local *maximum* vector in D for J, we find the necessary condition

$$\delta^2 J(x^*; \Delta x) \leq 0 \qquad \text{for every vector } \Delta x \text{ in } \mathfrak{X}. \tag{8.2.4}$$

These necessary conditions (8.2.3) and (8.2.4) can sometimes be used to distinguish between the cases of a minimum vector and a maximum vector. However, most actual extremum problems which arise in practice involve certain constraints of various types, and the corresponding necessary conditions which involve the second variations are somewhat more complicated and more difficult to apply. For this reason these conditions which involve the second variation are not much used in practice (except in cases for which $\mathfrak{X}$ is a finite dimensional vector space, such as $\mathcal{R}_n$).

Exercises

1. Let $J = J(Y)$ be defined by $J(Y) = \int_0^1 [Y(t)^2 + Y(t)^3]\, dt$ for any function $Y = Y(t)$ of class $\mathcal{C}^0$ on the interval $[0, 1]$. Show that the constant functions $Y^*(t) = 0$ and $Y^{**}(t) = -\frac{2}{3}$ (for all $0 \leq t \leq 1$) are the only possible extremum vectors in $\mathcal{C}^0[0, 1]$ for J, and show that Y^* satisfies the condition (8.2.3) while Y^{**} satisfies (8.2.4).

2. Let J, Y^*, and Y^{**} be as in Exercise 1. Prove directly that $J(Y^* + \Delta Y) \geq J(Y^*)$ for all continuous functions $\Delta Y = \Delta Y(t)$ such that $\max_{0 \leq t \leq 1} |\Delta Y(t)| \leq 1$. Similarly, prove that $J(Y^{**} + \Delta Y) \leq J(Y^{**})$ for all $\Delta Y = \Delta Y(t)$ such that $\max_{0 \leq t \leq 1} |\Delta Y(t)| \leq 2$.

3. Let $K = K(Y)$ be defined by $K(Y) = \int_0^1 Y(t)^3\, dt$ for any function $Y = Y(t)$ in the normed vector space $\mathcal{C}^0[0, 1]$ equipped with the norm $\|Y\| = \max_{0 \leq t \leq 1} |Y(t)|$. Show that the zero vector $Y^* = 0$ satisfies the necessary conditions $\delta K(Y^*; \Delta Y) = 0$ and $\delta^2 K(Y^*; \Delta Y) \geq 0$ for every vector ΔY in $\mathcal{C}^0[0, 1]$ but that Y^* is *not* a local minimum vector for K.

8.3. Sufficient Conditions for a Local Extremum

As before, we let J be a functional defined on an open subset D of a normed vector space $\mathfrak{X}$, and we consider a vector x^* in D for which the first variation of J vanishes as

$$\delta J(x^*; \Delta x) = 0 \qquad \text{for every vector } \Delta x \text{ in } \mathfrak{X}. \tag{8.3.1}$$

We also assume that the second variation of J is nonnegative at x^* as

$$\delta^2 J(x^*; \Delta x) \geq 0 \qquad \text{for every vector } \Delta x \text{ in } \mathfrak{X}.$$

Hence we tentatively view x^* as a candidate which *may* be a local *minimum* vector in D for J [see (8.2.3)], although the example of Exercise 3 of Section 8.2 shows that x^* need not be such a local minimum vector. We seek suitable additional conditions which will be sufficient to guarantee that x^* is, in fact, a local minimum vector in D for J.

First, suppose that it is known that the second variation of J is actually *positive* at x^*, with

$$\delta^2 J(x^*; \Delta x) > 0 \qquad \text{for all nonzero vectors } \Delta x \text{ in } \mathfrak{X}. \tag{8.3.2}$$

As in Section 8.2, with (8.3.1) we have

$$J(x^* + \epsilon\, \Delta x) - J(x^*) = \frac{\epsilon^2}{2}\delta^2 J(x^*; \Delta x) + R_2(x^*; \Delta x; \epsilon),$$

where

$$\operatorname*{limit}_{\epsilon \to 0} \frac{R_2(x^*; \Delta x; \epsilon)}{\epsilon^2} = 0, \tag{8.3.3}$$

and it might be thought now that the condition (8.3.2) would guarantee that the vector x^* must actually be a local minimum vector for J. Indeed, this is true if $\mathfrak{X}$ is a finite dimensional vector space such as $\mathfrak{R}_n$. However, it need *not* be true in general, as is shown by the example of Exercise 1. This is perhaps not unexpected since the conditions (8.3.1), (8.3.2), and (8.3.3) are all largely independent of the particular norm being used on the vector space $\mathfrak{X}$, whereas the very notion of a *local* minimum vector x^* is in general dependent on the norm. (A vector $x^* + \Delta x$ may be close to x^* relative to one norm but *not* relative to another; see Section A2 of the Appendix.)

There are various strengthened conditions which involve the second variation of J at x^* and which are related to (8.3.2) and (8.3.3), and which if satisfied do imply that the vector x^* must be a local minimum vector for J. For example, if in addition to (8.3.1) J satisfies the condition

$$J(x^* + \Delta x) - J(x^*) = \tfrac{1}{2}\delta^2 J(x^*; \Delta x) + E_2(x^*; \Delta x),$$

where [compare with (8.3.3)]

$$\operatorname*{limit}_{\Delta x \to 0 \text{ in } \mathfrak{X}} \frac{E_2(x^*; \Delta x)}{\|\Delta x\|^2} = 0, \tag{8.3.4}$$

and if, moreover, there is some *positive constant* p for which the condition [compare with (8.3.2)]

$$\delta^2 J(x^*; \Delta x) \geq p\|\Delta x\|^2 \qquad \text{for all small vectors } \Delta x \text{ in } \mathfrak{X} \tag{8.3.5}$$

holds, then we find directly the inequality

$$J(x^* + \Delta x) - J(x^*) \geq \|\Delta x\|^2\left\{\frac{p}{2} + \frac{E_2(x^*; \Delta x)}{\|\Delta x\|^2}\right\}.$$

This inequality along with the condition (8.3.4) and the positiveness of the constant p now leads to the desired result:

$$J(x^* + \Delta x) - J(x^*) \geq 0 \qquad \text{for all sufficiently small vectors } \Delta x \text{ in } \mathfrak{X}.$$

Hence the conditions (8.3.4) and (8.3.5) along with (8.3.1) are sufficient to guarantee that x^* is a local minimum vector for J relative to the given norm. [It is clear that the validity of the conditions (8.3.4) and (8.3.5) will in general depend on the particular norm being used on the vector space $\mathfrak{X}$.] Unfortunately, these conditions are not widely applicable in practice since most extremum problems which arise in practice involve various constraints of a wide variety of different types. Moreover, the corresponding sufficient conditions based on the second variation are usually more complicated for such extremum problems with constraints. (See Exercise 2.) In fact, the actual verification that these sufficient conditions are satisfied in any particular case is often just as difficult as a direct verification that x^* is a minimum vector in the particular case. For this reason, throughout this book we have emphasized direct methods that can be used to determine whether or not a given candidate x^* is actually a solution to a given extremum problem. We shall not pursue any further the general study of sufficiency conditions based on the second and higher variations. We should mention, however, that there is a well-developed sufficiency theory for the classical problem of minimizing or maximizing a functional J of the form

$$J(Y) = \int_{t_0}^{t_1} F(t, Y(t), Y'(t))\, dt$$

subject only to the fixed end-point constraints $Y(t_0) = y_0$ and $Y(t_1) = y_1$.† In part this theory studies conditions under which it can be shown that (8.3.4) and (8.3.5) hold relative to certain norms, while in part the theory involves a study of certain direct methods that can be used to determine whether $J(Y)$ is maximized or minimized at a given potential extremum function $Y = Y^*$. One aspect of this theory involves a certain function $E = E(x, y, z_0, z)$, known as the *Weierstrass excess function*, which is defined by

$$E(x, y, z_0, z) = F(x, y, z) - [F(x, y, z_0) + F_z(x, y, z_0)(z - z_0)]$$

for any suitable numbers x, y, z_0, and z, where F is the integrand function appearing in the definition of the functional J. This Weierstrass excess function E is just the difference between the value of $F(x, y, z)$ and the first two terms of the Taylor expansion of F with respect to z about the point z_0. Hence E measures the vertical distance between the graph of the function $F(x, y, z)$, considered as a function of z alone, and the tangent to this graph at z_0. We

†See, for example, I. M. Gelfand and S. V. Fomin, *Calculus of Variations*, trans. Richard A. Silverman (Englewood Cliffs, N.J.: Prentice-Hall, Inc. 1963), and Hans Sagan, *Introduction to the Calculus of Variations* (New York: McGraw-Hill Book Company, 1969).

simply want to call attention to the existence of these results here; we refer the interested reader to the previously referenced book by Gelfand and Fomin and to the book by Sagan for the actual statement of the results.

Exercises

1. Let $J = J(Y)$ be defined by $J(Y) = \int_0^1 [Y(t)^2 - Y'(t)^4]\,dt$ for any vector Y in the vector space $\mathfrak{X}$ consisting of all continuously differentiable functions $Y(t)$ on $0 \leq t \leq 1$ satisfying the conditions $Y(0) = 0$, $Y(1) = 0$. Show that the first variation of J vanishes at the zero vector in $\mathfrak{X}$, $\delta J(0; \Delta Y) = 0$ for all vectors ΔY in $\mathfrak{X}$. Show also that the second variation of J is positive at the zero vector, $\delta^2 J(0; \Delta Y) > 0$ for all nonzero vectors ΔY in $\mathfrak{X}$. Finally, show that the zero vector is *not* a local minimum vector in $\mathfrak{X}$ for J relative to either of the norms $\|Y\|_1 = \sqrt{\int_0^1 Y(t)^2\,dt}$ or $\|Y\|_2 = \max_{0 \leq t \leq 1} |Y(t)|$. *Hint:* For the last result, consider the vectors $\Delta Y_1, \Delta Y_2, \Delta Y_3, \ldots$ given as $\Delta Y_n(t) = (\sin 2\pi n t)/\sqrt{n}$ for $n = 1, 2, 3, \ldots$.

2. Let x^* be a (candidate for a) local extremum vector in $D[K_i = k_i$ for $i = 1, 2, \ldots, m]$ for J, where $J, K_1, K_2, \ldots, K_m$ are given functionals defined on an open subset D of a normed vector space $\mathfrak{X}$, and where the constraint set $D[K_i = k_i$ for $i = 1, 2, \ldots, m]$ consists of all vectors x in D which satisfy the simultaneous constraints $K_i(x) = k_i$ for $i = 1, 2, \ldots, m$. Here $k_1, k_2, \ldots, k_m$ may be any given real numbers. Assume that the functionals involved all have first variations which are weakly continuous near x^*, and assume that the given constraints are independent at x^* in the sense that the condition (3.6.2) *fails* to hold. Then the Euler-Lagrange multiplier theorem guarantees that there will be certain multipliers $\lambda_1, \lambda_2, \ldots, \lambda_m$ such that

$$\delta J(x^*; \Delta x) = \sum_{i=1}^{m} \lambda_i\, \delta K_i(x^*; \Delta x) \qquad \text{for all vectors } \Delta x \text{ in } \mathfrak{X}. \tag{8.3.6}$$

Let the functional $\mathcal{J}$ be defined by $\mathcal{J}(x) = J(x) - \sum_{i=1}^{m} \lambda_i K_i(x)$ for any vector x in D, where $\lambda_1, \lambda_2, \ldots, \lambda_m$ are the same Euler-Lagrange multipliers which appear in (8.3.6). *Show that x^* must actually be a local minimum vector in $D[K_i = k_i$ for $i = 1, 2, \ldots, m]$ for J* if x^* is in the constraint set and (8.3.6) holds and if in addition the two conditions

$$\mathcal{J}(x^* + \Delta x) - \mathcal{J}(x^*) = \tfrac{1}{2}\delta^2 \mathcal{J}(x^*; \Delta x) + \mathcal{E}_2(x^*; \Delta x),$$

with [compare with (8.3.4)] $\operatorname{limit}_{\Delta x \to 0 \text{ in } \mathfrak{X}} [\mathcal{E}_2(x^*; \Delta x)/\|\Delta x\|^2] = 0$, and [compare with (8.3.5)]

$\delta^2 \mathcal{J}(x^*; \Delta x) \geq p\|\Delta x\|^2$ for all small vectors Δx for which $x^* + \Delta x$ is in the given constraint set

hold. Here p may be any fixed positive constant. *Hint:* $\mathcal{J}(x^* + \Delta x) - \mathcal{J}(x^*) = J(x^* + \Delta x) - J(x^*)$ holds for any vector $x^* + \Delta x$ in the constraint set.

Appendix

A1. The Cauchy and Schwarz Inequalities

We first prove Cauchy's inequality (1.4.5), which can be written as

$$(x \cdot y)^2 \leq (x \cdot x)(y \cdot y) \tag{A1.1}$$

for any n-tuples $x = (x_1, x_2, \ldots, x_n)$ and $y = (y_1, y_2, \ldots, y_n)$ in $\mathcal{R}_n$, where the *inner product* (dot product) $x \cdot y$ is defined by

$$x \cdot y = \sum_{i=1}^{n} x_i y_i$$

for any two vectors $x = (x_1, x_2, \ldots, x_n)$ and $y = (y_1, y_2, \ldots, y_n)$.

Since the inequality (A1.1) clearly holds if $y \cdot y = 0$ (why?), we shall consider further only the case $y \cdot y > 0$. We begin with the obvious inequality $z \cdot z \geq 0$ which holds for any n-tuple $z = (z_1, z_2, \ldots, z_n)$. If we take $z = x + ty$ for any fixed real number t, we then find $(x + ty) \cdot (x + ty) \geq 0$, where $(x + ty) \cdot (x + ty) = x \cdot x + 2tx \cdot y + t^2 y \cdot y$, as follows directly from the definition of the inner product. Hence

$$x \cdot x + 2tx \cdot y + t^2 y \cdot y \geq 0 \tag{A1.2}$$

holds for any number t. The desired inequality (A1.1) now follows easily from this last inequality if we take for t the value (recall that $y \cdot y > 0$) $t = -x \cdot y/y \cdot y$. We leave the details as an exercise for the reader.

The same method of calculation can be used to prove Schwarz's inequality (1.4.7), which can be written as

$$(\phi \cdot \psi)^2 \leq (\phi \cdot \phi)(\psi \cdot \psi) \tag{A1.3}$$

for any pair of continuous functions $\phi = \phi(x)$ and $\psi = \psi(x)$ on the fixed interval $a \leq x \leq b$, where in this case the inner product $\phi \cdot \psi$ is defined by

$$\phi \cdot \psi = \int_a^b \phi(x)\psi(x)\, dx$$

for any two continuous functions $\phi = \phi(x)$ and $\psi = \psi(x)$. In this case we find, as before, that [compare with (A1.2)]

$$\phi \cdot \phi + 2t\phi \cdot \psi + t^2\psi \cdot \psi \geq 0$$

for any number t, and the desired result (A1.3) can be shown to follow if we take $t = -\phi \cdot \psi/\psi \cdot \psi$. (Again we need only consider the case $\psi \cdot \psi > 0$.) We leave the details to the reader.

A2. An Example on Normed Vector Spaces

We give an example of a fixed vector space for which different norms can lead to strikingly different normed vector spaces. We let $\mathfrak{X}_1$ be the normed vector space given by the vector space $\mathcal{C}^0[a, b]$ with norm $\|\cdot\|_1$ defined by formula (1.4.6) as

$$\|\phi\|_1 = \sqrt{\int_a^b \phi(x)^2\, dx}$$

for any function ϕ in $\mathcal{C}^0[a, b]$, and we let $\mathfrak{X}_2$ be the normed vector space given by the same vector space $\mathcal{C}^0[a, b]$ but with norm $\|\cdot\|_2$ defined by formula (1.4.8) as

$$\|\phi\|_2 = \max_{a \leq x \leq b} |\phi(x)|$$

for any continuous function ϕ on $[a, b]$. Here $[a, b]$ is any given fixed closed interval. Even though $\mathfrak{X}_1$ and $\mathfrak{X}_2$ are constructed from the same vector space $\mathcal{C}^0[a, b]$, we shall see that they are quite different as normed vector spaces. In fact, there are vectors in $\mathcal{C}^0[a, b]$ with extremely *small* norm in $\mathfrak{X}_1$ but with extremely *large* norm in $\mathfrak{X}_2$.

For example, the function ϕ_k defined on the interval [0, 2] by (see Figure

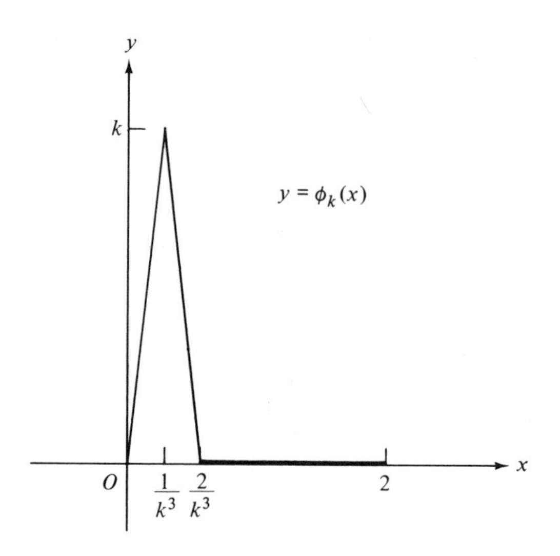

Figure A.1

A.1)

$$\phi_k(x) = \begin{cases} k^4 x & \text{for } 0 \le x \le k^{-3} \\ -k^4 x + 2k & \text{for } k^{-3} \le x \le 2k^{-3} \\ 0 & \text{for } 2k^{-3} \le x \le 2 \end{cases}$$

is of class $\mathcal{C}^0$ on [0, 2] for each integer $k = 1, 2, 3, \ldots$, and a calculation using the definition of the norm $\| \cdot \|_1$ shows in this case that

$$\| \phi_k \|_1 = \sqrt{\frac{2}{3k}},$$

while clearly (why?)

$$\| \phi_k \|_2 = k.$$

Hence

$$\lim_{k \to \infty} \| \phi_k \|_1 = 0,$$

while

$$\lim_{k \to \infty} \| \phi_k \|_2 = +\infty,$$

so that for large integers k the vectors ϕ_k are, on the one hand, extremely large in $\mathfrak{X}_2$ and, on the other hand, very small in $\mathfrak{X}_1$.

This situation, with a given sequence of vectors in a vector space $\mathfrak{X}$ becoming *unbounded* in one norm while becoming small in another norm, cannot occur in finite dimensional vector spaces such as $\mathcal{R}_n$ since all norms are "equivalent" on any finite dimensional space.†

†See Richard E. Williamson, Richard H. Crowell, and Hale F. Trotter, *Calculus of Vector Functions* (Englewood Cliffs, N.J.: Prentice-Hall, Inc., 1968), pp. 407–410.

A3. An Integral Inequality

We shall give one of the standard proofs for the inequality [see (1.5.9)]

$$\left|\int_a^b f(x)\,dx\right| \leq \int_a^b |f(x)|\,dx \tag{A3.1}$$

which holds for any continuous function f on the interval $[a, b]$, a $< b$. The crucial fact that we shall use about the integral is the following: *The integral of a nonnegative function is nonnegative*; i.e.,

$$\int_a^b g(x)\,dx \geq 0 \qquad \text{if } g(x) \geq 0 \quad \text{for } a \leq x \leq b. \tag{A3.2}$$

We first consider the function g defined by

$$g(x) = |f(x)| - f(x) \qquad \text{for } a \leq x \leq b.$$

Since $g(x) \geq 0$, in this case we can use (A3.2) to find that

$$0 \leq \int_a^b g(x)\,dx = \int_a^b |f(x)|\,dx - \int_a^b f(x)\,dx$$

or that

$$\int_a^b f(x)\,dx \leq \int_a^b |f(x)|\,dx. \tag{A3.3}$$

On the other hand, if we define g by

$$g(x) = |f(x)| + f(x) \qquad \text{for } a \leq x \leq b,$$

we find similarly that

$$-\int_a^b |f(x)|\,dx \leq \int_a^b f(x)\,dx. \tag{A3.4}$$

The desired inequality (A3.1) follows directly from the two results (A3.3) and (A3.4), as the reader can easily verify.

A4. A Fundamental Lemma of the Calculus of Variations

Most of the early work done in the calculus of variations was somewhat vague with respect to the fundamental concepts involved. This situation began to change in 1879 with the appearance of an important paper by Paul

Du Bois-Reymond.† In this paper Du Bois-Reymond proved the following basic lemma, and he also obtained the results given below in Section A5 of the Appendix.

Lemma (*Du Bois-Reymond*). *Let $f = f(x)$ be any given continuous real-valued function on the closed interval* $[a, b]$, *and suppose that for some nonnegative integer n* ($n = 0, 1, 2, \ldots$)

$$\int_a^b f(x)h(x)\,dx = 0 \tag{A4.1}$$

holds for all functions h of class $\mathfrak{C}^n$ on $[a, b]$ *which vanish at the end points along with their derivatives of order up to and including order n,*

$$h^{(k)}(a) = h^{(k)}(b) = 0 \qquad \text{for } k = 0, 1, 2, \ldots, n.$$

Then f must necessarily vanish identically on $[a, b]$*;* i.e., $f(x) = 0$ *for all* $a \leq x \leq b$.

Proof. We shall show that $f(x)$ must vanish for each interior point x in the open interval (a, b). Then by continuity f must vanish on $[a, b]$. We argue by contradiction. Suppose that there is some interior point x^*, $a < x^* < b$, for which $f(x^*) \neq 0$. Then either $f(x^*) > 0$ or $f(x^*) < 0$, and for definiteness we shall assume that

$$f(x^*) > 0.$$

(The other case can be reduced to this case by considering $-f$.) Since f is assumed to be continuous, it follows in this case that f is everywhere positive in some small interval containing x^*, say

$$f(x) > 0 \qquad \text{for all} \qquad \alpha < x < \beta, \tag{A4.2}$$

for some suitable numbers α and β satisfying $a < \alpha < x^* < \beta < b$. We now define a function h^* by

$$h^*(x) = \begin{cases} (x-\alpha)^{n+1}(\beta - x)^{n+1} & \text{for } \alpha \leq x \leq \beta \\ 0 & \text{for } a \leq x < \alpha \text{ and} \\ & \quad \text{for } \beta < x \leq b. \end{cases}$$

It is easy to check that this function h^* is of class $\mathfrak{C}^n$ on $[a, b]$, while h^* clearly vanishes at the end points $x = a$ and $x = b$ along with all its derivatives. Hence h^* is an admissible function which may be used in the relation (A4.1),

†Paul Du Bois-Reymond, "Erläuterungen zu den Anfangsgründen der Variationsrechnung," *Mathematische Annalen*, **15** (1879), 283–314.

yielding

$$\int_{\alpha}^{\beta} f(x)(x-\alpha)^{n+1}(\beta-x)^{n+1}\,dx = 0 \tag{A4.3}$$

since h^* vanishes outside the interval (α, β). But this result (A4.3) is certainly *not possible*. Indeed, $x-\alpha > 0$ and $\beta - x > 0$ for x in (α, β), and therefore $(x-\alpha)^{n+1}(\beta-x)^{n+1} > 0$ for all x in (α, β), from which we find with (A4.2) that

$$f(x)(x-\alpha)^{n+1}(\beta-x)^{n+1} > 0$$

for all x in (α, β). This result is clearly incompatible with (A4.3), and therefore the original assumption that $f(x^*) \neq 0$ is impossible. This completes the proof of the lemma.

A5. Du Bois-Reymond's Derivation of the Euler-Lagrange Equation

If $Y = Y(x)$ is any local extremum vector in $D[K_0 = y_0, K_1 = y_1]$ for the functional J of (4.1.1), then equation (4.1.6) implies that

$$\int_{x_0}^{x_1} [F_Y(x, Y(x), Y'(x))\,\Delta Y(x) + F_{Y'}(x, Y(x), Y'(x))\,\Delta Y'(x)]\,dx = 0 \tag{A5.1}$$

for all continuously differentiable functions $\Delta Y = \Delta Y(x)$ which vanish at the end points $x = x_0$ and $x = x_1$. We recall that Y is assumed to be continuously differentiable, so that the two functions given by $F_Y(x, Y(x), Y'(x))$ and $F_{Y'}(x, Y(x), Y'(x))$ may be assumed to be *continuous* for $x_0 \leq x \leq x_1$.

We wish to derive the Euler-Lagrange equation (4.1.11) from (A5.1) *without* assuming that the function $F_{Y'}(x, Y(x), Y'(x))$ is differentiable. The method of Lagrange involves an integration by parts as given by equation (4.1.8), which requires the differentiability of this function. Du Bois-Reymond discovered the following method which eliminates this additional requirement. We define a function $g = g(x)$ by

$$g(x) = \int_{x_0}^{x} F_Y(\xi, Y(\xi), Y'(\xi))\,d\xi.$$

By the fundamental theorem of calculus it follows that this function g is continuously differentiable with

$$g'(x) = F_Y(x, Y(x), Y'(x)), \tag{A5.2}$$

so that

$$\int_{x_0}^{x_1} F_Y(x,\ Y(x)\ Y'(x))\,\Delta Y(x)\,dx = \int_{x_0}^{x_1} g'(x)\,\Delta Y(x)\,dx$$

holds. But now we can integrate this last integral by parts and find that

$$\int_{x_0}^{x_1} F_Y(x,\ Y(x),\ Y'(x))\,\Delta Y(x)\,dx = -\int_{x_0}^{x_1} g(x)\,\Delta Y'(x)\,dx$$

for any continuously differentiable function $\Delta Y = \Delta Y(x)$ which vanishes at the end points x_0 and x_1. Hence equation (A5.1) can be rewritten in the form

$$\int_{x_0}^{x_1} [-g(x) + F_{Y'}(x,\ Y(x),\ Y'(x))]\,\Delta Y'(x)\,dx = 0 \tag{A5.3}$$

for all functions ΔY of class $\mathcal{C}^1$ on $[x_0, x_1]$ which satisfy the conditions

$$\Delta Y(x_0) = 0, \qquad \Delta Y(x_1) = 0. \tag{A5.4}$$

We shall now show that the condition (A5.3) implies that the function $-g + F_{Y'}$ *must be everywhere constant.* Of course if this is true, then this constant value, say c, must be given as the following average value:

$$\frac{1}{x_1 - x_0}\int_{x_0}^{x_1} [-g(x) + F_{Y'}(x,\ Y(x),\ Y'(x))]\,dx = c. \tag{A5.5}$$

Since

$$\int_{x_0}^{x_1} c\,\Delta Y'(x)\,dx = c[\Delta Y(x_1) - \Delta Y(x_0)] = 0$$

holds for any constant c and for any function ΔY which satisfies (A5.4), we can modify (A5.3) and obtain

$$\int_{x_0}^{x_1} [-g(x) + F_{Y'}(x,\ Y(x),\ Y'(x)) - c]\,\Delta Y'(x)\,dx = 0, \tag{A5.6}$$

which must still hold for all functions ΔY of class $\mathcal{C}^1$ which satisfy (A5.4). If we can take $\Delta Y'$ in (A5.6) to be equal to the function appearing in square brackets there, we would find that

$$\int_{x_0}^{x_1} [-g(x) + F_{Y'}(x,\ Y(x),\ Y'(x)) - c]^2\,dx = 0,$$

which would indeed lead to the asserted result that $-g + F_{Y'}$ is everywhere constant. Hence we try to choose some particular admissible function $\Delta Y = \Delta Y(x)$ such that

$$\Delta Y'(x) = -g(x) + F_{Y'}(x,\ Y(x),\ Y'(x)) - c. \tag{A5.7}$$

Since we want ΔY to vanish at $x = x_0$, we take

$$\Delta Y(x) = \int_{x_0}^{x} [-g(\xi) + F_{Y'}(\xi, Y(\xi), Y'(\xi)) - c]\, d\xi, \tag{A5.8}$$

which then satisfies (A5.7) and also satisfies the requirement $\Delta Y(x_0) = 0$. However, we also require the condition $\Delta Y(x_1) = 0$, and this will hold for (A5.8) if and only if the constant c is defined by (A5.5). (The reader should verify this fact.) Hence with this special choice of the constant c we can obtain a suitable admissible function ΔY in (A5.8), and the above argument then leads to the asserted result

$$-g(x) + F_{Y'}(x, Y(x), Y'(x)) = c \tag{A5.9}$$

for all $x_0 \leq x \leq x_1$.

But now the last result leads directly to the Euler-Lagrange equation. Indeed, it now follows from (A5.9) that the function $F_{Y'}(x, Y(x), Y'(x))$ is automatically continuously differentiable for the extremum function $Y = Y(x)$ since the function $g(x) + c$ is continuously differentiable and

$$F_{Y'}(x, Y(x), Y'(x)) = g(x) + c.$$

Hence we can differentiate this last equation on both sides and use (A5.2) to obtain the Euler-Lagrange equation (4.1.11).

Exercise. Let $f = f(x)$ be any given continuous real-valued function on the interval $a \leq x \leq b$ which is known to satisfy the condition

$$\int_{a}^{b} f(x)h''(x)\, dx = 0 \tag{A5.10}$$

for all functions h of class $\mathcal{C}^2$ on $[a, b]$ which vanish at the end points along with their first derivatives,

$$h(a) = h(b) = h'(a) = h'(b) = 0. \tag{A5.11}$$

Prove that f must be a constant plus a linear function; i.e., $f(x) = c_0 + c_1 x$ for some suitable constants c_0 and c_1. *Hints*: For any constants c_0 and c_1, show that it is possible to modify the condition (A5.10) so as to obtain the condition

$$\int_{a}^{b} [f(x) - (c_0 + c_1 x)]h''(x)\, dx = 0, \tag{A5.12}$$

which must still hold for all functions h of class $\mathcal{C}^2$ which satisfy (A5.11). Now show that it is possible to find a *particular* function h which satisfies

(A5.11) and which also satisfies the relation

$$h''(x) = f(x) - (c_0 + c_1 x)$$

for all x in $[a, b]$. The desired result should then follow easily. Are the constants c_0 and c_1 uniquely determined?

A6. A Useful Result from Calculus

We shall give here a proof of [see (4.4.14)]

$$\frac{d}{d\epsilon}\int_{x_0}^{\xi(\epsilon)} f(x;\epsilon)\,dx = f(\xi(\epsilon);\epsilon)\xi'(\epsilon) + \int_{x_0}^{\xi(\epsilon)} \frac{\partial}{\partial\epsilon} f(x;\epsilon)\,dx \tag{A6.1}$$

for any continuously differentiable function $\xi = \xi(\epsilon)$ depending on ϵ and for any smooth function $f = f(x; \epsilon)$ depending on x and ϵ. Specifically, we shall assume that $f = f(x; \epsilon)$ is continuous with respect to the first argument x (for each fixed value of ϵ) and differentiable with respect to the second argument ϵ (for each fixed value of x). Moreover, we shall assume that the resulting derivative $\partial f(x; \epsilon)/\partial\epsilon$ is Riemann integrable with respect to x over the interval between x_0 and $\xi(\epsilon)$, and we assume also that the quotient

$$\frac{f(x;\epsilon+\alpha) - f(x;\epsilon)}{\alpha} \tag{A6.2}$$

is uniformly bounded for all x between x_0 and $\xi(\epsilon)$ as the number α tends toward $\alpha = 0$. [This quotient tends to the derivative $\partial f(x; \epsilon)/\partial\epsilon$ as $\alpha \longrightarrow 0$, for each fixed x.] There are other conditions on f which also ensure the validity of (A6.1), but the present conditions suit our needs.

We shall denote the integral

$$\int_{x_0}^{\xi(\epsilon)} f(x;\epsilon)\,dx$$

as $h = h(\epsilon)$, i.e.,

$$h(\epsilon) = \int_{x_0}^{\xi(\epsilon)} f(x;\epsilon)\,dx, \tag{A6.3}$$

and we wish to prove that h is differentiable with respect to ϵ, with $h'(\epsilon)$ given by the expression on the right-hand side of equation (A6.1). To this end we shall calculate the difference quotient

$$\frac{h(\epsilon+\alpha) - h(\epsilon)}{\alpha}$$

for an arbitrary small nonzero number α. From (A6.3) we find after a brief calculation that

$$\frac{h(\epsilon+\alpha)-h(\epsilon)}{\alpha} = \frac{1}{\alpha}\int_{\xi(\epsilon)}^{\xi(\epsilon+\alpha)} f(x;\epsilon)\,dx + \int_{x_0}^{\xi(\epsilon+\alpha)} \frac{f(x;\epsilon+\alpha)-f(x;\epsilon)}{\alpha}\,dx. \tag{A6.4}$$

Since

$$\frac{d}{d\epsilon}\int_{x_0}^{\xi(\epsilon)} f(x;\epsilon)\,dx = h'(\epsilon) = \operatorname*{limit}_{\alpha\to 0}\frac{h(\epsilon+\alpha)-h(\epsilon)}{\alpha},$$

it is clear now that (A6.1) will follow from (A6.4) upon passing to the limit $\alpha \longrightarrow 0$ *provided that*

$$\operatorname*{limit}_{\alpha\to 0}\frac{1}{\alpha}\int_{\xi(\epsilon)}^{\xi(\epsilon+\alpha)} f(x;\epsilon)\,dx = f(\xi(\epsilon);\epsilon)\,\xi'(\epsilon) \tag{A6.5}$$

and

$$\operatorname*{limit}_{\alpha\to 0}\int_{x_0}^{\xi(\epsilon+\alpha)} \frac{f(x;\epsilon+\alpha)-f(x;\epsilon)}{\alpha}\,dx = \int_{x_0}^{\xi(\epsilon)} \frac{\partial}{\partial\epsilon}f(x;\epsilon)\,dx \tag{A6.6}$$

hold. Hence we must now prove the validity of these last two relations.

We consider (A6.5) first. In this case we find that

$$\begin{aligned}\frac{1}{\alpha}\int_{\xi(\epsilon)}^{\xi(\epsilon+\alpha)} f(x;\epsilon)\,dx &= \frac{f(\xi(\epsilon);\epsilon)}{\alpha}\int_{\xi(\epsilon)}^{\xi(\epsilon+\alpha)} dx + \int_{\xi(\epsilon)}^{\xi(\epsilon+\alpha)} \frac{f(x;\epsilon)-f(\xi(\epsilon);\epsilon)}{\alpha}\,dx \\ &= f(\xi(\epsilon);\epsilon)\frac{\xi(\epsilon+\alpha)-\xi(\epsilon)}{\alpha} \\ &\quad + \int_{\xi(\epsilon)}^{\xi(\epsilon+\alpha)} \frac{f(x;\epsilon)-f(\xi(\epsilon);\epsilon)}{\alpha}\,dx,\end{aligned}$$

and since $\xi'(\epsilon) = \operatorname{limit}_{\alpha\to 0}\,[\xi(\epsilon+\alpha)-\xi(\epsilon)]/\alpha$, it is clear from the last equation that (A6.5) will hold if and only if

$$\operatorname*{limit}_{\alpha\to 0}\int_{\xi(\epsilon)}^{\xi(\epsilon+\alpha)} \frac{f(x;\epsilon)-f(\xi(\epsilon);\epsilon)}{\alpha}\,dx = 0 \tag{A6.7}$$

holds. We can now appeal to the (uniform) continuity of f with respect to x and to the differentiability of ξ with respect to ϵ to prove the last result. Indeed, if δ is any given positive number, then there is a corresponding positive number α_0 such that

$$|\,f(x;\epsilon)-f(\xi(\epsilon);\epsilon)| \le \frac{\delta}{1+|\xi'(\epsilon)|}$$

for all x satisfying $|x - \xi(\epsilon)| \leq \alpha_0(1 + |\xi'(\epsilon)|)$, and such that

$$\left|\frac{\xi(\epsilon + \alpha) - \xi(\epsilon)}{\alpha}\right| \leq 1 + |\xi'(\epsilon)|$$

for all α satisfying $0 < |\alpha| \leq \alpha_0$. Using these estimates, we easily find that

$$\left|\int_{\xi(\epsilon)}^{\xi(\epsilon+\alpha)} \frac{f(x;\epsilon) - f(\xi(\epsilon);\epsilon)}{\alpha}\,dx\right| \leq \frac{\delta}{1 + |\xi'(\epsilon)|}\left|\frac{1}{\alpha}\int_{\xi(\epsilon)}^{\xi(\epsilon+\alpha)} dx\right|$$
$$= \frac{\delta}{1 + |\xi'(\epsilon)|} \cdot \frac{|\xi(\epsilon + \alpha) - \xi(\epsilon)|}{|\alpha|} \leq \delta$$

for all $0 < |\alpha| \leq \alpha_0$. This proves the validity of (A6.7) and therefore completes the proof of (A6.5).

We turn now to the proof of (A6.6). Since

$$\int_{x_0}^{\xi(\epsilon+\alpha)} \frac{f(x;\epsilon + \alpha) - f(x;\epsilon)}{\alpha}\,dx = \int_{x_0}^{\xi(\epsilon)} \frac{f(x;\epsilon + \alpha) - f(x;\epsilon)}{\alpha}\,dx$$
$$+ \int_{\xi(\epsilon)}^{\xi(\epsilon+\alpha)} \frac{f(x;\epsilon + \alpha) - f(x;\epsilon)}{\alpha}\,dx$$

holds, it is clear that (A6.6) will follow from

$$\lim_{\alpha \to 0} \int_{x_0}^{\xi(\epsilon)} \frac{f(x;\epsilon + \alpha) - f(x;\epsilon)}{\alpha}\,dx = \int_{x_0}^{\xi(\epsilon)} \frac{\partial}{\partial\epsilon} f(x;\epsilon)\,dx \tag{A6.8}$$

and

$$\lim_{\alpha \to 0} \int_{\xi(\epsilon)}^{\xi(\epsilon+\alpha)} \frac{f(x;\epsilon + \alpha) - f(x;\epsilon)}{\alpha}\,dx = 0.$$

The proof of the last result follows easily from the boundedness of the quotient given in (A6.2) along with the differentiability of $\xi = \xi(\epsilon)$, and we leave the details in this case to the reader. Finally, then, we need only prove the result (A6.8). But this result follows now directly from our stated hypotheses on f and the *Arzela bounded convergence theorem*.† This completes the proof of (A6.1).

A7. The Construction of a Certain Function

We show one way to construct a function $h = h(x)$ of class $\mathcal{C}^2$ on a given interval $x_0 \leq x \leq x_1$ subject to the requirements

$$h(x_0) = h'(x_0) = 0 \tag{A7.1}$$

†See Tom M. Apostol, *Mathematical Analysis* (Reading, Mass.: Addison-Wesley Publishing Company, Inc., 1957), p. 405.

and

$$h(x_1) = 0, \qquad h'(x_1) = 1. \tag{A7.2}$$

[Some such function is required in order to derive (4.9.19) from (4.9.18).] We assume without loss that $x_0 < x_1$.

We simply set $h(x) = A(x - x_0)^3 + B(x - x_0)^5$ for suitable constants A and B to be determined. No matter how we choose A and B, it is clear that the conditions of (A7.1) are automatically satisfied. Hence we use the two conditions of (A7.2) to determine the two constants A and B. In this way we find that

$$A = -\frac{1}{2(x_1 - x_0)^2}$$

$$B = \frac{1}{2(x_1 - x_0)^4}.$$

The resulting function clearly satisfies all the stated requirements. We leave the details to the reader.

A8. The Fundamental Lemma for the Case of Several Independent Variables

We shall prove here a result for functions of several independent variables which is analogous to the lemma proved in Section A4 of the Appendix for functions of one variable. Such a result is needed in deriving the Euler-Lagrange equation (4.10.15) from the condition (4.10.14). For simplicity we consider only the case of two independent variables here.

We let R denote any bounded open region in the (x, y)-plane with a piecewise smooth boundary curve ∂R, and we denote by $\mathcal{C}^n(R + \partial R)$ the vector space of all functions $\phi = \phi(x, y)$ which have continuous *partial* derivatives of all orders up to and including nth order on $R + \partial R$. The values of the partial derivatives of such a function on the boundary ∂R may be taken to be the limits of those values upon approach to the boundary from within R.

Lemma. *Let $f = f(x, y)$ be any continuous real-valued function on the closed region $R + \partial R$, and suppose that for some nonnegative integer n ($n =$ 0, 1, 2, . . .)*

$$\iint_R f(x, y)h(x, y)\, dx\, dy = 0 \tag{A8.1}$$

holds for all functions h in $\mathcal{C}^n(R + \partial R)$ which vanish on the boundary ∂R along with their partial derivatives of order up to and including order n. Then f must *necessarily vanish identically on $R + \partial R$; i.e., $f(x, y) = 0$ for all points (x, y) in $R + \partial R$.*

Proof. It will be enough to prove the result $f(x, y) = 0$ for every (interior) point (x, y) in the open region R (why?). We argue by contradiction. Suppose that $f(x^*, y^*) > 0$ for some interior point (x^*, y^*) in R. [The case $f(x^*, y^*) < 0$ can be reduced to this case by considering $-f$.] Then by continuity

$$f(x, y) > 0 \tag{A8.2}$$

holds for all points (x, y) in some ball $B_\rho(x^*, y^*)$ centered at (x^*, y^*) and contained in R (for some radius $\rho > 0$), where

$$B_\rho(x^*, y^*) = \{(x, y): (x - x^*)^2 + (y - y^*)^2 < \rho^2\}.$$

We now define a function $h^* = h^*(x, y)$ on $R + \partial R$ by

$$h^*(x, y) = \begin{cases} (\rho^2 - [(x - x^*)^2 + (y - y^*)^2])^{n+1} & \text{for } (x, y) \text{ in } B_\rho(x^*, y^*) \\ 0 & \text{for } (x, y) \text{ in } R + \partial R \text{ but } \textit{not} \text{ in } B_\rho(x^*, y^*). \end{cases}$$

It is easy to check that this function h^* is of class $\mathcal{C}^n$ on $R + \partial R$, while h^* clearly vanishes on the boundary ∂R along with all its derivatives. Hence h^* is an admissible function which may be used in the relation (A8.1), yielding

$$\iint_{B_\rho(x^*, y^*)} f(x, y)h^*(x, y)\, dx\, dy = 0 \tag{A8.3}$$

since h^* vanishes outside the ball $B_\rho(x^*, y^*)$. But the last result is not possible. Indeed, it is easy to show that h^* is *positive* everywhere in $B_\rho(x^*, y^*)$, and then (A8.2) implies that

$$f(x, y)h^*(x, y) > 0$$

for all (x, y) in $B_\rho(x^*, y^*)$. This, however, is clearly incompatible with (A8.3), and therefore the original assumption that $f(x^*, y^*) > 0$ is impossible. Similarly, $f(x^*, y^*) < 0$ is impossible, and this completes the proof of the lemma.

A9. The Kinetic Energy for a Certain Model of an Elastic String

Here we shall give some justification for the definition (4.11.4) for the kinetic energy of motion of a vibrating elastic string. We shall use the same notation and terminology as in Section 4.11.

We first divide the interval $0 \leq x \leq l$ into n subintervals given as $[x_0, x_1], [x_1, x_2], \ldots, [x_{n-1}, x_n]$ with $0 = x_0 < x_1 < x_2 < \cdots < x_{n-1} < x_n = l$. Any such division of the interval $[0, l]$ into n subintervals gives at any time t a corresponding partition of the stretched string $\gamma = \gamma(t)$ into n pieces $\gamma_1, \gamma_2, \ldots, \gamma_n$, where the ith piece $\gamma_i = \gamma_i(t)$ is given parametrically as [recall (4.11.2)]

$$\gamma_i\colon \quad z = Z(x, t) \qquad \text{for } x_{i-1} \leq x \leq x_i$$

for $i = 1, 2, \ldots, n$.

We assume that initially the mass of the string is uniformly distributed when the string is in its equilibrium position at time $t = t_0$, so that the initial mass m_i of the ith piece of string is proportional to the ratio of the length of the ith piece to the total length of γ. Hence we find that

$$m_i = m\frac{x_i - x_{i-1}}{l} = \text{mass of } \gamma_i \text{ at time } t_0,$$

where m is the total mass of the string. On the other hand, we consider only transversal motions of the string where each piece vibrates up and down. Hence the mass m_i of the ith piece of string is constant during the entire motion, with

$$m_i = m\frac{x_i - x_{i-1}}{l} \qquad \text{for } t_0 \leq t \leq t_1$$

and for $i = 1, 2, \ldots, n$.

Now if n is large enough and if the length of each subinterval is small enough, then the ith portion of the string may be considered to *approximate* a *particle* with mass equal to m_i and with kinetic energy T_i given by the usual expression

$$T_i = \tfrac{1}{2} m_i v_i^2,$$

where the speed v_i of the ith *particle* is the rate of change of its displacement with respect to time, given as

$$v_i = \frac{\partial}{\partial t} Z(\bar{x}_i, t)$$

for some suitable point $\bar{x}_i$ between x_{i-1} and x_i. We can get an *approximation* to the total kinetic energy T of the entire stretched string by summing over all *n particles* as

$$\sum_{i=1}^{n} T_i = \frac{1}{2} \sum_{i=1}^{n} m_i v_i^2 = \frac{m}{2l} \sum_{i=1}^{n} (x_i - x_{i-1}) \left[\frac{\partial}{\partial t} Z(\bar{x}_i, t)\right]^2,$$

and it is reasonable to expect that this approximate value will be close to the exact value of T if n is large while each subinterval is small. Hence we are led to *define* the kinetic energy T of the stretched string to be the *limit* of the above approximate expression as $n \longrightarrow \infty$, i.e.,

$$T = \frac{m}{2l} \operatorname*{limit}_{n \to \infty} \sum_{i=1}^{n} (x_i - x_{i-1}) \left[\frac{\partial}{\partial t} Z(\bar{x}_i, t) \right]^2,$$

where it is understood here that the lengths of the subintervals are required to shrink toward zero in the limiting process. But if Z is continuously differentiable, then the limit appearing here is (by the definition of the Riemann integral) equal to the integral

$$\int_0^l \left[\frac{\partial Z(x, t)}{\partial t} \right]^2 dx,$$

which gives

$$T = \frac{m}{2l} \int_0^l \left[\frac{\partial Z(x, t)}{\partial t} \right]^2 dx.$$

It is this expression that we have taken in (4.11.4) to give the kinetic energy of motion for the vibrating string.

The basic assumption made in the above discussion is that the string must move transversally with each piece vibrating up and down. Although a real string will not satisfy this requirement *exactly*, it is nevertheless true that elastic strings undergoing *small* vibrations do *approximately* satisfy this requirement. The resulting simplified model of a vibrating string leads to theoretical results which are in close agreement with actual experimental results in the case of small vibrations.

A10. The Variation of an Initial Value Problem with Respect to a Parameter

We shall give the essential ideas in the derivation of (6.4.12) for the solution $X = X(t; U)$ of the differential equation

$$\frac{dX}{dt} = G(t, X, U(t)) \qquad \text{for } t > t_0 \tag{A10.1}$$

subject to the fixed initial condition

$$X = x_0 \qquad \text{at } t = t_0. \tag{A10.2}$$

Here $U = U(t)$ may be any given continuous (or piecewise continuous) control function defined for $t \geq t_0$, while the function $G = G(t, x, u)$ is assumed to be continuous with respect to all its variables and have continuous first-order partial derivatives with respect to x and u.

We wish to calculate the variation of $X = X(t; U^*)$ at any fixed control function U^* and in any fixed "direction" ΔU; i.e., we wish to calculate [see (6.4.9)]

$$\frac{\partial}{\partial \epsilon} X(t; U^* + \epsilon\, \Delta U)\bigg|_{\epsilon=0} = \operatorname*{limit}_{\epsilon \to 0} \frac{X(t; U^* + \epsilon\, \Delta U) - X(t; U^*)}{\epsilon} \tag{A10.3}$$

for any two given continuous functions $U^* = U^*(t)$ and $\Delta U = \Delta U(t)$.

Since $X = X(t; U^*)$ and $X = X(t; U^* + \epsilon\, \Delta U)$ both satisfy the same fixed initial condition (A10.2),

$$\frac{X(t_0; U^* + \epsilon\, \Delta U) - X(t_0; U^*)}{\epsilon} = 0$$

holds for any small nonzero number ϵ, and then (A10.3) implies that

$$\frac{\partial}{\partial \epsilon} X(t_0; U^* + \epsilon\, \Delta U)\bigg|_{\epsilon=0} = 0. \tag{A10.4}$$

On the other hand, our stated assumptions on the function G imply† that the expression $(d/dt)X(t; U^* + \epsilon\, \Delta U)$ is continuously differentiable with respect to the real variable ϵ near $\epsilon = 0$ (for each fixed t near $t = t_0$), with

$$\frac{\partial}{\partial \epsilon}\frac{d}{dt} X(t; U^* + \epsilon\, \Delta U) = \frac{d}{dt}\frac{\partial}{\partial \epsilon} X(t; U^* + \epsilon\, \Delta U). \tag{A10.5}$$

Since $X = X(t; U^* + \epsilon\, \Delta U)$ satisfies the differential equation (A10.1) with $U = U^* + \epsilon\, \Delta U$, we have

$$\frac{d}{dt} X(t; U^* + \epsilon\, \Delta U) = G(t, X(t; U^* + \epsilon\, \Delta U), U^*(t) + \epsilon\, \Delta U(t)),$$

and this equation can now be differentiated with respect to ϵ to find with (A10.5) that

$$\frac{d}{dt}\frac{\partial}{\partial \epsilon} X(t; U^* + \epsilon\, \Delta U) = \frac{\partial}{\partial \epsilon} G(t, X(t; U^* + \epsilon\, \Delta U), U^*(t) + \epsilon\, \Delta U(t)).$$

The required derivative of G with respect to ϵ can be calculated with the chain rule of differential calculus as

$$\begin{aligned}
&\frac{\partial}{\partial \epsilon} G(t, X(t; U^* + \epsilon\, \Delta U), U^*(t) + \epsilon\, \Delta U(t)) \\
&\quad = G_X(t, X(t; U^* + \epsilon\, \Delta U), U^*(t) + \epsilon\, \Delta U(t)) \frac{\partial}{\partial \epsilon} X(t; U^* + \epsilon\, \Delta U) \\
&\qquad + G_U(t, X(t; U^* + \epsilon\, \Delta U), U^*(t) + \epsilon\, \Delta U(t))\, \Delta U(t).
\end{aligned}$$

†See Garrett Birkhoff and Gian-Carlo Rota, *Ordinary Differential Equations* (Boston: Ginn and Company, 1962), pp. 123–124, for the method of proof of this result.

If we now set $\epsilon = 0$ in these last equations, we find that

$$\frac{d}{dt}Y(t) = A(t)Y(t) + B(t)\,\Delta U(t) \qquad \text{for } t > t_0, \tag{A10.6}$$

where $Y(t)$, $A(t)$, and $B(t)$ are defined by

$$Y(t) = \frac{\partial}{\partial \epsilon} X(t; U^* + \epsilon\,\Delta U)\Big|_{\epsilon=0} \tag{A10.7}$$

$$A(t) = G_X(t, X(t; U^*), U^*(t)) = \frac{\partial}{\partial x} G(t, x, u)\Big|_{\substack{x=X(t;\,U^*)\\ u=U^*(t)}}$$

and

$$B(t) = G_U(t, X(t; U^*), U^*(t)) = \frac{\partial}{\partial u} G(t, x, u)\Big|_{\substack{x=X(t;\,U^*)\\ u=U^*(t)}}.$$

The solution of the differential equation (A10.6) is given as†

$$Y(t) = e^{\int_{t_0}^{t} A(s)\,ds}\left[Y(t_0) + \int_{t_0}^{t} e^{-\int_{t_0}^{\tau} A(s)\,ds} B(\tau)\,\Delta U(\tau)\,d\tau\right].$$

Moreover, at time $t = t_0$ we can use (A10.4) and (A10.7) to find that $Y(t_0) = 0$. Hence (A10.7) and these last results imply the desired equation

$$\frac{\partial}{\partial \epsilon} X(t; U^* + \epsilon\,\Delta U)\Big|_{\epsilon=0} = \int_{t_0}^{t} e^{\int_{\tau}^{t} A(s)\,ds} B(\tau)\,\Delta U(\tau)\,d\tau,$$

which agrees with (6.4.12).

†See, for example, Chapter 20 of George B. Thomas, Jr., *Calculus and Analytic Geometry*, (Reading, Mass.: Addison-Wesley Publishing Company, Inc., 1968), or any elementary text on differential equations.

Subject Index

J

K

L

M

N

Author Index

D

E

F

G

H

I

J

K

L

M

N

O

P

R

S

T

V

W

Y

Z